Probability and Mathematical Statistics (Continued)

WILLIAMS • Diffusions, Markov Processes, and Martingales, Volume I: Foundations
ZACKS • Theory of Statistical Inference

Applied Probability and Statistics

ANDERSON, AUQUIER, HAUCK, OAKES, VANDAELE, and WEISBERG • Statistical Methods for Comparative Studies
ARTHANARI and DODGE • Mathematical Programming in Statistics
BAILEY • The Elements of Stochastic Processes with Applications to the Natural Sciences
BAILEY • Mathematics, Statistics and Systems for Health
BARNETT • Interpreting Multivariate Data
BARNETT and LEWIS • Outliers in Statistical Data
BARTHOLOMEW • Stochastic Models for Social Processes, *Second Edition*
BARTHOLOMEW and FORBES • Statistical Techniques for Manpower Planning
BECK and ARNOLD • Parameter Estimation in Engineering and Science
BELSLEY, KUH, and WELSCH • Regression Diagnostics: Identifying Influential Data and Sources of Collinearity
BENNETT and FRANKLIN • Statistical Analysis in Chemistry and the Chemical Industry
BHAT • Elements of Applied Stochastic Processes
BLOOMFIELD • Fourier Analysis of Time Series: An Introduction
BOX • R. A. Fisher, The Life of a Scientist
BOX and DRAPER • Evolutionary Operation: A Statistical Method for Process Improvement
BOX, HUNTER, and HUNTER • Statistics for Experimenters: An Introduction to Design, Data Analysis, and Model Building
BROWN and HOLLANDER • Statistics: A Biomedical Introduction
BROWNLEE • Statistical Theory and Methodology in Science and Engineering, *Second Edition*
BURY • Statistical Models in Applied Science
CHAMBERS • Computational Methods for Data Analysis
CHATTERJEE and PRICE • Regression Analysis by Example
CHERNOFF and MOSES • Elementary Decision Theory
CHOW • Analysis and Control of Dynamic Economic Systems
CHOW • Econometric Analysis by Control Methods
CLELLAND, BROWN, and deCANI • Basic Statistics with Business Applications, *Second Edition*
COCHRAN • Sampling Techniques, *Third Edition*
COCHRAN and COX • Experimental Designs, *Second Edition*
CONOVER • Practical Nonparametric Statistics, *Second Edition*
CORNELL • Experiments with Mixtures: Designs, Models and The Analysis of Mixture Data
COX • Planning of Experiments
DANIEL • Biostatistics: A Foundation for Analysis in the Health Sciences, *Second Edition*
DANIEL • Applications of Statistics to Industrial Experimentation
DANIEL and WOOD • Fitting Equations to Data: Computer Analysis of Multifactor Data, *Second Edition*
DAVID • Order Statistics, *Second Edition*
DEMING • Sample Design in Business Research
DODGE and ROMIG • Sampling Inspection Tables, *Second Edition*
DRAPER and SMITH • Applied Regression Analysis, *Second Edition*
DUNN • Basic Statistics: A Primer for the Biomedical Sciences, *Second Edition*
DUNN and CLARK • Applied Statistics: Analysis of Variance and Regression
ELANDT-JOHNSON • Probability Models and Statistical Methods in Genetics
ELANDT-JOHNSON and JOHNSON • Survival Models and Data Analysis

continued on back

Simulation and the
Monte Carlo Method

Simulation and the Monte Carlo Method

REUVEN Y. RUBINSTEIN
Technion, Israel Institute of Technology

John Wiley & Sons

New York Chichester Brisbane Toronto Singapore

Library of Congress Cataloging in Publication Data:

Rubinstein, Reuven Y.
Simulation and the Monte Carlo method.

(Wiley series in probability and mathematical statistics)
Includes bibliographies and index.
1. Monte Carlo method. 2. Digital computer
simulation. I. Title. II. Series.

QA298.R8 519.2′82 81-1873
ISBN 0-471-08917-6 AACR2

Printed in the United States of America

10 9 8 7 6 5 4

To my wife Rina
and to my friends Eitan Finkelstein and Alexandr Lerner—
Russian refuseniks.

Preface

In the last 15 years more than 3000 articles on simulation and the Monte Carlo method have been published. There is real need for a book providing detailed treatment of the statistical aspect of these topics. This book attempts to fill this need, at least partially. I hope it will make the users of simulation and the Monte Carlo method more knowledgeable about these topics.

It is assumed that the readers are familiar with the basic concepts of probability theory, mathematical statistics, integral and differential equations, and that they have an elementary knowledge of vector and matrix operators. Sections 6.5, 6.6, 7.3, and 7.6 require more sophistication in probability, statistics, and stochastic processes; they can be omitted for a first reading.

Since most complex simulations are implemented on digital computers, a rudimentary acquaintance with computer programming will probably be an asset to the readers of this book, though no computer programs are included.

Chapter 1 describes concepts such as systems, models, and the ideas of Monte Carlo and simulation. A discussion of these concepts seems necessary as there is no uniform terminology in the literature. Instead of giving rigid definitions, I try to make clear what I mean when I use these terms. In addition to the terminology, some examples and ideas of simulation and Monte Carlo methods are given.

Chapter 2 deals with several alternative methods for generating random and pseudorandom numbers on a computer, as well as several statistical methods for testing the "randomness" of pseudorandom numbers.

Chapter 3 describes methods for generating random variables and random vectors from different probability distributions.

Chapter 4 provides a basic treatment of Monte Carlo integration, and Chapter 5 provides a solution of linear, integral, and differential equations by Monte Carlo methods. It is shown that, in order to find a solution by Monte Carlo methods, we must choose a proper distribution and present the problem in terms of its expected value. Then, taking a sample from this distribution, we can estimate the expected value. In addition, variance reduction techniques (importance sampling, control variates, stratified sampling antithetic variates, etc.) are discussed.

Chapter 6 deals with simulating regenerative processes and in particular with estimating some output parameters of the steady-state distribution associated with these processes. Simulation results for several practical problems are presented, and variance reduction techniques are given as well.

Chapter 7 discusses random search methods, which are also related to Monte Carlo methods. In this chapter I describe how random search methods can be successfully applied for solving complex optimization problems.

The final version of this book was written during my 1980 summer visit at IBM Thomas J. Watson Research Center. I express my gratitude to the Computer Sciences Department for their hospitality and for providing a rich intellectual environment.

A number of people have contributed corrections and suggestions for improvement of the earlier draft of the manuscript, especially P. Feigin, I. Kreimer, O. Maimon, H. Nafetz, G. Samorodnitsky, and E. Yaschin from Technion, Israel Institute of Technology, and P. Heidelberger and S. Lavenberg of IBM Thomas J. Watson Research Center. It is a pleasure to acknowledge my debt to them. I would also like to express my indebtedness to Beatrice Shube of John Wiley & Sons and to Eliezer Goldberg of Technion for their efficient editorial guidance. Many thanks to Marylou Dietrich of IBM and to Eva Gaster of Technion for their excellent typing.

Finally, I thank the following authors and publishers for granting permission for publication of the cited material:

Pages 12–17 based on *Handbook of Operations Research, Foundations and Fundamentals*, Edited by Joseph T. Modern and Salah E. Elmagraby, Von Nostrand Reinhold Company, 1978, pp. 570–573.

Pages 23–25 based on D. E. Knuth, *The Art of Computer Programming: Seminumerical Algorithms*, Vol. 2, Addisson-Wesley, Reading, Massachusetts, 1969, pp. 155–156.

Pages 199–208 based on Y. R. Rubinstein, Selecting the best stable stochastic system, in *Stochastic Processes and their Applications*, 1980. (to appear)

Pages 253–255 based on Y. R. Rubinstein and I. Weisman, The Monte Carlo method for global optimization, *Cahiers du Centre d'Etudes de Recherche Operationelle*, **21**, No. 2, 1979, pp. 143–149.

Pages 248–251 based on Y. R. Rubinstein, and A. Kornovsky, Local and integral properties of a search algorithm of the stochastic approximation type. *Stochastic Processes Appl.*, **6**, 1978, 129–134.

REUVEN Y. RUBINSTEIN

Haifa, Israel
March 1981

Contents

Simulation and the
Monte Carlo Method

CHAPTER 1

Systems, Models, Simulation, and the Monte Carlo Methods

In this chapter we discuss the concepts of systems, models, simulation and Monte Carlo methods. This discussion seems necessary in the absence of a unified terminology in the literature. We do not give rigid definitions, however, but explain what we mean when using the above-mentioned terms.

1.1 SYSTEMS

By a system we mean a set of related entities sometimes called *components* or *elements*. For instance, a hospital can be considered as a system, with doctors, nurses, and patients as elements. The elements have certain characteristics, or *attributes*, that have logical or numerical values. In our example an attribute can be, for instance, the number of beds, the number of X-ray machines, skill, quantity, and so on. A number of activities (relations) exist among the elements, and consequently the elements interact. These activities cause changes in the system. For example, the hospital has X-ray machines that have an operator. If there is no operator, the doctors cannot have X-rays of the patients taken.

We consider both internal and external relationships. The internal relationships connect the elements within the system, while the external relationships connect the elements with the environment, that is, with the world outside the system. For instance, an internal relationship is the relationship or interaction between the doctors and nurses, or between

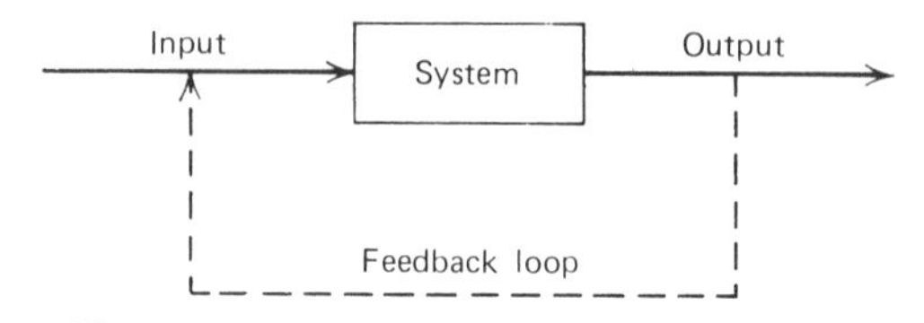

Fig. 1.1.1 Graphical representation of a system.

the nurses and the patients. An external relationship is, for example, the way in which the patients are delivered to the emergency room. We can represent a system by a diagram, as in Fig. 1.1.1.

The system is influenced by the environment through the input it receives from the environment. When a system has the capability of reacting to changes in its own state, we say that the system contains feedback. A nonfeedback, or open-loop, system lacks this characteristic. For an example of feedback consider a waiting line; when there are more than a certain number of patients, the hospital can add more staff to handle the increased workload.

The attributes of the system elements define its *state*. In our example the number of patients waiting for a doctor describe the system's state. When a patient arrives at or leaves the hospital, the system moves to a new state. If the behavior of the elements cannot be predicted exactly, it is useful to take random observations from the probability distributions and to average the performance of the objective. We say that a system is in *equilibrium* or in the *steady state* if the probability of being in some state does not vary in time. There are still actions in the system, that is, the system can still move from one state to another, but the *probabilities* of its moving from one state to another are fixed. These fixed probabilities are limiting probabilities that are realized after a long period of time, and they are independent of the state in which the system started. A system is called *stable* if it returns to the steady state after an external shock in the system. If the system is not in the steady state, it is in a *transient* state.

We can classify systems in a variety of ways. There are *natural* and *artificial systems*, *adaptive* and *nonadaptive systems*. An adaptive system reacts to changes in its environment, whereas a nonadaptive system does not. Analysis of an adaptive system requires a description of how the environment induces a change of state.

Suppose that over a period of time the number of patients increases. If the hospital adds more staff to handle the increased workload, we say that the hospital is an adaptive system.

1.2 MODELS

The first step in studying a system is building a model. The importance of models and model-building has been discussed by Rosenbluth and Wiener [32], who wrote:

> No substantial part of the universe is so simple that it can be grasped and controlled without abstraction. Abstraction consists in replacing the part of the universe under consideration by a model of similar but simpler structure. Models... are thus a central necessity of scientific procedure.

A scientific model can be defined as an abstraction of some real system, an abstraction that can be used for prediction and control. The purpose of a scientific model is to enable the analyst to determine how one or more changes in various aspects of the modeled system may affect other aspects of the system or the system as a whole.

A crucial step in building the model is constructing the objective function, which is a mathematical function of the decision variables.

There are many types of models. Churchman et al. [4] and Kiviat [18] described the following kinds:

1 **Iconic models** Those that pictorially or visually represent certain aspects of a system.

2 **Analog models** Those that employ one set of properties to represent some other set of properties that the system being studied possesses.

3 **Symbolic models** Those that require mathematical or logical operations and can be used to formulate a solution to the problem at hand.

In this book, however, we are concerned only with symbolic models (which are also called *abstract* models), that is, we deal with models consisting of mathematical symbols or flowcharts. All other models (iconic, analog, verbal, physical, etc.), although no less important, are excluded from this book.

There are many advantages by using mathematical models. According to Fishman [8] they do the following:

1 Enable investigators to organize their theoretical beliefs and empirical observations about a system and to deduce the logical implications of this organization.

2 Lead to improved system understanding.

3 Bring into perspective the need for detail and relevance.

4 Expedite the analysis.

5 Provide a framework for testing the desirability of system modifications.

6 Allow for easier manipulation than the system itself permits.
7 Permit control over more sources of variation than direct study of a system would allow.
8 Are generally less costly than the system.

An additional advantage is that a mathematical model describes a problem more concisely than, for instance, a verbal description does.

On the other hand, there are at least three reservations in Fishman's monograph [8], which we should always bear in mind while constructing a model.

First, there is no guarantee that the time and effort devoted to modeling will return a useful result and satisfactory benefits. Occasional failures occur because the level of resources is too low. More often, however, failure results when the investigator relys too much on method and not enough on ingenuity; the proper balance between the two leads to the greatest probability of success.

The second reservation concerns the tendency of an investigator to treat his or her particular depiction of a problem as the best representation of reality. This is often the case after much time and effort have been spent and the investigator expects some useful results.

The third reservation concerns the use of the model to predict the range of its applicability without proper qualification.

Mathematical models can be classified in many ways. Some models are *static*, other are *dynamic*. Static models are those that do not explicitly take time-variation into account, whereas dynamic models deal explicitly with time-variable interaction. For instance, Ohm's law is an example of a static model, while Newton's law of motion is an example of a dynamic model.

Another distinction concerns *deterministic* versus *stochastic* models. In a deterministic model all mathematical and logical relationships between the elements are fixed. As a consequence these relationships completely determine the solutions. In a stochastic model at least one variable is random.

While building a model care must be taken to ensure that it remains a valid representation of the problem.

In order to be useful, a scientific model necessarily embodies elements of two conflicting attributes—realism and simplicity. On the one hand, the model should serve as a reasonably close approximation to the real system and incorporate most of the important aspects of the system. On the other hand, the model must not be so complex that it is impossible to understand and manipulate. Being a formalism, a model is necessarily an abstraction.

Often we think that the more details a model includes the better it resembles reality. But adding details makes the solution more difficult and converts the method for solving a problem from an analytical to an approximate numerical one.

In addition, it is not even necessary for the model to approximate the system to indicate the measure of effectiveness for all various alternatives. All that is required is that there be a high *correlation* between the prediction by the model and what would actually happen with the real system. To ascertain whether this requirement is satisfied or not, it is important to test and establish control over the solution.

Usually, we begin testing the model by re-examining the formulation of the problem and revealing possible flaws. Another criterion for judging the validity of the model is determining whether all mathematical expressions are dimensionally consistent. A third useful test consists of varying input parameters and checking that the output from the model behaves in a plausible manner. The fourth test is the so-called *retrospective* test. It involves using historical data to reconstruct the past and then determining how well the resulting solution would have performed if it had been used. Comparing the effectiveness of this hypothetical performance with what actually happened then indicates how well the model predicts the reality. However, a disadvantage of retrospective testing is that it uses the same data that guided formulation of the model. Unless the past is a true replica of the future, it is better not to resort to this test at all.

Suppose that the conditions under which the model was built change. In this case the model must be modified and control over the solution must be established. Often, it is desirable to identify the critical input parameters of the model, that is, those parameters subject to changes that would affect the solution, and to establish systematic procedures to control them. This can be done by *sensitivity analysis*, in which the respective parameters are varied over their ranges to determine the degree of variation in the solution of the model.

After constructing a mathematical model for the problem under consideration, the next step is to derive a solution from this model. There are *analytic* and *numerical* solution methods.

An analytic solution is usually obtained directly from its mathematical representation in the form of formula.

A numerical solution is generally an approximate solution obtained as a result of substitution of numerical values for the variables and parameters of the model. Many numerical methods are iterative, that is, each successive step in the solution uses the results from the previous step. Newton's method for approximating the root of a nonlinear equation can serve as an example.

Two special types of numerical methods are simulation and the Monte Carlo methods. The following section discusses these.

1.3 SIMULATION AND THE MONTE CARLO METHODS

Simulation has long been an important tool of designers, whether they are simulating a supersonic jet flight, a telephone communication system, a wind tunnel, a large-scale military battle (to evaluate defensive or offensive weapon systems), or a maintenance operation (to determine the optimal size of repair crews).

Although simulation is often viewed as a "method of last resort" to be employed when everything else has failed, recent advances in simulation methodologies, availability of software, and technical developments have made simulation one of the most widely used and accepted tools in system analysis and operations research.

Naylor et al. [28] define simulation as follows:

> Simulation is a numerical technique for conducting experiments on a digital computer, which involves certain types of mathematical and logical models that describe the behavior of business or economic system (or some component thereof) over extended periods of real time.

This definition is extremely broad, however, and can include such seemingly unrelated things as economic models, wind tunnel testing of aircraft, war games, and business management games.

Naylor et al. [28] write:

> The fundamental rationale for using simulation is man's unceasing quest for knowledge about the future. This search for knowledge and the desire to predict the future are as old as the history of mankind. But prior to the seventeenth century the pursuit of predictive power was limited almost entirely to the purely deductive methods of such philosophers as Plato, Aristotle, Euclid, and others.

Simulation deals with both *abstract* and *physical* models. Some simulation with physical and abstract models might involve participation by real people. Examples include link-trainers for pilots and military or business games. Two types of simulation involving real people deserve special mention. One is *operational gaming*, the other *man-machine* simulation.

The term "operational gaming" refers to those simulations characterized by some form of conflict of interest among players or human decision-makers within the framework of the simulated environment, and the experimenter, by observing the players, may be able to test hypotheses concerning the behavior of the individuals and/or the decision system as a whole.

In operational gaming a computer is often used to collect, process, and produce information that human players, usually adversaries, need to make decisions about system operation. Each player's objective is to perform as well as possible. Moreover, each player's decisions affect the information that the computer provides as the game progresses through simulated time. The computer can also play an active role by initiating predetermined or random actions to which the players respond.

War games and business management games are commonly discussed in operational gaming literature (see, e.g., Morgenthaler [23] and Shubik [38]).

Military gaming is essentially a training device for military leaders; it enables them to test the effects of alternative strategies under simulated war conditions. For example, the Naval Electronic Warfare Simulator, developed in the 1950s, consisted of a large analog computer designed primarily to assess ship damage and to provide information to two opposite forces regarding their respective effectiveness in a naval engagement [14, pp. 15, 16]. The exercise, which is one form of simulation gaming, has been used as an educational device for naval fleet officers in the final stages of their training.

Business games are also a type of educational tool, but for training managers or business executives rather than military leaders.

> A business game is a contrived situation which imbeds players in a simulated business environment, where they must make management-type decisions from time to time, and their choices at one time generally affect the environmental conditions under which subsequent decisions must be made. Further, the interaction between decisions and environment is determined by a refereeing process which is not open to argument from the players [30, pp. 7, 8].

In man-machine simulation there is no need for gaming. While interacting with the computer real people in the laboratory perform the data reduction and analysis.

The following two examples are drawn from Fishman [8]:

The Rand Systems Research Laboratory employed simulation to generate stimuli for the study of information processing centers [14, p. 16]. The principal features of a radar site were reproduced in the laboratory, and by carefully controlling the synthetic input to the system and recording the behavior of the human detectors it was possible to examine the relative effectiveness of various man-machine combinations and procedures.

In 1956 Rand established the Logistics System Laboratory under U.S. Air Force sponsorship [10]. The first study in this laboratory involved simulation of two large logistics systems in order to compare their effectiveness under different management and resource utilization policies. Each system consisted of men and machines, together with policy rules for the use of such resources in simulated stress situations such as war. The simulated environment required a specified number of aircraft in flying and alert states, while the system's capability to meet these objectives was limited by malfunctioning parts, procurement and transportation delays, and the like. The human participants represented management personnel, while higher echelon policies in the utilization of resources were simulated on the computer. The ultimate criteria of the effectiveness of each system were the number of operationally ready aircraft and the dollar cost of maintaining this number.

Although the purpose of the first study in this laboratory was to test the feasibility of introducing new procedures into an existing air force logistics system and to compare the modified system with the original one, the second laboratory problem had quite a different objective. Its purpose was to improve the design of the operational control system through the use of simulation.

Naylor et al. [28] describe many situations where simulation can be successfully used. We mention some of them.

First, it may be either impossible or extremely expensive to obtain data from certain processes in the real world. Such processes might involve, for example, the performance of large-scale rocket engines, the effect of proposed tax cuts on the economy, the effect of an advertising campaign on total sales. In this case we say that the simulated data are necessary to *formulate hypotheses* about the system.

Secondly, the observed system may be so complex that it cannot be described in terms of a set of mathematical equations for which analytic solutions are obtainable. Most economic systems fall into this category. For example, it is virtually impossible to describe the operation of a business firm, an industry, or an economy in terms of a few simple equations. Simulation has been found to be an extremely effective tool for dealing with problems of this type. Another class of problems that leads to similar difficulties is that of large-scale queueing problems involving multiple channels that are either parallel or in series (or both).

Thirdly, even though a mathematical model can be formulated to describe some system of interest, it may not be possible to obtain a solution to the model by straightforward analytic techniques. Again, economic systems and complex queueing problems provide examples of this type of difficulty. Although it may be conceptually possible to use a set of mathematical equations to describe the behavior of a dynamic system operating under conditions of uncertainty, present-day mathematics and computer technology are simply incapable of handling a problem of this magnitude.

Fourth, it may be either impossible or very costly to perform validating experiments on the mathematical models describing the system. In this case we say that the simulation data can be used to test alternative hypotheses.

In all these cases simulation is the only practical tool for obtaining relevant answers.

Naylor et al. [28] have suggested that simulation analysis might be appropriate for the following reasons:

1 Simulation makes it possible to study and experiment with the complex internal interactions of a given system whether it be a firm, an industry, an economy, or some subsystem of one of these.

2 Through simulation we can study the effects of certain informational, organizational, and environmental changes on the operation of a system by making alterations in the model of the system and observing the effects of these alterations on the system's behavior.

3 Detailed observation of the system being simulated may lead to a better understanding of the system and to suggestions for improving it, suggestions that otherwise would not be apparent.

4 Simulation can be used as a pedagogical device for teaching both students and practitioners basic skills in theoretical analysis, statistical analysis, and decision making. Among the disciplines in which simulation has been used successfully for this purpose are business administration, economics, medicine, and law.

5 Operational gaming has been found to be an excellent means of stimulating interest and understanding on the part of the participant, and is particularly useful in the orientation of persons who are experienced in the subject of the game.

6 The experience of designing a computer simulation model may be more valuable than the actual simulation itself. The knowledge obtained in designing a simulation study frequently suggests changes in the system being simulated. The effects of these changes can then be tested via simulation before implementing them on the actual system.

7 Simulation of complex systems can yield valuable insight into which variables are more important than others in the system and how these variables interact.

8 Simulation can be used to experiment with new situations about which we have little or no information so as to prepare for what may happen.

9 Simulation can serve as a "preservice test" to try out new policies and decision rules for operating a system, before running the risk of experimenting on the real system.

10 Simulations are sometimes valuable in that they afford a convenient way of breaking down a complicated system into subsystems, each of which may then be modeled by an analyst or team that is expert in that area [23, p. 373].

11 Simulation makes it possible to study dynamic systems in either real time, compressed time, or expanded time.

12 When new components are introduced into a system, simulation can be used to help foresee bottlenecks and other problems that may arise in the operation of the system [23, p. 375].

Computer simulation also enables us to *replicate* an experiment. Replication means rerunning an experiment with selected changes in parameters or operating conditions being made by the investigator. In addition, computer simulation often allows us to induce correlation between these random number sequences to improve the statistical analysis of the output of a simulation. In particular a negative correlation is desirable when the results of two replications are to be summed, whereas a positive correlation is preferred when the results are to be differenced, as in the comparison of experiments.

Simulation does not require that a model be presented in a particular format. It permits a considerable degree of freedom so that a model can bear a close correspondence to the system being studied. The results obtained from simulation are much the same as observations or measurements that might have been made on the system itself. To demonstrate the principles involved in executing a discrete simulation, an example of simulating a machine shop is given in Section 1.4. Many programming systems have been developed, incorporating simulation languages. Some of them are general-purpose in nature, while others are designed for specific types of systems. FORTRAN, ALGOL, and PL/1 are examples of general-purpose languages, while GPSS, SIMSCRIPT, and SIMULA are examples of special simulation languages.

Simulation is indeed an invaluable and very versatile tool in those problems where analytic techniques are inadequate. However, it is by no means ideal. Simulation is an imprecise technique. It provides only *statistical estimates* rather than exact results, and it only compares alternatives rather than generating the optimal one. Simulation is also a *slow* and *costly* way to study a problem. It usually requires a large amount of time and great expense for analysis and programming. Finally, simulation yields only *numerical data* about the performance of the system, and sensitivity analysis of the model parameters is very expensive. The only possibility is to conduct series of simulation runs with different parameter values.

We have defined simulation as a technique of performing *sampling experiments* on the model of the system. This general definition is often called simulation in a *wide sense*, whereas simulation in a *narrow sense*, or *stochastic simulation*, is defined as experimenting with the model over time; it includes sampling stochastic variates from probability distribution [19]. Therefore stochastic simulation is actually a statistical sampling experiment with the model. This sampling involves all the problems of statistical design analysis.

Because sampling from a particular distribution involves the use of random numbers, stochastic simulation is sometimes called *Monte Carlo simulation*. Historically, the Monte Carlo method was considered to be a technique, using random or pseudorandom numbers, for solution of a model. Random numbers are essentially independent random variables uniformly distributed over the unit interval [0, 1]. Actually, what are available at computer centers are arithmetic codes for generating sequences of *pseudorandom digits*, where each digit (0 through 9) occurs with approximately equal probability (likelihood). Consequently, the sequences can model successive flips of a fair ten-side die. Such codes are called *random number generators*. Grouped together, these generated digits yield pseudorandom numbers with any required number of elements. We discuss random and pseudorandom numbers in the next chapter.

One of the earliest problems connected with Monte Carlo method is the famous Buffon's needle problem. The problem is as follows. A needle of length l units is thrown randomly onto a floor composed of parallel planks of equal width d units, where $d > l$. What is the probability that the needle, once it comes to rest, will cross (or touch) a crack separating the planks on the floor? It can be shown that the probability of the needle hitting a crack is $P = 2l/\pi d$, which can be estimated as the ratio of the number of throws hitting the crack to the total number of throws.

In the begining of the century the Monte Carlo method was used to examine the Boltzmann equation. In 1908 the famous statistician Student used the Monte Carlo method for estimating the correlation coefficient in his t-distribution.

The term "Monte Carlo" was introduced by von Neumann and Ulam during World War II, as a code word for the secret work at Los Alamos; it was suggested by the gambling casinos at the city of Monte Carlo in Monaco. The Monte Carlo method was then applied to problems related to the atomic bomb. The work involved direct *simulation* of behavior concerned with random neutron diffusion in fissionable material. Shortly thereafter Monte Carlo methods were used to evaluate complex multidimensional integrals and to solve certain integral equations, occurring in physics, that were not amenable to analytic solution.

The Monte Carlo method can be used not only for solution of stochastic problems, but also for solution of deterministic problems. A deterministic problem can be solved by the Monte Carlo method if it has the same formal expression as some stochastic process. In Chapter 4 we show how the Monte Carlo method can be used for evaluating multidimensional integrals and some parameters of queues and networks. In Chapter 5 the Monte Carlo method is used for solution of certain integral and differential equations.

Another field of application of the Monte Carlo methods is sampling of random variates from probability distributions, which Morgenthaler [23] calls model sampling. Chapter 3 deals with sampling from various distributions.

The Monte Carlo method is now the most powerful and commonly used technique for analyzing complex problems. Applications can be found in many fields from radiation transport to river basin modeling. Recently, the range of applications has been broadening, and the complexity and computational effort required has been increasing, because realism is associated with more complex and extensive problem descriptions.

Finally, we mention some differences between the Monte Carlo method and simulation:

1 In the Monte Carlo method time does not play as substantial a role as it does in stochastic simulation.

2 The observations in the Monte Carlo method, as a rule, are independent. In simulation, however, we experiment with the model over time so, as a rule, the observations are serially correlated.

3 In the Monte Carlo method it is possible to express the response as a rather simple function of the stochastic input variates. In simulation the response is usually a very complicated one and can be expressed explicitly only by the computer program itself.

1.4 A MACHINE SHOP EXAMPLE

This example is quoted from Gordon [11, pp. 570–573]. For better understanding of the example an important distinction to be made is whether an entity is permanent or temporary. Permanent entities can be compactly and efficiently represented in tables, while temporary entities will be volatile records and are usually handled by the list processing technique described later.

Consider a simple machine shop (or a single stage in the manufacturing process of a more complex machine shop). The shop is to machine five types of parts. The parts arrive at random intervals and are distributed randomly among the different types. There are three machines, all equally able to machine any part. If a machine is available at the time a part arrives, machining begins immediately. If all machines are busy upon arrival, the part will wait for service. On completion of machining the part will be dispatched to a certain destination, depending on its type. The progress of the part is not followed after it is dispatched from the shop. However, a count of the number of parts dispatched to each destination is kept.

Clearly, there are two types of elements in the system: parts and machines. There will be a stream of temporary elements, that is, the parts that enter and leave the system. There is no point in representing the different types of parts as different elements; rather, the type is an attribute of the parts. As indicated before, it is simpler to consider the group of machines as a single permanent element, having as attributes the number of machines and a count of the number currently busy. The activities causing changes in the system are the generation of parts, waiting, machining, and departing.

(a) System Image A set of numbers is needed to record the state of the system at any time. This set of numbers is called the *system image*, since it reflects the state of the system. The simulation proceeds by deciding, from the system image, when the next event is due to occur and what type of event it will be; testing whether it can be executed; and executing the changes to the image implied by the event.

The image must have a number representing clock time, and this number is advanced, in uneven steps, with the succession of events in the system. For each part record, there are four numbers to represent the part type, the arrival time, the machining time, and the time the part will next be involved in an event. The first three of these items are random variates derived by the methods described in Chapters 3 and 4. The next event time, in general, depends on the state of the system, and must be derived as the simulation proceeds.

The organization used for the system image is illustrated in Fig. 1.4.1. There are four frames in this figure, representing successive states of the system. The frames are read from left to right and from top to bottom. The frame in the top left corner is the initial state. The description of the system image is made in terms of that particular frame.

	Part type	Machine time	Arrival time	Next event time
Next arrival	2	75	1002	1002
Waiting parts	4	52	992	—
	3	84	976	—
Parts being machined	3	43	972	1040
	1	21	936	1017
	2	62	896	1003
Clock time				1000

	1	2	3	4	5
Counters	12	22	20	31	15

	Part type	Machine time	Arrival time	Next event time
Next arrival	1	68	1018	1018
Waiting parts	2	75	1002	—
	4	52	992	—
	3	84	976	—
Parts being machined	3	43	972	1040
	1	21	936	1017
	2	62	896	1003
Clock time				1002

	1	2	3	4	5
Counters	12	22	20	31	15

	Part type	Machine time	Arrival time	Next event time
Next arrival	1	68	1018	1018
Waiting parts	2	75	1002	—
	4	52	992	—
Parts being machined	3	84	976	1087
	3	43	972	1040
	1	21	936	1017
Clock time				1003

	1	2	3	4	5
Counters	12	23	20	31	15

	Part type	Machine time	Arrival time	Next event time
Next arrival	1	68	1018	1018
Waiting parts	2	75	1002	—
Parts being machined	3	84	976	1087
	4	52	992	1069
	3	43	972	1040
Clock time				1017

	1	2	3	4	5
Counters	13	23	20	31	15

Fig. 1.4.1 Machine shop example.

The top line of the system image represents the part due to enter the system next. As shown here, it is a type 2 part, will require 75 minutes of machining, and is due to arrive at time 1002. This, of course, is also its next event time.

Below the next arrival listing is an open-ended list of the parts that have arrived and are now waiting for service. Currently, there are two waiting parts. As indicated, they are listed in order of arrival. Because the waiting parts are delayed, it is not possible to predict a next event time for them. It is necessary to see whether there is a waiting part when a machine finishes, and to offer service to the first part in the waiting line.

The next rows of numbers represent the parts now being machined, in this case limited to three. Once machining begins, the time to finish can be derived and entered as the next event time. Three parts are occupying the machines at this time and they have been listed in the order in which they will finish. Finally, a number represents the clock time, here set to an initial value of 1000, and there are five counters showing how many parts of each type have been completed. Note that it is not customary to precalculate all the random variates. Instead, each is calculated at the time it is needed, so a simulation program continually switches between the examination and manipulation of the system image and the subroutines that calculate the random variates.

(b) The Simulation Process Looking now at the system image in Fig. 1.4.1, assume all events that can be executed up to time 1000 have been processed. It is now time to begin one more cycle. The first step is to find the next potential event by scanning all the event times. Because of the ordering of the parts being machined, it is, in fact, necessary only to compare the time of the next arrival with the first listed time in the machining section. With the numbers shown in frame 1, the next event is the arrival of a part at time 1002, so the clock is updated to this time in the second frame.

The arriving part finds all machines busy and must join the waiting line. The successor to the part just arrived is generated and inserted as the next future arrival, due to arrive at time 1018. Another cycle can now begin. The next event is the completion of machining a part at time 1003. The third frame of Fig. 1.4.1 shows the state of the system at the end of this event. The clock is updated to 1003 and the finished part is removed from the system, after incrementing by 1 the counter for that part type. There is a waiting part, so machining is started on the first part in the waiting line, and its next event time, derived from the machining time of 84, is calculated as 1087. In this case the new part for machining has the largest finish time, and it joins the end of the waiting line. The records in the waiting line and the machine segment are all moved down one line. There is then another completion at 1017 that, as before, leads to a counter being incremented and service being offered to the first part in the waiting line. In this case, however, the machining time is short enough for the new part to finish ahead of one whose machining started earlier, so, instead of being the last listed part, the new part becomes the second in the list. This is shown in the last frame of Fig. 1.4.1.

(c) Statistics Gathering The purpose of the simulation, of course, is to learn something about the system. In this case only the counts of the number of completed parts by type have been kept. Depending upon the purpose of the simulation study, other statistics could be gathered. Simulation language programs include routines for collecting certain typical statistics. Among the commonly used types of statistics are the following:

1 Counts Counts give the number of elements of a given type or the number of times some event occurred.

2 Utilization of equipment This can be counted in terms of the fraction of time the equipment is in use or in terms of the average number of units in use.

3 Distributions This means distributions of random variates, such as processing times and response times, together with their means and standard deviations.

(d) List Processing In the machine shop example it was convenient to describe the records as though they were located in one of three places, corresponding to whether they represented parts that were arriving, waiting, or being processed. The simulation was described in terms of moving the records from one place to the next, possibly with some resorting. A computer program that used this approach would be very inefficient because of the large amount of data movement involved. Much better control and efficiency are obtained by using *list processing*. With this technique each record consists of a number of contiguous words (or bytes), some of which are reserved for constructing a list of the records. Each record contains, in a standard position, the address of the next record in the list. This is called a pointer. A special word, called a header, located in a known position, contains a pointer to the first record in the list. The last record in the list has an end-of-list symbol in place of its pointer. If the list happens to be empty, the end-of-list symbol appears in the header.

The pointers, beginning from the header, place the records in a specific order, and allow a program to search the records by following the chain of pointers. These lists, in fact, are usually called chains. There may be another set of pointers tracing through the chain from end to beginning so that a program can move along the chain in either direction. It is also possible for a record to be on more than one chain, simply by reserving pointer space for each possible chain.

Removing or adding a record, or reorganizing the order of a chain now becomes a matter of manipulating pointers. To remove C from a chain of the records $A, B, C, D, \ldots$, the pointer of B is redirected to D. If the record is being discarded, its storage space would probably be returned to another chain from which it can be reassigned later. To put the record Z between B and C, the pointer of B is directed to Z and the pointer of Z is set to indicate C. Reordering a chain consists of a series of removals and insertions.

As can be seen, list processing does not require that records be physically moved. It therefore provides an efficient way of transferring records from one category to another by moving them on and off chains, and it can easily manage lists that are constantly changing size; these are two properties that are very desirable in simulation programming. Therefore list processing is used in the implementation of all major discrete system simulation languages, including the GPSS and SIMSCRIPT simulation programs.

REFERENCES

1 Ackoff, R. L., Towards a system of systems concepts, *Manage. Sci.*, **17**, 1971, 661–671.

2 Burt, J. M., D. P. Graver, and M. Perlas, Simple stochastic networks: Some problems and procedures, *Nav. Res. Logist. Quart.*, **17**, 1970, 439–459.

3 Chorafas, D. N., *Systems and Simulation*, Academic, New York, 1965.

4 Churchman, C. W., R. L. Ackoff, and E. L. Arnoff, *Introduction to Operations Research*, Wiley, New York, 1959.

5 Emshoff, J. R. and R. L. Sisson, *Design and Use of Computer Simulation Models*, Macmillan, New York, 1970.

6 Ermakov, J. M., *Monte Carlo Method and Related Questions*, Nauka, Moskow, 1976 (in Russian).

7 Evans, G. W., G. F. Wallace, and G. L. Sutherland, *Simulation Using Digital Computers*, Prentice-Hall, Englewood Cliffs, New Jersey, 1967.

8 Fishman, G. S., *Concepts and Methods in Discrete Event Digital Simulation*, Wiley, New York, 1973.

9 Fishman, G. S., *Principles of Discrete Event Simulation*, Wiley, New York, 1978.

10 Geisler, M. A., The use of man-machine simulation for support planning, *Nav. Res. Logist. Quart.*, **7**, 1960, 421–428.

11 Gordon, G., *System Simulation*, Prentice-Hall, Englewood Cliffs, New Jersey, 1969.

12 *Handbook of Operations Research, Foundations and Fundamentals*, edited by J. J. Modern and S. E. Elmagraby, Van Nostrand Reinhold, New York, 1978.

13 Hammersley, I. M. and D. C. Handscomb, *Monte Carlo Methods*, Wiley, New York; Methuen, London, 1964.

14 Harman, H. H., Simulation: A survey, Report SP-260, System Development Corporation, Santa Monica, California, 1961.

15 Hillier, F. S. and G. J. Lieberman, *Introduction to Operations Research*, Holden-Day, San Francisco, California 1968, Chapter 14.

16 Hollingdale, S. H. (Ed.), *Digital Simulation in Operations Research*, American Elsevier, New York, 1967.

17 IBM Corporation, *Bibliography on Simulation*, Form No. 320-0924, 112 East Post Road, White Plains, New York, 1966.

18 Kiviat, P. J., *Digital Computer Simulation: Modeling Concepts*, Report RM-5378-PR, The Rand Corporation, Santa Monica, California, 1967.

19 Kleinen, J. P. C., *Statistical Techniques in Simulation*, Part 1, Marcel Decker, New York, 1974.

20 Lewis, P. A. W., Large-Scale Computer-Aided Statistical Mathematics, Naval Postgraduate School, Monterey, California, in *Proc. Computer Science and Statistics: 6th Annual Symp. Interface*, Western Periodical Co., Hollywood, California, 1972.

21 Lucas, H. C., Performance evaluation and monitoring, *Comput. Surv.*, **3**, 1971, 79–91.

22 Maisel, H. and G. Gnugnoli, *Simulation of Discrete Stochastic Systems*, Science Research Associates, Palo Alto, California, 1972.

23 Morgenthaler, G. W., The theory and application of simulation in Operations research, in *Progress in Operations Research*, edited by R. L. Ackoff, Wiley, New York, 1961.

24 McLeod, J. (Ed.), *Simulation*, McGraw-Hill, New York, 1968.

25 McMillan, C., Jr., and R. Conzales, *Systems Analysis: A Computer Approach to Decision Models*, Revised ed., Richard D. Ervin, Homewood, Illinois, 1965.

26 Mikailov, G. A., *Some Problems in the Theory of the Monte-Carlo Method*, Nauka, Novosibirsk, U.S.S.R., 1974 (in Russian).

27 Mize, J. H. and J. G. Cox, *Essentials of Simulation*, Prentice-Hall, Englewood Cliffs, New Jersey, 1968.

28 Naylor, T. J., J. L. Balintfy, D. S. Burdick, and K. Chu, *Computer Simulation Techniques*, Wiley, New York, 1966.

29 Naylor, T. J., *Computer Simulation Experiments with Models of Economic Systems*, Wiley, New York, 1971.

30 *Proc. Conf. Business Games*, sponsored by the Ford Foundation and School of Business Administration, Tulane University, April 26–28, 1961.

31 Reitman, J., *Computer Simulation Applications: Discrete-Event Simulation for the Synthesis and Analysis of Complex Systems*, Wiley, New York, 1971.

32 Rosenbluth, A. and N. Wiener, The role of models in science, *Philos. Sci.*, **XII**, No. 4, Oct. 1945, 316–321.

33 Smith, J., *Computer Simulation Models*, Hafner, New York, 1968.

34 Sobol, J. M., *Computational Method of Monte Carlo*, Nauka, Moskow, 1973 (in Russian).

35 Shreider, Y. A. (Ed.), *Method of Statistical Testing: Monte Carlo Method*, Elsevier, Amsterdam, 1964.

36 Stephenson, R. E., *Computer Simulation for Engineers*, Harcourt Brace Jovanovitch, New York, 1971.

37 Shubik, M., On gaming and game theory, *Manage. Sci.*, Professional Series, **18**, 1972, 37–53.

38 Shubik, M., *A Preliminary Bibliography on Gaming*, Department of Administrative Sciences, Yale University, New Haven, Connecticut, 1970.

39 Shubik, M., Bibliography on simulation, gaming, artificial intelligence and allied topics, *J. Amer. Stat. Assoc.*, **55**, 1960, 736–751.

40 Tocher, K. D., *The Art of Simulation*, D. Van Nostrand, Princeton, New Jersey, 1963.

41 Yakowitz, S. J., *Computational Probability and Simulation*, Addison-Wesley, Reading, Massachusetts, 1977.

CHAPTER 2

Random Number Generation

2.1 INTRODUCTION

In this chapter we are concerned with methods of generating random numbers on digital computers. The importance of the random numbers in the Monte Carlo method and simulation has been discussed in Chapter 1. The emphasis in this chapter is mainly on the properties of numbers associated with uniform random variates. The term *random number* is used instead of *uniform random number*. Many techniques for generating random numbers have been suggested, tested, and used in recent years. Some of these are based on random phenomena, others on deterministic recurrence procedures.

Initially, *manual methods* were used, including such techniques as coin flipping, dice rolling, card shuffling, and roulette wheels. It was believed that only mechanical (or electronic) devices could yield "truly" random numbers. These methods were too slow for general use, and moreover, sequences generated by them could not be reproduced. Shortly following the advent of the computer it became possible to obtain random numbers with its aid. One method of generating random numbers on a digital computer consists of preparing a table and storing it in the memory of the computer. In 1955 the RAND Corporation published [46] a well known table of a million random digits that may be used in forming such a table. The advantage of this method is reproducibility; its disadvantage is its lack of speed and the risk of exhausting the table.

In view of these difficulties, John von Neumann [56] suggested the *mid-square* method, using the arithmetic operations of a computer. His idea was to take the square of the preceding random number and extract the middle digits; for example, if we are generating four-digit numbers and arrive at 5232, we square it, obtain 27,373,824; the next number consists of the middle four digits—namely, 3738—and the procedure is repeated. This raises a logical question: how can such sequences, defined in a completely deterministic way, be random? The answer is that they are not really random, but only seem so, and are in fact referred to as *pseudorandom* or *quasi-random*; still we call them *random*, with the appropriate reservation. Von Neumann's method likewise proved slow and awkward for statistical analysis; in addition the sequences tend to cyclicity, and once a zero is encountered the sequence terminates.

We say that the random numbers generated by this or any other method are "good" ones if they are uniformly distributed, statistically independent, and reproducible. A good method is, moreover, necessarily fast and requires minimum memory capacity. Since all these properties are rarely, if ever, realized, some compromise must be found. The congruential methods for generating pseudorandom numbers, discussed in the next section, were designed specifically to satisfy as many of these requirements as possible.

2.2 CONGRUENTIAL GENERATORS

The most commonly used present-day method for generating pseudorandom numbers is one that produces a nonrandom sequence of numbers according to some recursive formula based on calculating the residues modulo of some integer m of a linear transformation. It is readily seen from this definition that each term of the sequence is available in advance, before the sequence is actually generated. Although these processes are completely deterministic, it can be shown [31] that the numbers generated by the sequence appear to be uniformly distributed and statistically independent. Congruential methods are based on a fundamental congruence relationship, which may be expressed as [32]

$$X_{i+1} = (aX_i + c)(\bmod m), \qquad i = 1, \ldots, n, \tag{2.2.1}$$

where the *multiplier* a, the *increment* c, and the *modulus* m are nonnegative integers. The modulo notation $(\bmod m)$ means that

$$X_{i+1} = aX_i + c - mk_i, \tag{2.2.2}$$

where $k_i = [(aX_i + c)/m]$ denotes the largest positive integer in $(aX_i + c)/m$.

Given an initial starting value X_0 (also called the *seed*), (2.2.2) yields a congruence relationship (modulo m) for any value i of the sequence $\{X_i\}$.

Generators that produce random numbers according to (2.2.1) are called *mixed congruential generators*. The random numbers on the unit inverval (0, 1) can be obtained by

$$U_i = \frac{X_i}{m}. \tag{2.2.3}$$

Clearly, such a sequence will repeat itself in at most m steps, and will therefore be periodic. For example, let $a = c = X_0 = 3$ and $m = 5$; then the sequence obtained from the recursive formula $X_{i+1} = 3X_i + 3(\text{mod}\, 5)$ is $X_i = 3, 2, 4, 0, 3$.

It follows from (2.2.2) that $X_i < m$ for all i. This inequality means that the period of the generator cannot exceed m, that is, the sequence X_i contains at most m distinct numbers (the period of the generator in the example is 4, while $m = 5$).

Because of the deterministic character of the sequence, the entire sequence recurs as soon as any number is repeated. We say that the sequence "gets into a loop," that is, there is a cycle of numbers that is repeated endlessly. It is shown [31] that all sequences having the form $X_{i+1} = f(X_i)$ "get into a loop." We want, of course, to choose m as large as possible to ensure a sufficiently large sequence of distinct numbers in a cycle.

Let p be the period of the sequence. When p equals its maximum, that is, when $p = m$, we say that the random number generator has a *full period*. It can be shown [31] that the generator defined in (2.2.1) has a full period, m, if and only if:

1 c is *relatively prime* to m, that is, c and m have no common divisor.
2 $a \equiv 1(\text{mod}\, g)$ for every prime factor g of m.
3 $a \equiv 1(\text{mod}\, 4)$ if m is a multiple of 4.

Condition 1 means that the greatest common divisor of c and m is unity. Condition 2 means that $a = g[a/g] + 1$. Let g be a prime factor of m; then denoting $K = [a/g]$, we may write

$$a = 1 + gk. \tag{2.2.4}$$

Condition 3 means that

$$a = 1 + 4[a/4] \tag{2.2.5}$$

if $m/4$ is an integer.

Greenberger [19] showed that the correlation coefficient between X_i and X_{i+1} lies between the values

$$\frac{1}{a} - \left(\frac{6c}{am}\right)\left(1 - \frac{c}{m}\right) \pm \frac{a}{m},$$

and that its upper bound is achieved when $a = m^{1/2}$ irrespective of the value of c.

Since most computers utilize either a binary or a decimal digit system, we select $m = 2^\beta$ or $m = 10^\beta$, respectively where β denotes the word-length of the particular computer. We discuss both cases separately in the following.

For a binary computer we have from condition 1 that $m = 2^\beta$ guarantees a full period. It follows also from (2.2.1) that, for $m = 2^\beta$, the parameter c must be odd and

$$a = 1(\text{mod}\, 4), \tag{2.2.6}$$

which can be achieved by setting

$$a = 2^r + 1, \qquad r \geq 2.$$

It is noted in the literature [25, 35, 44] that good statistical results can be achieved while choosing $m = 2^{35}$, $a = 2^7 + 1$, and $c = 1$.

For a decimal computer $m = 10^\beta$. In order to generate a sequence with a full period, c must be a positive number not divisible by $g = 2$ or $g = 5$, and the multiplier a must satisfy the condition $a \equiv 1(\text{mod}\, 20)$, or alternatively, $a = 10^r + 1$, $r > 1$.

Satisfactory statistical results have been achieved [1] by choosing $a = 101$, $c = 1$, $r \geq 4$. In this case X_0 had little or no effect on the statistical properties of the generated sequences.

The second widely used generator is the *multiplicative generator*

$$X_{i+1} = aX_i(\text{mod}\, m), \tag{2.2.7}$$

which is a particular case of the mixed generator (2.2.1) with $c = 0$.

It can be shown [1, 2, 5, 31] that, generally, a *full* period cannot be achieved here, but a *maximal* period can, provided that X_0 is relatively prime to m and a meets certain congruence conditions.

For a binary computer we again choose $m = 2^\beta$ and it is shown [31] that the maximal period is achieved when $a = 8r \pm 3$. Here r is any positive integer.

The procedure for generating pseudorandom numbers on a binary computer* can be written as:

1 Choose any odd number as a starting value X_0.

2 Choose an integer $a = 8r \pm 3$, where r is any positive integer. Choose a close to $2^{\beta/2}$ (if $\beta = 35$, $a = 2^{17} + 3$ is a good selection).

3 Compute X_1, using fixed point integer arithmetic. This product will consist of 2β bits from which the high-order β bits are discarded, and the low-order β bits represent X_1.

4 Calculate $U_1 = X_1 / 2^\beta$ to obtain a uniformly distributed variable.

*This procedure and the one that follows are reproduced almost verbatim from Ref. 31.

5 Each successive random number X_{i+1} is obtained from the low-order bits of the product aX_i.

For a decimal computer $m = 10^\beta$. It is shown in Ref. 49 that the maximal period is achieved when $a = 200r \pm p$, where r is any positive integer and p is any of the following 16 numbers: (3, 11, 13, 19, 21, 27, 29, 37, 53, 59, 61, 67, 69, 77, 83, 91). The procedure for generating random numbers on a decimal computer can be written as:

1 Choose any odd integer not divisible by 5 as a starting value X_0.
2 Choose an integer $a = 200r \pm p$ for a constant multiplier, where r is any integer and p is any of the values 3, 11, 13, 19, 21, 27, 29, 37, 53, 59, 61, 67, 69, 77, 83, 91. Choose a close to $10^{\beta/2}$. (If $\beta = 10$, $a = 100{,}000 \pm 3$ is a good selection.)
3 Compute aX_0 using fixed point integer arithmetic. This product will consist of 2β digits, from which the high-order β digits are discarded, and the low-order digits are the value of X_1. Integer multiplication instructions automatically discard the high-order β digits.
4 The decimal point must be shifted β digits to the left to convert the random number (which is an integer) into a uniformly distributed variate defined over the unit interval $U_1 = X_1/10^\beta$.
5 Each successive random number X_{i+1} is obtained from the low-order digits of the product aX_i.

Another type of generator in which X_{i+1} depends on more than one of the preceding values is the additive congruential generator [17]

$$X_{i+1} \equiv X_i + X_{i-k} (\operatorname{mod} m), \qquad k = 1, 2, \ldots, i-1. \tag{2.2.8}$$

In the particular case $k = 1$ we obtain the well known Fibonacci sequence, which behaves like sequences produced by the multiplicative congruential method with $a = (1 + \sqrt{5})/2$. Unfortunately, a Fibonacci sequence is not satisfactorily random, but its statistical properties improve as k increases.

RESUME: We have seen that a sequence of pseudorandom numbers produced by a congruential generator is completely defined by the numbers X_0, a, c, and m. In order to obtain satisfactory statistical results our choice must be based on the following six principles*:

1 The number X_0 may be chosen arbitrarily. If the program is run several times and a different source of random numbers is desired each time, set X_0 equal to the last value attained by X on the preceding run, or (if more convenient) set X_0 equal to the current date and time.

*These six principles are reproduced by permission from Knuth [31, pp. 155–156].

2 The number m should be large. It may conveniently be taken as the computer's word length, since this makes the computation of $(aX + c)$ $(\text{mod}\, m)$ quite efficient. The computation of $(aX + c)(\text{mod}\, m)$ must be done exactly, with no roundoff error.

3 If m is a power of 2 (i.e., if a binary computer is being used), pick a so that $a(\text{mod}\, 8) = 5$. If m is a power of 10 (i.e., if a decimal computer is being used), choose a so that $a(\text{mod}\, 200) = 21$. This choice of a, together with the choice of c given below, ensures that the random number generator will produce all m different possible values of X before it starts to repeat.

4 The multiplier a should be larger than $\sqrt{m}$, preferably larger than $m/100$, but smaller than $m - \sqrt{m}$. The best policy is to take some haphazard constant to be the multiplier, such as $a = 3{,}141{,}592{,}621$ (which satisfies both of the conditions in 3).

5 The constant c should be an odd number when m is a power of 2 and, when m is a power of 10, should also not be a multiple of 5.

6 The least significant (right-hand) digits of X are not very random, so decisions based on the number X should always be primarily influenced by the most significant digits. It is generally better to think of X as a random fraction X/m between 0 and 1, that is, to visualize X with a decimal point at its left, than to regard X as a random integer between 0 and $m - 1$. To compute a random integer between 0 and $k - 1$, we would multiply by k and truncate the result.

Finally, we present in this section the IBM System/360 Uniform Random Number Generator, a multiplicative congruential generator that utilizes the full word size, which is equal to 32 bits with 1 bit reserved for algebraic sign. Therefore an obvious choice for m is 2^{31}.

A pure congruential generator $(c = 0)$ with $m = 2^k$ $(k > 0)$ can have a maximum period length of $m/4$. Thus the maximum period length is $2^{31}/4 = 2^{29}$. The period length also depends on the starting value. When the modulus m is prime, the maximum possible period length is $m - 1$. The largest prime less than or equal to 2^{31} is $2^{31} - 1$. Hence, if we choose $m = 2^{31} - 1$, the uniform random number generators will have a maximum period length of $m - 1 = 2^{31} - 2$, which is only the upper bound on the period length. The maximum period length depends on the choice of the multiplier. Note that the conditions ensuring a maximum period length do not necessarily guarantee good statistical properties for the generator, although the choice of the particular multiplier 7^5 does satisfy some known conditions regarding the statistical performance of the generated sequence. The System/360 Generator can be described as follows. Choose any

$X_0 > 0$. For $n > 1$,

$$X_n = 7^5 X_{n-1} (\text{mod } 2^{31} - 1) = 16{,}807 X_{n-1} (\text{mod } 2^{31} - 1).$$

The random numbers are (see (2.2.3)) $U_n = X_n / (2^{31} - 1)$.

The results of the statistical tests of the System/360 Uniform Random Number Generator indicate that it is very satisfactory. Versions of this generator are used in the IBM SL/MATH package, the IBM version of APL, the Naval Postgraduate School random number generator package LLRANDOM, and the International Mathematics and Statistics Library (IMSL) package. The generator is also used in the simulation programming language SIMPL/I. The assembly language subroutines GGL1 and GGL2 of IBM Corporation (1974) also implement this generator, as well as the FORTRAN subroutine GGL.

2.3 STATISTICAL TESTS OF PSEUDORANDOM NUMBERS

In this section we describe some statistical tests for checking independence and uniformity of a sequence of pseudorandom numbers produced by a computer program. As mentioned earlier, a sequence of pseudorandom numbers is completely deterministic, but insofar as it passes the set of statistical tests, it may be treated as one of "truly" random numbers, that is, as a sample from $\mathcal{U}(0, 1)$. Our object in this section is to provide some idea of these tests rather than present rigorous proofs. For a more detailed discussion of this topic the reader is referred to Fishman [11] and Knuth [31].

2.3.1 Chi-Square Goodness-of-Fit Test

The chi-square goodness-of-fit test, proposed by Pearson in 1900, is perhaps the best known of all statistical tests.

Let $X_1, \ldots, X_N$ be a sample drawn from a population with unknown cumulative distribution function (c.d.f.) $F_X(x)$. We wish to test the null hypothesis

$$H_0 : F_X(x) = F_0(x), \qquad \text{for all } x,$$

where $F_0(x)$ is a completely specified c.d.f., against the alternative

$$H_1 : F_X(x) \neq F_0(x), \qquad \text{for some } x.$$

Assume that the N observations have been grouped into k mutually exclusive categories, and denote by N_j and Np_j^0 the observed number of trial outcomes and the expected number for the jth category, $j = 1, \ldots, k$, respectively, when H_0 is true.

The test criterion suggested by Pearson uses the following statistic:

$$Y = \sum_{j=1}^{k} \frac{\left(N_j - Np_j^0\right)^2}{N \cdot p_j^0}, \qquad \sum_{j=1}^{k} N_j = N, \tag{2.3.1}$$

which tends to be small when H_0 is true and large when H_0 is false. The exact distribution of the random variable Y is quite complicated, but for large samples its distribution is approximately chi-square with $k-1$ degrees of freedom [15].

Under the H_0 hypothesis we expect

$$P\left(Y > \chi_{1-\alpha}^2\right) = \alpha, \tag{2.3.2}$$

where α is the significant level, say 0.05 or 0.1; the quantile $\chi_{1-\alpha}^2$ that corresponds to probability $1-\alpha$ is given in the tables of chi-square distribution.

When testing for uniformity we simply divide the interval [0, 1] into k nonoverlapping subintervals of length $1/k$ so that $Np_j^0 = N/k$. In this case we have

$$Y = \frac{k}{N} \sum_{j=1}^{k} \left(N_j - \frac{N}{k}\right)^2, \tag{2.3.3}$$

and (2.3.2) can again be applied for testing random number generators.

To ensure the asymptotical properties of Y it is often recommended in the literature to choose $N > 5k$ and $k > 1000$, where $k = 2^\beta$ and $k = 10^\beta$ for a binary and a decimal computer, respectively.

2.3.2 Kolmogorov-Smirnov Goodness-of-Fit Test

Another test well known in statistical literature is the one proposed by Kolmogorov and developed by Smirnov.

Let $X_1, \ldots, X_N$ again denote a random sample from unknown c.d.f. $F_X(x)$. The *sample cumulative distributive function*, denoted by $F_N(x)$, is defined as

$$\begin{aligned} F_N(x) &= \frac{1}{N}(\text{number of } X_i \text{ less than or equal to } x) \\ &= \frac{1}{N} \sum_{i=1}^{N} I_{(-\infty, x)}(X_i), \end{aligned}$$

where $I_{(-\infty,x)}(X)$ is the indicator random variable (r.v.) that is,

$$I_{(-\infty,x)}(X) = \begin{cases} 1, & \text{if } -\infty < X \leq x \\ 0, & \text{otherwise.} \end{cases} \tag{2.3.4}$$

For fixed x, $F_N(x)$ is itself an r.v., since it is a function of the sample.

Let us show that $F_N(x)$ has the same distribution as the sample mean of a Bernoulli distribution, namely

$$P\left[F_N(x)=\frac{k}{N}\right]=\binom{N}{k}[F_x(x)]^k[1-F_x(x)]^{N-k}. \tag{2.3.5}$$

Denote $V_i = I_{(-\infty,x)}(X_i)$; then V_i has a Bernoulli distribution with parameter $P(V_i=1)=P(X_i \le x)=F_X(x)$. Since $\Sigma_{i=1}^N V_i$ has a binomial distribution with parameters N and $F_X(x)$, and since $F_N(x)=(1/N)\ \Sigma_{i=1}^N V_i$, the result follows immediately.

From (2.3.5) we see that

$$E[F_N(x)]=\sum_{k=0}^{N}\frac{k}{N}\binom{N}{k}[F_X(x)]^k[1-F_X(x)]^{N-k}=F_X(x) \tag{2.3.6}$$

and

$$\operatorname{var} F_N(x)=\frac{1}{N}F_X(x)[1-F_X(x)]. \tag{2.3.7}$$

Equations (2.3.6) and (2.3.7) show that, for fixed x, $F_N(x)$ is an unbiased and consistent estimator of $F_X(x)$ irrespective of the form of $F_X(x)$. Since $F_N(x)$ is the sample mean of random variables $I_{(-\infty,x)}(X_i)$, $i=1,\ldots,N$, it follows from the central-limit theorem that $F_N(x)$ is asymptotically normally distributed with mean $F_X(x)$ and variance $(1/N)F_X(x)[1-F(x)]$. We are interested in estimating $F_X(x)$ for every x (or rather, for a fixed x) and in finding how close $F_N(x)$ is to $F_X(x)$ jointly over all values x.

The result

$$\lim_{N\to\infty} P\left[\sup_{-\infty<x<\infty}|F_N(x)-F_X(x)|>\varepsilon\right]=0 \tag{2.3.8}$$

is known as the *Glivenko-Cantelli theorem*, which states that for every $\varepsilon>0$ the step function $F_N(x)$ converges uniformly to the distribution function $F_X(x)$. Therefore for large N the deviation $|F_N(x)-F_X(x)|$ between the true function $F_X(x)$ and its statistical image $F_N(x)$ should be small for all values of x.

The random quantity

$$D_N=\sup_{-\infty<x<\infty}|F_N(x)-F_X(x)|, \tag{2.3.9}$$

which measures how far $F_N(x)$ deviates from $F_X(x)$ is called the *Kolmogorov-Smirnov one-sample statistic*. Kolmogorov and Smirnov proved that, for any continuous distribution $F_X(x)$,

$$\lim_{N\to\infty} P(\sqrt{N}\,D_N\le x)=\left[1-2\sum_{j=1}^{\infty}(-1)^{j-1}\exp(-2j^2x^2)\right]=H(x). \tag{2.3.10}$$

The function $H(x)$ has been tabulated and the approximation was found to be sufficiently close for practical applications, so long as N exceeds 35.

The c.d.f. $H(x)$ does not depend on the one from which the sample was drawn; that is, the limiting distribution of $\sqrt{N}\ D_N$ is *distribution-free*. This fact allows D_N to be broadly used as a statistic for goodness-of-fit.

For instance, assume that we have the random sample $X_1, \ldots, X_N$ and wish to test $H_0: F_X(x) = F_0(x)$ for all x where $F_0(x)$ is a completely specified c.d.f. (in our case $F_0(x)$ is the uniform distribution in the interval (0, 1)). If H_0 is true, which means that we have a good random number generator, then

$$\sqrt{N}\ D_N = \sqrt{N} \sup_{-\infty < x < \infty} |F_N(x) - F_X(x)| \tag{2.3.11}$$

is approximately distributed as the c.f.d. $H(x)$.

If H_0 is false, which means that we have a bad random number generator, then $F_N(x)$ will tend to be near the true c.d.f. $F_X(x)$ rather than near $F_0(x)$, and consequently $\sup_{-\infty < x < \infty} |F_N(x) - F_0(x)|$ will tend to be large. Hence a reasonable test criterion is to reject H_0 if $\sup_{-\infty < x < \infty} |F_N(x) - F_0(x)|$ is large.

The Kolmogorov-Smirnov goodness-of-fit test with significance level α rejects H_0 if and only if $\sqrt{N}\ D_N > x_{1-\alpha}$ where the quantile $x_{1-\alpha}$ is given in the tables of $H(x)$.

Before we leave the chi-square and Kolmogorov-Smirnov tests, a word is in order on the similarity and difference between them. The similarity lies in the fact that both of them indicate how well a given set of observations (pseudorandom numbers) fits some specified distribution (in our case the uniform distribution); the difference is that the Kolmogorov-Smirnov test applies to continuous (jumpless) c.d.f.'s and the chi-square to distributions consisting exclusively of jumps (since all the observations are divided into k categories). Still the chi-square test may be applied to a continuous $F_X(x)$, provided its domain is divided into k parts and the variables within each part are disregarded. This is essentially what we did earlier when testing whether or not the sequence obtained from the random number comes from the uniform distribution. When applying the chi-square test allowance must be made for its sensitivity to the number of classes and their widths, arbitrarily chosen by the statistician.

Another difference is that chi-square requires grouped data whereas Kolmogorov-Smirnov does not. Therefore when the hypothesized distribution is continuous Kolmogorov-Smirnov allows us to examine the goodness-of-fit for each of the n observations, instead of only for k classes, where $k \leq n$. In this sense Kolmogorov-Smirnov makes more complete use of the available data.

As regards the efficiency of the Kolmogorov-Smirnov and chi-square tests, at present too few theoretical results are available to allow meaningful judgment.

2.3.3 Cramer-von Mises Goodness-of-Fit Test [4]

This test, like the preceding two, belongs to the goodness-of-fit tests and its object is the same as theirs: for a given sample $X_1, \ldots, X_N$ from some unknown c.d.f. we wish to test the null hypothesis

$$H_0 : F_X(x) = F_0(x),$$

where $F_0(x)$ is a completely specified distribution, against the alternative

$$H_1 : F_X(x) \neq F_0(x)$$

for at least one value of x. Denote by $X_{(1)}, \ldots, X_{(N)}$ the order statistic and consider the following test statistic:

$$Y = \frac{1}{12N} + \sum_{i=1}^{N} \left[F_0(X_{(i)}) - \frac{2i-1}{2N} \right]^2. \tag{2.3.12}$$

In other words, the ordinate of $F_0(x)$ is found at each value in the random sample $X_{(i)}$, and from this is subtracted the quantity $(2i-1)/2N$, which is the average just before and just after the jump at $X_{(i)}$—that is, the average of $(i-1)/N$ and i/N. The difference is squared, so that positive differences do not cancel the negative ones, and the results are added together.

The quantities of Y are tabulated by using an asymptotic distribution function of Y as given by Anderson and Darling [2]. The Cramer-von Mises goodness-of-fit test, with significance level α, rejects H_0 if and only if $Y > y_{1-\alpha}$, where the quantity $y_{1-\alpha}$ can be found from the appropriate tables.

2.3.4 Serial Test [31]

The serial test is used to check the degree of randomness between successive numbers in a sequence and represents an extension of the chi-square goodness-of-fit test.

Let $X_1 = (U_1, \ldots, U_k)$, $X_2 = (U_{k+1}, \ldots, U_{2k})$, $\ldots$, $X_N = (U_{(N-1)k+1}, \ldots, U_{Nk})$ be a sequence of N k-tuples. We wish to test the hypothesis that the r.v.'s $X_1, X_2, \ldots, X_N$ are independent and uniformly distributed over the k-dimensional unit hypercube.

Dividing this hypercube into r^k elementary hypercubes, each with volume $1/r^k$, and denoting by $V_{j_1}, \ldots, _{j_k}$ the number of k-tuples falling within the element

$$\left(\frac{j_{i-1}}{r}, \frac{j_i}{r}\right), \qquad i = 1, \ldots, k; \quad j_i = 1, \ldots, r,$$

we have that the statistic

$$Y = \frac{r^k}{N} \sum_{j_1, \ldots, j_k = 1}^{r} \left(V_{j_1, \ldots, j_k} - \frac{N}{r^k} \right)^2, \tag{2.3.13}$$

has an asymptotical chi-square distribution with $r^k - 1$ degrees of freedom. Since there are r^k hypercubes within which X_i may fall, the question of available space arises. If $k = 3$ and $r = 1000$, the serial test requires $1000^3 = 10^7$ counters—a problematic requirement in terms of both storage and search. In these circumstances the test is rarely used for $k > 2$.

2.3.5 The-Up-and-Down Test [43]

For this test the magnitude of each element is compared with that of its immediate predecessor in the given sequence. If the next element is larger, we have a run-up: if smaller, a run-down. We thus observe whether the sequence increases or decreases and for how long. A decision concerning the pseudorandom number generator may then be based on the number and length of the runs.

For example, the following seven-term sequence 0.2 0.4 0.1 0.3 0.6 0.7 0.5 consists of a run-up of length 1, followed by a run-down of length 1, followed by a run-up of length 3, and finally a run-down of length 1, and may be characterized by the binary symbol as 1 0 111 0, where 1 denotes a run-up and 0 a run-down. More generally, suppose there are N terms, say $X_1 < X_2 < \cdots < X_N$ when arranged in order of magnitude; the time-ordered sequence of observations represents a permutation of these N numbers. There are $N!$ permutations, each of them representing a possible set of sample observations. Under the null hypothesis each of these alternatives is equally likely to occur. The test of randomness, using runs-up and runs-down for the sequence $X_1, \ldots, X_N$ of dimension N, is based on the derived sequence of dimension $N - 1$, whose ith element is 0 or 1 depending on whether $X_{i+1} - X_i$, $i = 1, \ldots, N - 1$, is negative or positive. A large number of long runs should not occur in a "truly" random sample. The test rejects the null hypothesis if there are at least r runs of length t or more, where both r and t are determined by the desired significance level.

The means, variances, and covariances of the numbers of runs of length t or more are given in Levene and Wolfowitz [34].

The expected numbers of occurrences of runs in a "truly" random sample are [43]

$$\frac{2N-1}{3} \quad \text{for total runs}$$

$$\frac{N+1}{12} \quad \text{for runs of length 1}$$

$$\frac{11N-14}{12} \quad \text{for runs of length 2}$$

. . . .

$$\frac{2\left[(k^2+3k+1)N-(k^3+3k^2-k-4)\right]}{(k+3)!} \quad \text{for runs of length } k,$$

for $k < N-1$

$$\frac{2}{N!} \quad \text{for runs of length } N-1.$$

Tables of the exact probabilities of at least r runs of the length t or more are available in Olmstead [44] for $n \geq 14$, from which the appropriate critical region can be found.

A test of randomness can also be based on the total number of runs, whether up or down, irrespective of their lengths. The hypothesis of randomness is rejected when the total number of runs is small. Levene [33] has shown that the r.v.

$$Z = \frac{U-(2N-1)/3}{\left[(16N-29)/90\right]^{1/2}} \tag{2.3.14}$$

has a standard normal distribution, so that for large N the test of significance can be readily done.

2.3.6 Gap Test [31]

The gap test is concerned with the randomness of the digits in a sequence of numbers. Let $U_1, \ldots, U_N$ be such a sequence. We say that any subsequence $U_j, U_{j+1}, \ldots, U_{j+r}$ of $r+1$ numbers represents a *gap* of length r if U_j and U_{j+r} lie between α and β $(0 \leq \alpha < \beta \leq 1)$ but U_{j+i}, $i = 1, \ldots, r-1$, does not. For a "true" sequence of random numbers the probability of obtaining a gap of length r is given in Ref. 44 and is equal to

$$P(r) = (0.9)^r(0.1). \tag{2.3.15}$$

A chi-square goodness-of-fit test based on the comparison of the expected and actual numbers of gaps of length r may again be used.

2.3.7 Maximum Test [35]

Let $Y_j^k = \max\{U_{(j-1)k+1}, \ldots, U_{jk}\}$, $j = 1, \ldots, N$, be a sequence of N k-tuples. It is shown in Ref. 35 that, if the sequence $U_1, \ldots, U_{Nk}$ is from $\mathcal{U}(0, 1)$, then $Y_1^k, \ldots, Y_N^k$ is also from $\mathcal{U}(0, 1)$. To check whether or not $U_1, \ldots, U_{Nk}$ is a "true" sequence of random numbers, we can apply the chi-square or the Kolmogorov-Smirnov test to the sequence $\{Y_j^k, j = 1, \ldots, N\}$.

The reader might ask: "How many tests do we need to check the random number generator?" and also "Which of them should we choose?" In fact, more computer time may be spent testing random numbers than generating them.

Another question that arises is: "What should be done with the sequence of numbers if it passes most of the tests but fails one of them?" These questions, as well as many others, must be solved by the statistician.

EXERCISES

1 Consider a sequence

$$X_{i+1} = f(X_i),$$

where $X_1, X_2, \ldots$ are integers, $0 \le X_i < m$, and $0 \le f(X_i) < m$.

(a) Show that the sequence is ultimately periodic, in the sense that there exist numbers λ and μ for which the values $X_0, X_1, \ldots, X_\mu, \ldots, X_{\mu+\lambda-1}$ are distinct, but $X_{n+\lambda} = X_n$ when $n \ge \mu$. Find the maximum and minimum possible values of μ and λ.

(b) Show that there exists an $n > 0$ such that $X_n = X_{2n}$; the smallest such value of n lies in the range $\mu \le n \le \mu + \lambda$, and the value of X_n is unique in the sense that, if $X_i = X_{2i}$ and $X_r = X_{2r}$, then $X_r = X_i$ (hence $r - i$ is a multiple of λ).

From Knuth [31].

2 Prove that the middle-square method using $2n$-digit numbers to the base β has the following disadvantage: if ever a number X, whose most significant n digits are zero, appears, then the succeeding numbers will get smaller and smaller until zero occurs repeatedly. From Knuth [31].

3 A sequence generated as in exercise 1 must begin to repeat after at most m values have been generated. Suppose we generalize the method so that X_{i+1} depends on X_{i-1} as well as on X_i; formally, let $f(x,y)$ be a function such that, if $0 \le x,y < m$, then $0 \le f(x,y) < m$. The sequence is constructed by selecting X_0 and X_1 arbitrarily, and then letting

$$X_{i+1} = f(X_i, X_{i-1}), \qquad \text{for } i > 0.$$

Show that the maximum period conceivably attainable in this case is m^2. From Knuth [31].

4 Given the two conditions that c is odd and $a(\text{mod})4 = 1$, prove that they are necessary and sufficient to guarantee the maximum length period in the sequence

$$X_{i+1} = aX_i - c(\text{mod}\, m)$$

when $m = a^e$, $e \neq 2$. From Knuth [31].

5 Prove that the sequence

$$X_{i+1} = aX_i - c(\text{mod}\, m),$$

with $m = 10^e$, $e > 3$, and c not a multiple of 2 and not a multiple of 5, will have a full period if and only if $a(\text{mod } 20) = 1$. From Knuth [31].

6 Show that the random function

$$S_n(x) = \sum_{i=1}^{n} \frac{I(x - X_i)}{n}, \qquad \text{where } I(t) = \begin{cases} 1, & \text{if } t \geq 0 \\ 0, & \text{if } t < 0 \end{cases}$$

is the empirical distribution function of a sample $X_1, X_2, \ldots, X_n$; this should be done by showing that $S_n(x) = F_n(x)$ for all x.

7 Let $F_n(x)$ be the empirical distribution function for a random sample of size n from $\mathcal{U}(0, 1)$. Define

$$X_n(t) = \sqrt{n}\,[F_n(t) - t]$$

$$Z_n(t) = (t+1)X_n\left(\frac{t}{t+1}\right), \qquad \text{for } 0 \leq t \leq 1.$$

Prove that $\text{var}[X_n(t)] \leq \text{var}[Z_n(t)]$ for all $0 \leq t \leq 1$ and all n.

8 Find the minimum sample size N required such that

$$P(D_N < 0.05) \geq 0.95.$$

9 A random sample of size 10 is obtained:

$X_1 = 0.503$ $X_2 = 0.621$ $X_3 = 0.447$ $X_4 = 0.203$ $X_5 = 0.710$

$X_6 = 0.480$ $X_7 = 0.320$ $X_8 = 0.581$ $X_9 = 0.551$ $X_{10} = 0.386$.

For a level of significance $\alpha = 0.05$ test, the null hypothesis

$$F_X(x) = F_0(x), \qquad \text{for all } x,$$

where $F_0(x)$ is from uniform distribution, that is,

$$F_0(x) = \begin{cases} 0, & \text{if } x < 0 \\ x, & \text{if } 0 \leq x < 1 \\ 1, & \text{if } x \geq 1 \end{cases}$$

using:

(a) The Kolmogorov-Smirnov test.
(b) The Cramer-von Mises test.

REFERENCES

1 Allard, J. L., A. R. Dobell, and T. E. Hull, Mixed congruential random number generators for decimal machines, *J. Assoc. Comp. Mach.*, **10**, 1966 131–141.

2 Anderson, T. W. and D. A. Darling, Asymptotic theory of certain "goodness of fit" criteria based on stochastic processes. *Ann. Math. Stat.*, **23**, 1952, 193–212.

3 Barnett, V. D., The behavior of pseudo-random sequences generated on computers by the multiplicative congruential method, *Math. Comp.*, **16**, 1969, 63–69.

4 Conover, W. J., *Practical Nonparametric Statistics*, Wiley, New York, 1971.

5 Coveyou, R. R., Serial correlation in the generation of pseudo-random numbers, *J. Assoc. Comp. Mach.*, **7**, 1960, 72–74.

6 Coveyou, R. R. and R. D. MacPherson, Fourier analysis of uniform random number generators, *J. Assoc. Comp. Mach.*, **14**, 1967, 100–119.

7 Dieter, U., Pseudo-random numbers: The exact distribution of pairs, *Math. Comp.*, **25**, 1971, 855–883.

8 Dieter, U. and J. Ahrens, An exact determination of serial correlations of pseudo-random numbers, *Numer. Math.*, **17**, 1971, 101–123.

9 Downham, D. Y., The runs up and down test, *Comp. J.*, **12**, 1969, 373–376.

10 Downham, D. Y. and F. D. K. Roberts, Multiplicative congruential pseudorandom number generators, *Comp. J.* , **10** 1967, 74–77.

11 Fishman, G., *Principles of Discrete Event Simulation*, Wiley, New York, 1978.

12 Forsythe, G. E., Generation and testing of random digits, *U.S. Nat. Bur. Stand. Appl. Math. Ser.*, No. 12, pp. 34–5, 1951.

13 Franklin, J. N. Deterministic simulation of random processes, *Math. Comp.*, **17**, 1963, 28–59.

14 Franklin, J. N., Numerical simulation of stationary and non-stationary Gaussian random processes, *Soc. Indust. Appl. Math. Rev.*, **7**, 1965, 68–80.

15 Gibbons, J. D., *Nonparametric Statistical Inference*, McGraw-Hill, Tokio; Kogakusha, 1971.

16 Gorenstein, S., Testing a random number generator, *Comm. Assoc. Comp. Mach.* **10**, 1967, 111–118.

17 Green, B. F., J. E. K. Smith, and L. Klem, Empirical tests of an additive random generator, *J. Assoc. Comp. Mach.*, **6**, 1959, 527–537.

18 Greenberger, M., Notes in a new pseudo-random number generator, *J. Assoc. Comp. Mach.*, **8**, 1961, 163–167.

19 Greenberger, M., An a priori determination of serial correlation in computer generated random numbers, *Math. Comp.*, **15**, 1961, 383–389.

20 Greenberger, M., Method in randomness, *Comm. Assoc. Comp. Mach.*, **8**, 1965, 177–179.

21 Gruenberger, F., Tests of random digits, *Math. Tab. Aids Comp.*, **5**, 1950, 244–245.

22 Gruenberger, F. and A. M. Mark, The d^2 test of random digits, *Math. Tab. Aids Comp.*, **5**, 1951, 109–110.

23 Hammer, P. C., The mid-square method of generating digits, *U.S. Nat. Bur. Stand. Appl. Math. Ser.*, No. 12, p. 33, 1951.

24 Hull, T. E. and A. R. Dobell, Random number generators, *Soc. Indust. Appl. Math. Rev.*, **4**, 1962, 230–254.

25 Hull, T. E. and A. R. Dobell, Mixed congruential random number generators for binary machines, *J. Assoc. Comp. Mach.*, **11**, 1964, 31–40.

26 Hutchinson, D. W., *A New Uniform Pseudo-Random Number Generator*, File 651, Department of Computer Sciences, University of Illinois, Urbana, Illinois, April 27, 1965.

27 Hutchinson, D. W., A new uniform pseudorandom number generator, *Comm. Assoc. Comput. Mach.*, **9**, 1966, 432–433.

28 IBM Corporation, *Random Number Generation and Testing*, Form C20-8011, 1959.

29 IBM Corporation, *General Purpose Simulation System/360 User's Manual*, GH 20-0326, White Plains, New York, January 1970.

30 Jansson, B., *Random Number Generators*, Almquist and Wiskell, Stockholm, 1966.

31 Knuth, D. E., *The Art of Computer Programming*: *Seminumerical Algorithms*, Vol. 2, Addison-Wesley, Reading, Massachusetts, 1969.

32 Lehmer, D. H., Mathematical methods in large-scale computing units, *Ann. Comp. Lab. Harvard Univ.*, **26**, 1951, 141–146.

33 Levene, M., On the power function of tests of randomness based on runs up and down, *Ann. Math. Stat.*, **23**, 1952, 34–56.

34 Levene, M. and T. Wolfowitz, The covariance matrix of runs up and down, *Ann. Math. Stat.*, **15**, 1944, 58–69.

35 MacLaren, M. D. and G. Marsaglia, Uniform random number generators, *J. Assoc. Comp. Mach.*, **12**, 1965, 83–89.

36 Marsaglia, G., Random numbers fall mainly in the planes, *Proc. Nat. Acad. Sci.*, **61**, Sept. 1968, 25–28.

37 Marsaglia, G. The structure of linear congruential sequences in *Applications of Number Theory to Numerical Analysis*, edited by S. K. Zaremba, Academic, New York, 1972.

38 Mood, A. M., F. A. Graybill, and D. C. Boes, *Introduction to the Theory of Statistics*, 3rd ed., McGraw-Hill, New York, 1974.

39 Moore, P. G., A sequential test for randomness, *Biometrika*, **40**, 1953, 111–115.

40 Moshman, J., The generation of pseudo-random numbers on a decimal calculator, *J. Assoc. Comp. Mach.*, **1**, 1954, 88–91.

41 Moshman, J., Random number generation in *Mathematical Methods for Digital Computers*, Vol. 2, edited by A. Ralston and H. S. Wilf, Wiley, New York, 1967, 249–263.

42 Nance, R. and C. Overstreet, Bibliography on random number generation, *Comp. Rev.*, **13**, 1972, 495–508.

43 Naylor, T. et al., *Computer Simulation Techniques*, Wiley, New York, 1966.

44 Olmstead, P. S., Distribution of sample arrangements for runs up and down, *Ann. Math. Stat.*, **17**, 1946, 24–33.

45 Owen, D. B., *Handbook of Statistical Tables*, Addison-Wesley, Reading, Massachusetts, 1962.

46 Page, E. S., Pseudo-random elements for computers, *Appl. Stat.*, **8**, 1959, 124–131.

47 Rand Corporation, *A Million Random Digits with 1000,000 Normal Deviates*, Free Press, Clencoe, Illinois, 1955.

48 Rotenberg, A., A new pseudo-random number generator, *J. Assoc. Comp. Mach.*, **7**, 1960.

49 Taussky, O. and J. Todd, Generation and testing of pseudo-random numbers, in *Symposium on Monte Carlo Methods*, edited by H. A. Meyer, Wiley, New York, 1956, 15–28.

50 Tausworthe, R. S., Random number generated by linear recurrence modulo two, *Math. Comp.*, **19**, 1965, 201–209.

51 Thompson, W. E., ERNIE—A mathematical and statistical analysis, *J. Roy. Stat. Soc., A*, **122**, 1959, 301–333.

52 Tippett, L. H. C., Random sampling numbers, in *Tracts for Computers*, No. XV, Cambridge University Press, New York, 1925.

53 Tocher, K. D., *The Art of Simulation*, English Universities Press, London, 1963.

54 Tootill, J. P. R., W. D. Robinson, and A. G. Adams, The runs up-and-down performance of Tausworthe pseudo-random number generators, *J. Assoc. Comp. Mach.*, **18**, 1971, 381–399.

55 Van Gelder, A., Some new results in pseudo-random number generation, *J. Assoc. Comp. Mach.*, **14**, 1967, 785–792.

56 Von Neumann, J., Various techniques used in connection with random digits, *U.S. Nat. Bur. Stand. Appl. Math. Ser.*, No. 12, 36–38, 1951.

57 Westlake, W. J., A uniform random number generator based on the combination of two congruential generators, *J. Assoc. Comp. Mach.*, **14**, 1967, 337–340.

58 Whittlesey, J., A comparison of the correlational behavior of random number generators, *Comm. Assoc. Comp. Mach.*, **11**, 1968, 641–644.

CHAPTER 3

Random Variate Generation

3.1 INTRODUCTION

In this chapter we consider some procedures for generating random variates (r.v.'s) from different distributions. These procedures are based on the following three methods: inverse transform method, composition method, and acceptance-rejection method, which are described, respectively, in Sections 3.2, 3.3, and 3.4. Some generalizations on von Neumann's acceptance-rejection method are given in Section 3.4.3. Several techniques for generating random vectors are the subject of Section 3.5. Sections 3.6 and 3.7 describe generation of random variates from most widely used continuous and discrete distributions, respectively.

The notations and mode of algorithm presentation are similar to those in Fishman [12] and are used here to provide uniformity with other works in the field of random variate generation.

For convenience we refer to sampling from a particular distribution by placing the name of the distribution of type of random variate before the word generation. For example, exponential generation denotes sampling from an exponential distribution.

For simplicity U is a uniform deviate with probability density function (p.d.f.)

$$f_U(u) = \begin{cases} 1, & 0 \le u \le 1 \\ 0, & \text{otherwise,} \end{cases}$$

V is a standard exponential deviate with p.d.f.

$$f_V(v) = \begin{cases} e^{-v}, & 0 < v < \infty \\ 0, & \text{otherwise,} \end{cases}$$

and Z is a standard normal deviate with p.d.f.

$$f_Z(z) = \frac{1}{\sqrt{2\pi}} e^{-z^2/2}, \qquad -\infty < z < \infty.$$

X usually denotes the random variable with p.d.f. $f_X(x)$ from which we wish to generate a value.

3.2 INVERSE TRANSFORM METHOD

Let X be a random variable with cumulative probability distribution function (c.d.f.) $F_X(x)$. Since $F_X(x)$ is a nondecreasing function, the inverse function $F_X^{-1}(y)$ may be defined for any value of y between 0 and 1 as: $F_X^{-1}(y)$ is the smallest x satisfying $F_X(x) \geq y$, that is,

$$F_X^{-1}(y) = \inf\{x : F_X(x) \geq y\}, \qquad 0 \leq y \leq 1. \tag{3.2.1}$$

Let us prove that if U is uniformly distributed over the interval (0, 1), then (Fig. 3.2.1)

$$X = F_X^{-1}(U) \tag{3.2.2}$$

has cumulative distribution function $F_X(x)$.

The proof is straightforward:

$$P(X \leq x) = P[F_X^{-1}(U) \leq x] = P[U \leq F_X(x)] = F_X(x). \tag{3.2.3}$$

So to get a value, say x, of a random variable X, obtain a value, say u, of a random variable U, compute $F_X^{-1}(u)$, and set it equal to x.

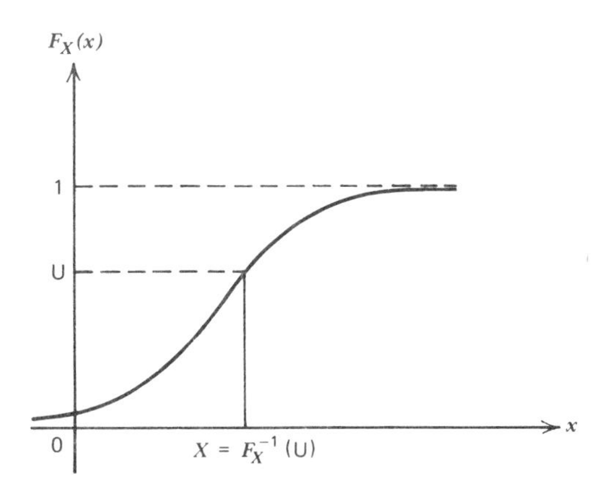

Fig. 3.2.1 Inverse probability integral transformation method.

The algorithm IT-1

1. Generate U from $\mathcal{U}(0, 1)$.
2. $X \leftarrow F_X^{-1}(U)$.
3. Deliver X.

Example 1 Generate an r.v. with p.d.f.

$$f_X(x) = \begin{cases} 2x, & 0 \le x \le 1 \\ 0, & \text{otherwise.} \end{cases} \tag{3.2.4}$$

The c.d.f. is

$$F_X(x) = \begin{cases} 0, & x < 0 \\ \int_0^x 2x\,dx = x^2, & 0 \le x \le 1 \\ 1, & x > 1. \end{cases}$$

Applying (3.2.2), we have

$$X = F_X^{-1}(U) = U^{1/2}, \qquad 0 \le u \le 1.$$

Therefore to generate a variate X from the p.d.f. (3.2.4) we generate from $\mathcal{U}(0, 1)$ and then take a square root from U.

Example 2 Generate an r.v. from the uniform distribution $\mathcal{U}(a, b)$, that is,

$$f_X(x) = \begin{cases} \dfrac{1}{b-a}, & a \le x \le b \\ 0, & \text{otherwise.} \end{cases}$$

The c.d.f. is

$$F_X(x) = \begin{cases} 0, & x < a \\ \dfrac{x-a}{b-a}, & a \le x \le b \\ 1, & x > b. \end{cases}$$

and $X = F_X^{-1}(U) = a + (b - a)U$

Example 3 Let $X_1, \ldots, X_n$ be independent and identically distributed (i.i.d) r.v.'s distributed $F_X(x)$. Define $Y_n = \max(X_1, \ldots, X_n)$ and $Y_1 = \min(X_1, \ldots, X_n)$. Generate Y_n and Y_1. The distributions of Y_n and Y_1 are, respectively [23],

$$F_{Y_n}(y) = [F_X(y)]^n$$

and

$$F_{Y_1}(y) = 1 - [1 - F_X(y)]^n.$$

Applying (3.2.2), we get

$$Y_n = F_X^{-1}(U^{1/n})$$

and

$$Y_1 = F_X^{-1}(1 - U^{1/n}).$$

In the particular case where $X = U$ we have

$$Y_n = U^{1/n}$$

and

$$Y_1 = 1 - U^{1/n}.$$

To apply this method $F_X(x)$ must exist in a form for which the corresponding inverse transform can be found analytically. Distributions in this group are exponential, uniform, Weibull, logistic, and Cauchy. Unfortunately, for many probability distributions it is either impossible or extremely difficult to find the inverse transform, that is, to solve

$$U = \int_{-\infty}^{x} f_X(t)\,dt$$

with respect to x.

Even in the case when F_X^{-1} exists in an explicit form, the inverse transform method is not necessarily the most efficient method for generating random variates.

Example 4 Generate a random variable from the piece-wise constant p.d.f. (Fig. 3.2.2)

$$f_X(x) = \begin{cases} C_i, & x_{i-1} \le x \le x_i;\ i = 1, 2, \ldots, n \\ 0, & \text{otherwise} \end{cases}$$

where $C_i \geq 0$, $a = x_0 < x_1 < \cdots < x_{n-1} < x_n = b$. Denote $P_i = \int_{x_{i-1}}^{x_i} f_X(x)\,dx$, $i = 1, \ldots, n$, and $F_i = \sum_{j=1}^{i} P_j$, $F_0 = 0$; then

$$F_X(x) = \sum_{j=1}^{i-1} P_j + \int_{x_{i-1}}^{x} C_i\,dx = F_{i-1} + C_i(x - x_{i-1}),$$

$$\text{where } i = \max_j \{j : x_{j-1} \leq x\}.$$

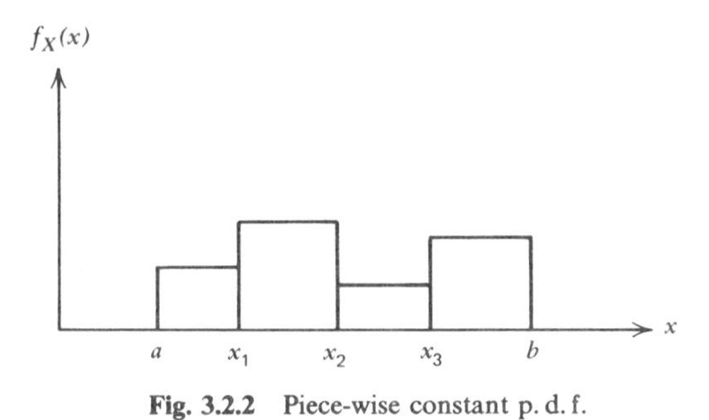

Fig. 3.2.2 Piece-wise constant p. d. f.

Now solving $F_X(X) = U$ with respect to X, we obtain

$$X = x_{i-1} + \frac{U - F_{i-1}}{C_i}, \qquad \text{where } F_{i-1} \leq U < F_i.$$

To carry out the method:

1 Generate U from $\mathcal{U}(0, 1)$.

2 Find i from

$$\sum_{j=1}^{i-1} P_j < U \leq \sum_{j=1}^{i} P_j, \qquad i = 1, \ldots, n.$$

3

$$X \leftarrow x_{i-1} + \frac{U - \sum_{j=1}^{i-1} P_j}{C_i}.$$

4 Deliver X.

Example 5 Let $f_X(x)$ be represented as

$$f_X(x) = \sum_{i=1}^{n} f_i(x), \qquad f_i \geq 0. \tag{3.2.5}$$

Denote

$$P_i = \int_{\infty}^{\infty} f_i(x)\, dx,$$

$$F_0 = 0, \qquad F_i = \sum_{j=1}^{i} P_j, \qquad i = 1, 2, \ldots, n$$

and

$$\phi_i(x) = \int_{-\infty}^{x} f_i(x)\,dx.$$

Let us prove that

$$X = \phi_i^{-1}(U - F_{i-1}), \qquad \text{where } F_{i-1} \le U < F_i \tag{3.2.6}$$

It is easy to see that ϕ_i/P_i is a c.d.f. and that $(U - F_{i-1})/P_i$ is distributed $\mathcal{U}(0, 1)$ if $F_{i-1} < U \le F_i$. Therefore the r.v. $X = F_X^{-1}((U - F_{i-1})/P_i)$ has a p.d.f. f_i/P_i conditional to $F_{i-1} < U \le F_i$. Noticing that $X = \phi_i^{-1}(U - F_{i-1}) = F_X^{-1}((U - F_{i-1})/P_i)$, $f_X(x) = \sum_{i=1}^{n} (f_i/P_i)P_i$, the results follow immediately. To carry out the method:

1 $P_i \leftarrow \int_{-\infty}^{\infty} f_i(x)\,dx$, $i = 1, \ldots, n$.
2 $F_i \leftarrow \sum_{j=1}^{i} P_j$, $i = 1, \ldots, n$.
3 Generate U from $\mathcal{U}(0, 1)$.
4 Find i from $F_{i-1} < U \le F_i$, $F_0 \leftarrow 0$.
5 $\phi_i(x) \leftarrow \int_{-\infty}^{x} f_i(x)\,dx$, $i = 1, \ldots, n$.
6 $X \leftarrow \phi_i^{-1}(U - F_{i-1})$.
7 Deliver X.

As an example, let [22]

$$f_X(x) = \tfrac{3}{8}(1 + x^2), \qquad -1 \le x \le 1.$$

Assume $f_1(x) = \frac{3}{8}$, $f_2(x) = \frac{3}{8}x^2$, $-1 \le x \le 1$. Then $P_1 = \frac{3}{4}$, $P_2 = \frac{1}{4}$, $\phi_1(x) = \frac{3}{8}(x + 1)$, $\phi_2(x) = \frac{1}{8}(x^3 + 1)$, and

$$x = \begin{cases} \frac{8}{3}u - 1 & \text{if } u \le \frac{3}{4} \\ (8u - 1)^{1/3}, & \text{if } u > \frac{3}{4}. \end{cases}$$

3.3 COMPOSITION METHOD

This method is employed by Butler [7]. Refs. 11, 22, 29, and 35 exploit this method to great advantage.

In this technique $f_X(x)$, the p.d.f. of the distribution to be simulated, is expressed as a probability mixture of properly selected density functions.

Mathematically, let $g(x|y)$ be a family of one-parameter density functions, where y is the parameter identifying a unique $g(x)$. If a value of y is drawn from a continuous cumulative function $F_Y(y)$ and then if X is sampled from the $g(x)$ for that chosen y, the density function for X

$$f_X(x) = \int g(x|y)\,dF_Y(y). \tag{3.3.1}$$

If y is an integer parameter, then

$$f_X(x) = \sum_i P_i g(x|y=i) \tag{3.3.2}$$

where

$$\sum_i P_i = 1, \qquad P_i > 0;\ i = 1, 2, \ldots;\ P_i = P(y=i).$$

By using this technique some important distributions can be generated. This technique may be applied for generating complex distributions from simpler distributions that are themselves easily generated by the inverse transform technique or by the acceptance-rejection technique.

Another advantage of this technique is that we can sometimes find a decomposition (3.3.2) that assigns high probabilities P_i to p.d.f.'s from which sampling X is inexpensive and concomitantly assign low probabilities P_i to p.d.f.'s from which sampling X is expensive.

Example 1 Generate an r.v. from

$$f_X(x) = \tfrac{5}{12}\left[1 + (x-1)^4\right], \qquad 0 \le x \le 2,$$

which can be written

$$f_X(x) = \tfrac{5}{6} f_1(x) + \tfrac{1}{6} f_2(x), \qquad 0 \le x \le 2,$$

where

$$f_1(x) = \tfrac{1}{2}, \qquad f_2(x) = \tfrac{5}{2}(x-1)^4, \qquad 0 \le x \le 2.$$

Therefore

$$x = \begin{cases} 2u_2, & \text{if } u_1 < \frac{5}{6} \\ 1 + \sqrt[5]{2u_2}, & \text{if } u_1 \ge \frac{5}{6}. \end{cases}$$

Example 2 (Butler [7]) Generate an r.v. from

$$f_X(x) = n\int_1^\infty y^{-n} e^{-xy}\,dy.$$

Let

$$dF_Y(y) = \frac{n\,dy}{y^{n+1}}, \qquad 1 < y < \infty;\ n \ge 1$$

and $g(x|y) = ye^{-yx}$. A variate is now drawn from distribution $F_Y(y)$. Once this y is selected, it determines a particular $g(x) = ye^{-yx}$. The desired variate from $f_X(x)$ is then simply a variate generated from $g(x) = ye^{-yx}$. To carry out the composition method:

1 Generate U_1, U_2 from $\mathcal{U}(0, 1)$.
2 $Y \leftarrow U_1^{-1/n}$.
3 $X \leftarrow -(1/Y)\ln U_2$.
4 Deliver X.

Example 3 Generate an r.v. from

$$F_X(x) = \sum_{i=1}^{\infty} P_i x^i, \qquad 0 \le x \le 1,$$

where $\sum_{i=1}^{\infty} P_i = 1$, $P_i \ge 0$. The algorithm can be written directly:

1 Generate U_1 and U_2 from $\mathcal{U}(0, 1)$.
2 Find i from $\sum_{k=1}^{i-1} P_k \le U_1 \le \sum_{k=1}^{i} P_k$, where $\sum_{k=1}^{-1} P_k = 0$.
3 $X \leftarrow U_2^{1/i}$.
4 Deliver X.

3.4 ACCEPTANCE-REJECTION METHOD

This method is due to von Neumann [34] and consists of sampling a random variate from an appropriate distribution and subjecting it to a test to determine whether or not it will be acceptable for use.

3.4.1 Single-Variate Case

Let X to be generated from $f_X(x)$, $x \in I$. To carry out the method we represent $f_X(x)$ as

$$f_X(x) = Ch(x)g(x), \tag{3.4.1}$$

where $C \ge 1$, $h(x)$ is also a p.d.f., and $0 < g(x) \le 1$. Then we generate two random variates U and Y from $\mathcal{U}(0, 1)$ and $h(y)$, respectively, and test to see whether or not the inequality $U \le g(Y)$ holds:

1 If the inequality holds, then accept Y as a variate generated from $f_X(x)$.
2 If the inequality is violated, reject the pair U, Y and try again.

The theory behind this method is based on the following.

Theorem 3.4.1 Let X be a random variate distributed with the p.d.f. $f_X(x)$, $x \in I$, which is represented as

$$f_X(x) = Cg(x)h(x),$$

where $C \geq 1$, $0 < g(x) \leq 1$, and $h(x)$ is also a p.d.f. Let U and Y be distributed $\mathcal{U}(0, 1)$ and $h(y)$, respectively. Then

$$f_Y(x|U \leq g(Y)) = f_X(x). \tag{3.4.2}$$

Proof By Bayes' formula

$$f_Y(x|U \leq g(Y)) = \frac{P(U < g(Y)|Y = x)h(x)}{P(U \leq g(Y))}. \tag{3.4.3}$$

We can directly compute

$$P(U \leq g(Y)|Y = x) = P(U \leq g(x)) = g(x) \tag{3.4.4}$$

$$P(U \leq g(Y)) = \int P(U \leq g(Y|Y = x))h(x)\,dx \tag{3.4.5}$$

$$= \int g(x)h(x)\,dx = \int \frac{f_X(x)}{C}\,dx = \frac{1}{C}.$$

Upon substituting (3.4.4) and (3.4.5) into (3.4.3), we obtain

$$f_Y(x|U \leq g(Y)) = Cg(x)h(x) = f_X(x).$$

Q.E.D.

The efficiency of the acceptance-rejection method is determined by the inequality $U \leq g(Y)$ (see (3.4.5)). Since the trials are independent, the probability of success in each trial is $p = 1/C$. The number of trials X before a successful pair U, Y is found has a geometric distribution:

$$P_X(x) = p(1-p)^x, \qquad x = 0, 1, \ldots, \tag{3.4.6}$$

with the expected number of trials equal to C.

Algorithm AR-1 describes the necessary steps.

Algorithm AR-1

1. Generate U from $\mathcal{U}(0, 1)$.
2. Generate Y from the p.d.f. $h(y)$.
3. If $U \leq g(Y)$, deliver Y as the variate generated from $f_X(x)$.
4. Go to step 1.

For this method to be of practical interest the following criteria must be used in selecting $h(x)$:

1 It should be easy to generate an r.v. from $h(x)$.

2 The efficiency of the procedure $1/C$ should be large, that is, C should be close to 1 (which occurs when $h(x)$ is similar to $f_X(x)$ in shape). To illustrate this method (Fig. 3.4.1) let us choose C such that $f_X(x) \leq Ch(x)$ for all $x \in I$, where $C \geq 1$.
The problem then is to find a function $\phi(x) = Ch(x)$ such that $\phi(x) \geq f_X(x)$ and a function $h(x) = \phi(x)/C$, from which the r.v.'s can be easily generated.

The maximum efficiency is achieved when $f_X(x) = \phi(x)$, $\forall x \in I$. In this case $1/C = C = 1$, $g(x) = 1$, and there is no need for the acceptance-rejection method because $h(x) = f_X(x)$ (to generate a variate from $f_X(x)$ is the same as from $h(x)$).

There exist an infinite number of ways to choose $h(x)$ to satisfy (3.4.1). Many papers connected with choosing $h(x)$ have been written, and we consider some of them later.

In the particular case when $\phi(x) = M$, $a \leq x \leq b$, and

$$h(x) = \frac{1}{b-a}, \tag{3.4.7}$$

we obtain from (3.4.1)

$$C = M(b-a) \tag{3.4.8}$$

$$g(x) = \frac{f_X(x)}{M}, \qquad a \leq x \leq b. \tag{3.4.9}$$

Von Neumann [34] first considered the acceptance-rejection method for this particular case, and his algorithm can be described as follows.

Algorithm AR-2

1 Generate U_1 and U_2 from $\mathcal{U}(0, 1)$.

2 $Y \leftarrow a + U_2(b-a)$.

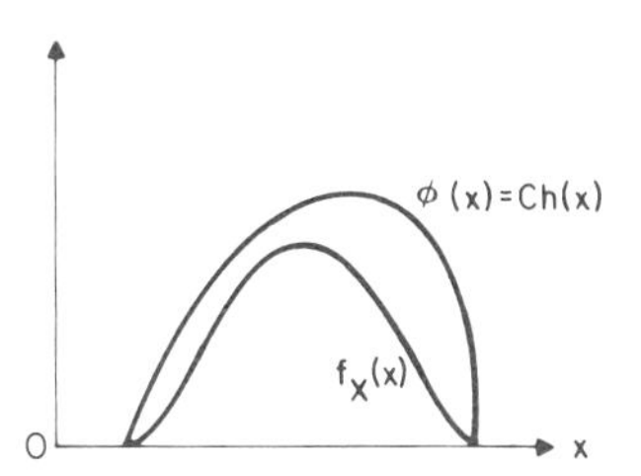

Fig. 3.4.1 Illustration of von Neumann's procedure.

3 If

$$U_1 \le g(Y) = \frac{f_X(Y)}{M} = \frac{f_X[a+(b-a)U_2]}{M},$$

deliver Y as the variate generated from $f_X(x)$.

4 Go to step 1.

We now consider three examples. The first two are related to Algorithm AR-2 and the third to Algorithm AR-1.

Example 1 Generate a random variate from

$$f_X(x) = 3x^2, \qquad 0 \le x \le 1.$$

Here $M = 3$, $a = 0$, and $b = 1$. To apply Algorithm AR-2:

1 Generate two uniform random variates U_1 and U_2 from $\mathcal{U}(0, 1)$.

2 Test to see if $U_1 \le U_2^2$.

3 If the inequality holds, accept U_2 as the variate generated from $f_X(x)$.

4 If the inequality is violated, reject U_1 and U_2 and repeat steps 1 through 3.

Example 2 Generate a random variate from

$$f_X(x) = \frac{2}{\pi R^2}\sqrt{R^2 - x^2}, \qquad -R \le x \le R.$$

Assume $M = 2/\pi R$; then Algorithm AR-2 is as follows:

1 Generate two uniform random variates U_1 and U_2 from $\mathcal{U}(0, 1)$.

2 Compute $Y = (2U_2 - 1)R$.

3 If $U_1 \le f_X(Y)/M$, which is equivalent to $(2U_2 - 1)^2 \le 1 - U_1^2$, then accept $Y = (2U_2 - 1)R$ as the variate generated from $f_X(x)$.

4 If the inequality is violated, reject U_1 and U_2 and repeat steps 1 through 3 again.

The expected number of trials $C = 4/\pi$ and the efficiency $1/C = \pi/4 = 0.785$.

Example 3 Generate a random variate from

$$f_X(x) = \frac{x^{\alpha-1}e^{-x}}{\Gamma(\alpha)}, \qquad 0 < \alpha < 1;\ x \ge 0.$$

To apply the acceptance-rejection method we use the inequality

$$x^{\alpha-1}e^{-x} \le \begin{cases} x^{\alpha-1}, & 0 \le x \le 1 \\ e^{-x}, & x > 1, \end{cases}$$

which is the same as

$$f_X(x) = \frac{x^{\alpha-1}e^{-x}}{\Gamma(\alpha)} \leq \phi(x) = Ch(x) = \begin{cases} \dfrac{x^{\alpha-1}}{\Gamma(\alpha)}, & 0 \leq x \leq 1 \\ \dfrac{e^{-x}}{\Gamma(\alpha)}, & x > 1. \end{cases}$$

Here

$$h(x) = \begin{cases} \dfrac{x^{\alpha-1}}{(1/\alpha) + (1/e)} & 0 \leq x \leq 1 \\ \dfrac{e^{-x}}{(1/e) + (1/\alpha)}, & 1 < x < \infty \end{cases}$$

$$C = \frac{1}{\Gamma(\alpha)}\left(\frac{1}{\alpha} + \frac{1}{e}\right),$$

and we obtain from (3.4.1)

$$g(x) = \begin{cases} e^{-x}, & 0 \leq x \leq 1 \\ x^{\alpha-1}, & 1 < x < \infty. \end{cases}$$

To generate a random variate from $f_X(x)$ we generate two random variates U and Y from $\mathcal{U}(0, 1)$ and $h(y)$, respectively, and then apply the acceptance rule $U \leq g(Y)$.

Note that the random variate Y can be easily generated by the inverse transform method. To apply Algorithm AR-1:

1 Generate U from $\mathcal{U}(0, 1)$.
2 Generate Y from $h(y)$.
3 If

$$U \leq \left\{ \begin{array}{ll} e^{-Y}, & 0 \leq Y \leq 1 \\ Y^{\alpha-1}, & 1 < Y < \infty, \end{array} \right\}$$

deliver Y as the variate generated from $f_X(x)$.
4 Go to step 1.

The probability of success is

$$\frac{1}{C} = \frac{\alpha + e}{\alpha e \Gamma(\alpha)},$$

and the mean number of trials is

$$C = \frac{\alpha e \Gamma(\alpha)}{\alpha + e}.$$

Let us assume that $h(x)$ is known up to the parameter β, that is, $h(x) = h(x, \beta)$. It is shown (see Michailov [22] and Tocher [33]) that the optimal β, which provides minimum to C, is achieved by

$$\min_{\beta} \max_{X} \frac{f_X(x)}{h(x, \beta)}. \tag{3.4.10}$$

3.4.2 Multivariate Case

Theorem 3.4.1 can easily be extended to the multivariate case. The proof of the following theorem will be left to the reader.

Theorem 3.4.2 Let $X = (X_1, \ldots, X_n)$ be a random vector distributed with the p.d.f. $f_X(x)$, $x = (x_1, \ldots, x_n) \in D$, where $D = \{(x_1, \ldots, x_n) : a_i \le x_i \le b_i, i = 1, \ldots, n\}$, and suppose $f_X(x) \le M$. Generate $U_1, \ldots, U_{n+1}$ from $\mathcal{U}(0, 1)$ and define $Y = (Y_1, \ldots, Y_n)$, where $Y_i = a_i + (b_i - a_i)U_i$, $i = 1, \ldots, n$. Then

$$P\left(Y_i \le x_i, i = 1, \ldots, n \,|\, U_{n+1} \le \frac{f_X(Y)}{M}\right)$$
$$= \int_{a_1}^{x_1} \cdots \int_{a_n}^{x_n} f_{X_1, \ldots, X_n}(x_1, \ldots, x_n)\, dx_1, \ldots, dx_n = F_X(x).$$

We can see that this theorem is an extension of von Neumann's method described in Algorithm AR-2 for the multivariate case.

Example 4 Generate a random vector uniformly distributed over the complex region G (Fig. 3.4.2). The algorithm is straightforward.

1 Generate a random vector Y uniformly distributed in Ω where Ω is a nice region (multidimensional rectangular, hypersphere, hyperellipsoid, etc.).

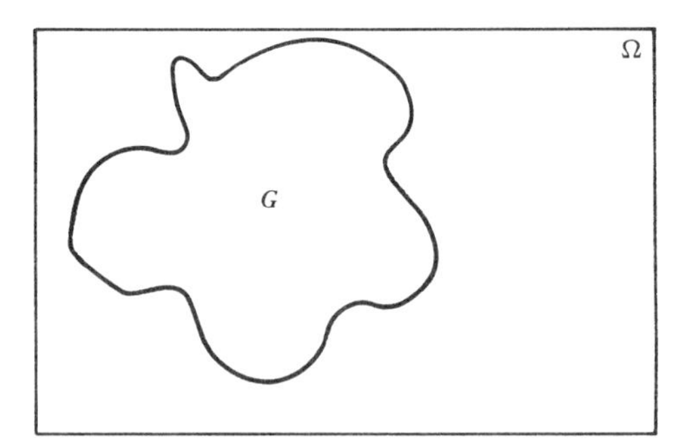

Fig. 3.4.2 Generating a random vector uniformly distributed over a complex area.

2 If $Y \in G$, accept Y as a variate uniformly distributed in G.
3 Go to step 1.

Example 5 Generate a random vector uniformly distributed on the surface of an n-dimensional unit sphere.

To generate a random vector uniformly distributed on the surface of an n-dimensional unit sphere, we simulate a random vector uniformly distributed in the n-dimensional hypercube $\{-1 \leq x_i \leq 1\}_{i=1}^{n}$ and then accept or reject the sample $(X_1, \ldots, X_n)$, depending on whether the point $(X_1, \ldots, X_n)$ is inside or outside the n-dimensional sphere.

The algorithm is as follows:

1 Generate $U_1, \ldots, U_n$ from $\mathfrak{U}(0, 1)$.
2 $X_1 \leftarrow 1 - 2U_1, \ldots, X_n \leftarrow 1 - 2U_n$, and $Y^2 \leftarrow \sum_{i=1}^{n} X_i^2$.
3 If $Y^2 < 1$, accept $Z = (Z_1, \ldots, Z_n)$, where $Z_i = (X_i / Y)$, $i = 1, \ldots, n$, as the desired vector.
4 Go to step 1.

The efficiency of the method is equal to the ratio

$$\frac{1}{C} = \frac{\text{volume of the sphere}}{\text{volume of the hypercube}} = \frac{1}{n2^{n-1}} \frac{\pi^{n/2}}{\Gamma(n/2)}.$$

For even n $(n = 2m)$

$$\frac{1}{C} = \frac{\pi^m}{m!2^{2m}} = \frac{1}{m!}\left(\frac{\pi}{2}\right)^m 2^{-m}$$

and

$$\lim_{m \to \infty} \frac{1}{C} = 0.$$

In other words, for n big enough the acceptance-rejection method is inefficient.

Remark To generate a random vector uniformly distributed inside an n-dimensional unit sphere, we have to rewrite only step 3 in the last algorithm as follows:

3 If $Y^2 \leq 1$, accept $Y = (Y_1, \ldots, Y_n)$ as the desired vector.

3.4.3 Generalization of von Neumann's Method

There are various modifications and generalizations of von Neumann's method [10, 29]. For simplicity consider the single variate case.

Consider a random vector $Y = (Y_1, Y_2)$ distributed $h_{Y_1Y_2}(y_1, y_2)$, $-\infty < y_1 < \infty$, $y_2 \in [0, M]$ and let $T(x)$ be an arbitrary continuous function such that $\sup_x T(x) = M$. Similarly to (3.4.2) let us find

$$f_{Y_1}(x \mid Y_2 \le T(Y_1))$$

which we denote $f_X(x)$. By Bayes' formula

$$\Pr(Y_1 \le x \mid Y_2 \le T(Y_1)) = \frac{P\{Y_1 \le x, Y_2 \le T(Y_1)\}}{P\{Y_2 \le T(Y_1)\}}$$

$$= \frac{\int_{-\infty}^{x} dy_1 \int_{-\infty}^{T(y_1)} h_{Y_1,Y_2}(y_1, y_2)\, dy_2}{\int_{y_1} dy_1 \int_{-\infty}^{T(y_1)} h_{Y_1,Y_2}(y_1, y_2)\, dy_2} = F_X(x). \tag{3.4.11}$$

Differentiating $F_X(x)$ with respect to x, we obtain

$$f_X(x) = f_{Y_1}(x \mid Y_2 \le T(Y_1))$$

$$= \frac{\int_{-\infty}^{T(x)} h_{Y_1Y_2}(x, y_2)\, dy_2}{\int_{y_1} dy_1 \int_{-\infty}^{T(y_1)} h_{Y_1Y_2}(y_1, y_2)\, dy_2}. \tag{3.4.12}$$

Theoretically, (3.4.12) offers an infinite number of possibilities for choosing h and T so as to define a proper $f_X(x)$. But, practically, this formula has no direct application for generating r.v.'s from $f_X(x)$.

Let Y_1 and Y_2 be independent. Consider some particular cases, as follows.

CASE 1 Let $h_{Y_1Y_2}(y_1, y_2) = h_{Y_1}(y_1) h_{Y_2}(y_2)$. Then

$$f_X(x) = \frac{h_{Y_1}(x) \int_{-\infty}^{T(x)} h_{Y_2}(y_2)\, dy_2}{\int_{y_1} h_{Y_1}(y_1)\, dy_1 \int_{-\infty}^{T(y_1)} h_{Y_2}(y_2)\, dy_2} \tag{3.4.13}$$

$$= \frac{h_{Y_1}(x) H_{Y_2}(T(x))}{\int_{y_1} h_{Y_1}(y_1) H_{Y_2}(T(y_1))\, dy_1},$$

where $H_Y(y)$ is the c.d.f. of Y.

The last formula can be written as

$$f_X(x) = C h_{Y_1}(x) H_{Y_2}(T(x)) \tag{3.4.14}$$

where

$$C^{-1} = \int_{y_1} h_{Y_1}(y_1) H_{Y_2}(T(y_1))\, dy_1 \tag{3.4.15}$$

is the efficiency of the method. Thus if Y_1 and Y_2 are independent and if $f_X(x)$ can be represented as (3.4.14), we have

$$f_{Y_1}(x \mid Y_2 \le T(Y_1)) = f_X(x).$$

We can see that (3.4.14) is similar to (3.4.1). When $g(x) = H_{Y_2}(T(x))$ both (3.4.1) and (3.4.14) coincide. In the particular case when $T(x) = x$ we obtain

$$f_X(x) = C h_{Y_1}(x) H_{Y_2}(x). \tag{3.4.16}$$

Algorithm AR-3 describes the acceptance-rejection method for case 1.

Algorithm AR-3

1 Generate Y_1 from $h_{y_1}(y)$.
2 Generate Y_2 from $h_{y_2}(y)$.
3 If $Y_2 \le T(Y_1)$, deliver Y_1.
4 Go to step 1.

Example 6 Generate a random variate from beta distribution

$$f_X(x) = \frac{\Gamma(\alpha+\beta)}{\Gamma(\alpha)\Gamma(\beta)} x^{\alpha-1}(1-x)^{\beta-1}, \qquad 0 \le x \le 1, \alpha > 0, \beta > 0. \tag{3.4.17}$$

Let us use (3.4.16), assuming

$$h_{Y_1}(x) = \beta(1-x)^{\beta-1}, \qquad 0 \le x \le 1 \tag{3.4.18}$$

$$H_{Y_2}(x) = x^{\alpha-1}, \qquad 0 \le x \le 1 \tag{3.4.19}$$

$$C = \frac{1}{\beta} \frac{\Gamma(\alpha+\beta)}{\Gamma(\alpha)\Gamma(\beta)}. \tag{3.4.20}$$

By the inverse transform method we have

$$Y_1 = 1 - U_1^{1/\beta}, \qquad Y_2 = U_2^{1/(\alpha-1)}, \tag{3.4.21}$$

and Algorithm AR-3 is as follows:

1 Generate U_1 and U_2 from $\mathcal{U}(0, 1)$.
2 $Y_1 \leftarrow 1 - U_1^{1/\beta}$.
3 $Y_2 \leftarrow U_2^{1/(\alpha-1)}$.
4 If $Y_2 \le Y_1$, deliver Y_1.
5 Go to step 1.

Example 7 Consider again the problem of generating a random variate from beta distribution (3.4.17). Let us make use of (3.4.14), assuming

$$h_{Y_1}(x) = \alpha x^{\alpha-1}, \qquad 0 \le x \le 1 \tag{3.4.22}$$

$$H_{Y_2}(T(x)) = (1-x)^{\beta-1}, \qquad 0 \le x \le 1 \tag{3.4.23}$$

$$C = \frac{1}{\alpha}\frac{\Gamma(\alpha+\beta)}{\Gamma(\alpha)\Gamma(\beta)}. \tag{3.4.24}$$

Here

$$T(x) = 1 - x. \tag{3.4.25}$$

By the inverse transform method $Y_1 = U_1^{1/\alpha}$, $Y_2 = U_2^{1/(\beta-1)}$, and Algorithm AR-3 is as follows:

1 Generate U_1 and U_2 from $\mathcal{U}(0, 1)$.
2 $Y_1 \leftarrow U_1^{1/\alpha}$.
3 $Y_2 \leftarrow U_2^{1/(\beta-1)}$.
4 If $Y_2 \le 1 - Y_1$, deliver Y_1.
5 Go to step 1.

Remark If $f_X(x)$ can be represented as $f_X(x) = Ch_{Y_1}(x)[1 - H_{Y_2}(T(x))]$, then it is easy to see that Algorithm AR-3 can be written as follows.

Algorithm AR-3′

1 Generate Y_1 from $h_{Y_1}(y)$.
2 Generate Y_2 from $h_{Y_2}(y)$.
3 If $Y_2 \ge T(Y_1)$, deliver Y_1.
4 Go to step 1.

CASE 2 Let $0 \le T(x) \le M$ and let Y_2 be from $\mathcal{U}(0, M)$, that is,

$$h_{Y_2}(y_2) = \begin{cases} \dfrac{1}{M}, & 0 \le y_2 \le M \\ 0, & \text{otherwise.} \end{cases} \tag{3.4.26}$$

Then it follows directly from (3.4.13) that

$$f_X(x) = \frac{h_{Y_1}(x)T(x)}{\int_{y_1} h_{Y_1}(y_1)T(y_1)\,dy_1} = C_1 h_{Y_1}(x)T(x), \tag{3.4.27}$$

where

$$C_1^{-1} = \int_{y_1} h_{Y_1}(y_1)T(y_1)\,dy_1. \tag{3.4.28}$$

The efficiency of the method is

$$C^{-1} = P(Y_2 \le T(Y_1)) = \int h_{Y_1}(y_1)\, dy_1 \int_0^{T(y_1)} h_{Y_2}(y_2)\, dy_2$$

$$= \int_{y_1} h_{Y_1}(y_1)\, dy_1 \frac{T(y_1)}{M} = \frac{1}{M} \int_{y_1} h_{Y_1}(y_1) T(y_1)\, dy_1 = \frac{1}{C_1 M}. \tag{3.4.29}$$

Substituting $C_1 = C/M$ in (3.4.1) and denoting $g(x) = T(x)/M$, we obtain

$$f_X(x) = C h_{Y_1}(x) g(x), \tag{3.4.30}$$

which is exactly (3.4.1). So case 2 corresponds to Algorithm AR-1.

Example 8 Consider again the problem of generating a random variate from beta distribution (3.4.17), representing $f_X(x)$ as in (3.4.30), that is, applying Algorithm AR-1 and taking into account that

$$g(x) = H_{Y_2}(x) = x^{\alpha-1} \tag{3.4.31}$$

and

$$g(x) = H_{Y_2}(T(x)) = (1-x)^{\beta-1}, \tag{3.4.32}$$

respectively, for both examples 6 and 7; Algorithm AR-1 for example 6 (see (3.4.17) through (3.4.21)) can be written as:

1. Generate U_1 and U_2 from $\mathcal{U}(0, 1)$.
2. $Y \leftarrow 1 - U_1^{1/\beta}$.
3. If $U_2 \le Y^{\alpha-1}$, deliver Y.
4. Go to step 1.

Similarly, for example 7 Algorithm AR-1 can be written as:

1. Generate U_1 and U_2 from $\mathcal{U}(0, 1)$.
2. $Y \leftarrow U_1^{1/\alpha}$.
3. If $U_2 \le (1-Y)^{\beta-1}$, deliver Y.
4. Go to step 1.

CASE 3 Let $a \le x \le b$, $0 \le T(x) \le M$, and let Y_1 and Y_2 be independent r.v.'s distributed $\mathcal{U}(a, b)$ and $\mathcal{U}(0, M)$, respectively. We immediately obtain from (3.4.14)

$$f_X(x) = T(x). \tag{3.4.33}$$

Rewriting $f_X(x)$ in the standard way (3.4.1)

$$f_X(x) = C h(x) g(x),$$

we have

$$C = M(b-a) \tag{3.4.34}$$

$$h(x) = \begin{cases} \dfrac{1}{b-a}, & a \leq x \leq b \\ 0, & \text{otherwise} \end{cases} \tag{3.4.35}$$

$$0 \leq g(x) = \frac{T(x)}{M} \leq 1. \tag{3.4.36}$$

Therefore case 3 corresponds to Algorithm AR-2.

We can easily see that Algorithm AR-3 generalizes both Algorithms AR-1 and *AR-2* in the sense that, when $h_{Y_2}(x)$ is distributed uniformly, we obtain Algorithm AR-1, and when both $h_{Y_1}(x)$ and $h_{Y_2}(x)$ are distributed uniformly, we obtain Algorithm AR-2. But (3.4.1) generalizes (3.4.14) in the sense that the c.d.f. $H_{Y_2}(T(x))$ is a particular case of $g(x)$, $0 \leq g(x) \leq 1$.

RESUME: Formula (3.4.1) generalizes 3.4.14. In the particular case when $g(x)$ can be represented as a c.d.f. $H_{Y_2}(T(x))$ from which a random variate Y_2 can be easily generated, Algorithm AR-3 generalizes Algorithm AR-1 and as a rule saves computation (CPU) time.

Formula (3.4.14) can be extended easily for the multivariate case.

Theorem 3.4.3 Let $Y = (Y_1, \dots, Y_n)$ be a random vector with p.d.f. $h_Y(x)$, $x = (x_1, \dots, x_n)$, and let W be a random variable with p.d.f. $h_W(w)$, $w \in [0, M]$. Let $T(x)$ be an arbitrary continuous function such that $\sup_x T(x) = M$. Then

$$f_{Y_1, \dots, Y_n}(x_1, \dots, x_n | W \leq T(Y)) = C h_Y(x) H_W(T(x)), \tag{3.4.37}$$

where

$$C^{-1} = \int h_Y(y) H_W(T(y))\, dy, \qquad y = (y_1, \dots, y_n). \tag{3.4.38}$$

The proof of this theorem is left for the reader.

3.4.4 Forsythe's Method

Forsythe's method is a rejection technique for sampling from a continuous distribution. The original idea is attributed to von Neumann [34]. Forsythe [15] described the method explicitly. Other descriptions are given by Ahrens and Dieter [2] and Fishman [12] with an application to different distributions. Our nomenclature follows that of Forsythe.

Suppose we wish to generate a random variable X from any p.d.f. of the form

$$f_X(x) = Ce^{-h(x)}, \qquad x \geq 0,$$

where

$$C = \left[\int_0^\infty e^{-h(x)}\, dx \right]^{-1} \tag{3.4.39}$$

and $h(x)$ is an increasing function of x over the range $[0, \infty]$. In the first stage of the method an interval is selected for x, and in the second stage the value of x is determined within the interval by a rejection.

For each $k = 1, 2, \ldots, K$ (K is defined below) pick g_k as large as possible subject to the constraints

$$h(g_k) - h(g_{k-1}) \leq 1, \qquad g_0 = 0. \tag{3.4.40}$$

Next compute

$$r_k = \int_0^{g_k} f_X(x)\, dx, \qquad k = 1, \ldots, K. \tag{3.4.41}$$

Here the number of intervals, K, is chosen as the least index such that r_k exceeds the largest number less than one that can be represented in a computer. (K may be chosen smaller if we set $r_k = 1$, and if we are willing to truncate the generated variable by reducing any value above g_k to the interval $[g_{k-1}, g_k)$. Finally, compute

$$d_k = g_k - g_{k-1}, \qquad k = 1, \ldots, K \tag{3.4.42}$$

and the function

$$G_k(x) = h(g_{k-1} + x) - h(g_{k-1}) \leq h(g_k) - h(g_{k-1}) \leq 1, \qquad 0 \leq x \leq d_k. \tag{3.4.43}$$

Now we present the algorithm. Steps 1 to 3 determine which interval $[g_{k-1}, g_k)$ the variable y will belong to. Steps 4 to 8 determine the value of y within that interval.

Algorithm F-1

1 Set $k \leftarrow 1$. Generate U from $\mathfrak{U}(0, 1)$.
2 If $U \leq r_k$, go to step 4 (the kth interval is selected).
3 If $U > r_k$, set $k \leftarrow k + 1$ and go back to step 1.
4 Generate another uniform deviate U and set $X = Ud_k$.
5 Set $t \leftarrow G_k(X)$.
6 Generate $U_1, U_2, \ldots, U_N$ where N is such that $t > U_1, t > U_2, \ldots, t > U_{N-1}$, but $t \leq U_N$ ($N = 1$ if $t \leq U_1$).
7 If N is even, reject X and return to step 1.
8 If N is odd, accept X.

The proof of the method is given in Forsythe [15] (see also Fishman [12, p. 400] and Ahrens and Dieter [2]).

Example 1 Exponential Distribution For $h(x) = x$, $f_X(x)$ is a standard exponential distribution and we have $g_k = k$, $d_k = 1$, and $r_k = 1 - e^{-k}$ for all k.

Example 2 Normal Distribution For $h(x) = x^2/2$, $f_X(x)$ corresponds to the positive half of the normal distribution and we have $g_0 = 0$, $g_1 = 1$, $g_k = (2k-1)^{1/2}$, $k \geq 2$. $d_1 = 1, d_2 = 3^{1/2} - 1, \ldots, d_k = (2k-1)^{1/2} - (2k-3)^{1/2}$, and $k \geq 2$. Also

$$G_k(x) = \frac{x^2}{2} + g_{k-1}x, \qquad k \geq 1.$$

The advantage of this method is that it provides a rejection technique for densities of the form (3.4.39) without the need for exponentiation. If $G_k(x)$ is easier to calculate than $e^{-h(x)}$, as it is for many members of the exponential family, the method can yield fast algorithms.

An important feature of the method is that it does not specify a unique algorithm, but rather a family of algorithms, subject to (3.4.40) being satisfied. The interval widths d_k can be chosen at will.

A disadvantage of the method is that it requires tables of the constants g_k, d_k, and r_k.

3.5 SIMULATION OF RANDOM VECTORS

3.5.1 Inverse Transform Method

Let $X = (X_1, \ldots, X_n)$ be a random vector to be generated from the given c.d.f. $F_X(x)$. We distinguish the following two cases.

CASE 1 The random variables $X_1, \ldots, X_n$ are independent. In this case the joint p.d.f. is

$$f_{X_1,\ldots,X_n}(x_1, \ldots, x_n) = \prod_{i=1}^{n} f_i(x_i), \tag{3.5.1}$$

where $f_i(x_i)$ is the marginal p.d.f. of the random variable X_i. It is easy to see that, in order to generate the random vector $X = (X_1, \ldots, X_n)$ from c.d.f. $F_X(x)$, we can apply the inverse-transform method

$$X_i = F_i^{-1}(U_i), \qquad i = 1, \ldots, n \tag{3.5.2}$$

to each variable separately.

Example 1 Let X_i be independent r.v.'s with the p.d.f.

$$f_i(x_i) = \begin{cases} \dfrac{1}{b_i - a_i}, & a_i \le x_i \le b_i,\ i = 1, \dots, n \\ 0, & \text{otherwise.} \end{cases}$$

To generate the random vector $X = (X_1, \dots, X_n)$ with the joint p.d.f.

$$f_{X_1, \dots, X_n}(x_1, \dots, x_n) = \begin{cases} \dfrac{1}{\prod\limits_{i=1}^{n} (b_i - a_i)}, & (x_1, \dots, x_n) \in D \\ 0, & \text{otherwise} \end{cases}$$

where $D = \{x_1, \dots, x_n) : a_i \le x_i \le b_i,\ i = 1, \dots, n\}$, we apply the inverse transform formula (3.5.2) and get $X_i = a_i + (b_i - a_i)U_i$, $i = 1, \dots, n$.

CASE 2 The random variables are dependent. In this case the joint c.d.f. is

$$f_{X_1, \dots, X_n}(x_1, \dots, x_n) = f_1(x_1) f_2(x_2 | x_1) \cdots f_n(x_n | x_1, \dots, x_{n-1}), \tag{3.5.3}$$

where $f_1(x_1)$ is the marginal p.d.f of X_1 and $f_k(x_k | x_1, \dots, x_{k-1})$ is the conditional p.d.f. of X_k given $X_1 = x_1, X_2 = x_2, \dots, X_{k-1} = x_{k-1}$.

Theorem 3.5.1 Let $U_1, \dots, U_n$ be independent uniformly distributed random variates from $\mathfrak{U}(0, 1)$. Then the vector $X = (X_1, \dots, X_n)$, which is obtained from the solution of the following system of equations

$$\begin{cases} F_1(X_1) = U_1 \\ F_2(X_2 | X_1) = U_2 \\ \vdots \qquad \vdots \\ F_n(X_n | X_1, \dots, X_{n-1}) = U_n, \end{cases} \tag{3.5.4}$$

is distributed according to $F_X(x)$. The proof of this theorem is similar to the proof of (3.2.2) and is left for the reader.

The procedure for generating random variates from (3.5.3) contains only two steps.

1 Generate n independent uniformly distributed variates from $\mathfrak{U}(0, 1)$.
2 Solve the system of equations (3.5.4) with respect to $X = (X_1, \dots, X_n)$.

There are $n!$ ordered combinations (possibilities) to represent the variables $X_1, \dots, X_n$ in vector X, and therefore $n!$ possibilities to generate X while solving (3.5.4). Thus for $n = 2$ and $n! = 2$ we can write $f_{X_1, X_2}(x_1, x_2)$

in two different ways:

1 $f_{X_1,X_2}(x_1,x_2)=f_1(x_1)f_2(x_2|x_1)$ (3.5.5)

2 $f_{X_1,X_2}(x_1,x_2)=f_2(x_2)f_1(x_1|x_2).$ (3.5.6)

The efficiency of simulation will generally depend on the order in which the random variates X_i, $i=1,\ldots,n$, are taken while forming the random vector X.

The following example, which is taken from Sobol [29], uses both formulas (3.5.5) and (3.5.6) for generating a two-variate random vector $X=(X_1,X_2)$ and shows the difference in their efficiency.

Example 1

$$f_{X_1,X_2}(x_1,x_2)=\begin{cases}6x_1, & \text{if } x_1+x_2\le 1,\ x_1\ge 0,\ x_2\ge 0\\ 0, & \text{otherwise.}\end{cases}$$

CASE 1

$$f_{X_1,X_2}(x_1,x_2)=f_1(x_1)f_2(x_2|x_1).$$

The marginal p.d.f. of the r.v. X_1 is

$$f_1(x_1)=\int_0^{1-x_1} f_{X_1X_2}(x_1,x_2)\,dx_2=6x_1(1-x_1),\qquad 0\le x_1\le 1.$$

The conditional p.d.f. of the r.v. X_2, given $X_1=x_1$, is

$$f_2(x_2|x_1)=\frac{f(x_1,x_2)}{f_1(x_1)}=\frac{1}{1-x_1},\qquad 0\le x_2\le 1-x_1.$$

The correspondent marginal and conditional distribution functions are, respectively,

$$F_1(x_1)=\int_0^{x_1} f_1(x_1)\,dx_1=3x_1^2-2x_1^3,\qquad 0\le x_1\le 1$$

$$F_2(x_2|x_1)=\int_0^{x_2} f_2(x_2|x_1)\,dx_2=x_2(1-x_1)^{-1},\qquad 0\le x_2\le 1-x_1,$$

and the system (3.5.4) is

$$\begin{cases}3X_1^2-2X_1^3=U_1\\ X_2(1-X_1)^{-1}=U_2.\end{cases}$$

CASE 2

$$f_{X_1,X_2}(x_1,x_2)=f_2(x_2)f_1(x_1|x_2).$$

The marginal and conditional p.d.f.'s are, respectively,

$$f_2(x_2) = \int_0^{1-x_2} f(x_1, x_2)\,dx_1 = 3(1-x_2)^2, \qquad 0 \le x_2 \le 1$$

$$f_1(x_1|x_2) = \frac{f(x_1, x_2)}{f_2(x_2)} = 2x_1(1-x_2)^{-2}, \qquad 0 \le x_1 \le 1 - x_2.$$

The corresponding marginal and conditional distribution functions are

$$F_2(x_2) = \int_0^{x_2} f_2(x_2)\,dx_2 = 1 - (1-x_2)^3, \qquad 0 \le x_2 \le 1$$

$$F_1(x_1|x_2) = \int_0^{x_1} f_1(x_1|x_2)\,dx_1 = x_1^2(1-x_2)^{-2}, \qquad 0 \le x_1 \le 1 - x_2$$

and the system (3.5.4) is

$$\begin{cases} 1 - (1-X_2)^3 = U_1 \\ X_1^2(1-X_2)^{-2} = U_2 \end{cases}.$$

Inasmuch as $1 - U$ is distributed in the same way as U, the last system can be written

$$\begin{cases} (1-X_2)^3 = U_1 \\ X_1^2 = U_2(1-X_2)^2. \end{cases}$$

Comparing both cases, we can see that the first system is rather difficult to solve (we would have to solve cubic and quadratic equations, respectively), while the second system has a trivial solution

$$X_2 = 1 - U_1^{1/3}$$
$$X_1 = U_1^{1/3} U_2^{1/2}.$$

Unfortunately, there is no way to find a priori the optimal order of representing the variates in the vector to minimize the CPU time.

Remark For independent r.v.'s the efficiency of simulation does not depend on the order in which the r.v.'s are taken in forming the random vector X.

An alternative method for generating random vectors is the acceptance-rejection method based on Theorem 3.4.3.

3.5.2 Multivariate Transformation Method

This method can sometimes be useful for generating both random variables and random vectors.

Suppose that we are given the joint p.d.f. $f_{X_1,\ldots,X_n}(x_1,\ldots,x_n)$ of the n-dimensional continuous random variable $(X_1,\ldots,X_n)$. Let

$$\kappa = \left\{(x_1,\ldots,x_n): f_{X_1,\ldots,X_n}(x_1,\ldots,x_n) > 0\right\}. \tag{3.5.7}$$

Again assume that the joint density of the random variables $Y_1 = g_1(X_1,\ldots,X_n),\ldots, Y_k = g_k(X_1,\ldots,X_n)$ is desired, where k is an integer satisfying $1 \leq k \leq n$. If $k < n$, we introduce additional new random variables $Y_{k+1} = g_{k+1}(X_1,\ldots,X_n),\ldots, Y_n = g_n(X_1,\ldots,X_n)$ for judiciously selected functions $g_{k+1},\ldots,g_n$; then we find the joint distribution of $Y_1,\ldots,Y_n$; finally, we find the desired marginal distribution of $Y_1,\ldots,Y_k$ from the joint distribution of $Y_1,\ldots,Y_n$. This use of possibly introducing additional random variables makes the transformation $y_1 = g_1(x_1,\ldots,x_n),\ldots,y_n = g_n(x_1,\ldots,x_n)$ a transformation from an n-dimensional space to an n-dimensional space. Henceforth we assume that we are seeking the joint distribution of $Y_1 = g_1(X_1,\ldots,X_n),\ldots,Y_n = g_n(X_1,\ldots,X_n)$ (rather than the joint distribution of $Y_1,\ldots,Y_k$) when we have given the joint probability density of $X_1,\ldots,X_n$.

We state our results for $n = 2$. The generalization for $n > 2$ is straightforward. Let $f_{X_1,X_2}(x_1,x_2)$ be given. Set $\kappa = \{(x_1,x_2): f_{X_1,X_2}(x_1,x_2) > 0\}$. We want to find the joint distribution of $Y_1 = g_1(X_1,X_2)$ and $Y_2 = g_2(X_1,X_2)$ for known functions $g_1(x_1,x_2)$ and $g_2(x_1,x_2)$. Now suppose that $y_1 = g_1(x_1,x_2)$ and $y_2 = g_2(x_1,x_2)$ defines a one-to-one transformation that maps κ onto, say, D. x_1 and x_2 can be expressed in terms of y_1 and y_2; so we can write, say, $x_1 = \varphi_1(y_1,y_2)$ and $x_2 = \varphi_2(y_1,y_2)$. Note that κ is a subset of the x_1x_2 plane and D is a subset of the y_1y_2 plane consisting of points (y_1,y_2) for which there exist a $(x_1,x_2) \in \kappa$ such that $(y_1,y_2) = [g_1(x_1,x_2), g_2(x_1,x_2)]$. The determinant

$$\begin{vmatrix} \dfrac{\partial x_1}{\partial y_1} & \dfrac{\partial x_1}{\partial y_2} \\[2ex] \dfrac{\partial x_2}{\partial y_1} & \dfrac{\partial x_2}{\partial y_2} \end{vmatrix}$$

is called the Jacobian of the transformation and is denoted by J. The above discussion permits us to state Theorem 3.5.2.

Theorem 3.5.2 Let X_1 and X_2 be jointly continuous random variables with density function $f_{X_1,X_2}(x_1,x_2)$. Set $\kappa = \{(x_1,x_2): f_{X_1,X_2}(x_1,x_2) > 0\}$. Assume that:

1 $y_1 = g_1(x_1,x_2)$ and $y_2 = g_2(x_1,x_2)$ defines a one-to-one transformation of κ onto D.

2 The first partial derivatives of $x_1 = \varphi_1(y_1, y_2)$ and $x_2 = \varphi_2(y_1, y_2)$ are continuous over D.

3 The Jacobian of the transformation is nonzero for $(y_1, y_2) \in D$. Then the joint density of $Y_1 = g_1(X_1, X_2)$ and $Y_2 = g_2(X_1, X_2)$ is given by

$$f_{Y_1, Y_2}(y_1, y_2) = |J| f_{X_1, X_2}(\varphi_1(y_1, y_2), \varphi_2(y_1, y_2)) I_D(y_1, y_2). \quad (3.5.8)$$

where

$$I_D(y_1, y_2) = I_\kappa(\varphi_1(y_1, y_2), \varphi_2(y_1, y_2)) = \begin{cases} 1, & \text{if } (y_1, y_2) \in D \\ 0, & \text{otherwise.} \end{cases}$$

The proof is essentially the derivation of the formulas for transforming variables in double integrals. For proof, the reader is referred to Neuts [25].

For the single variate case the transformation formula (3.5.8) becomes

$$f_Y(y) = f_X(g^{-1}(y)) \left| \frac{d(g^{-1}(y))}{dy} \right| I_\kappa(g^{-1}(y)) = f_X(x) \left| \frac{dx}{dy} \right| I_\kappa(x). \quad (3.5.9)$$

Here $f_X(x)$ is the given p.d.f., $f_Y(y)$ is the desired p.d.f., I_κ is the interval of x, and $Y = g(X)$. We can see that (3.5.9) is a particular case of (3.5.8).

Example 1 Let Z_1 and Z_2 be two independent standard normal random variables. Let $Y_1 = Z_1 + Z_2$ and $Y_2 = Z_1/Z_2$. Then

$$z_1 = \varphi_1(y_1, y_2) = \frac{y_1 y_2}{1 + y_2} \quad \text{and} \quad z_2 = \varphi_2(y_1, y_2) = \frac{y_1}{1 + y_2}$$

$$J = \begin{vmatrix} \dfrac{y_2}{1 + y_2} & \dfrac{y_1}{(1 + y_2)^2} \\ \dfrac{1}{1 + y_2} & -\dfrac{y_1}{(1 + y_2)^2} \end{vmatrix} = -\frac{y_1(y_2 + 1)}{(1 + y_2)^3} = -\frac{y_1}{(1 + y_2)^2}$$

$$f_{Y_1, Y_2}(y_1, y_2) = \frac{|y_1|}{(1 + y_2)^2} \frac{1}{2\pi} \exp\left\{ -\frac{1}{2} \left[\frac{(y_1 y_2)^2}{(1 + y_2)^2} + \frac{y_1^2}{(1 + y_2)^2} \right] \right\}$$

$$= \frac{1}{2\pi} \frac{|y_1|}{(1 + y_2)^2} \exp\left[-\frac{1}{2} \frac{(1 + y_2^2) y_1^2}{(1 + y_2)^2} \right].$$

To find the marginal distribution of, say Y_2, we must integrate out y_1, that is

$$f_{Y_2}(y_2) = \int_{-\infty}^{\infty} f_{Y_1, Y_2}(y_1, y_2)\, dy_1$$

$$= \frac{1}{2\pi} \frac{1}{(1+y_2)^2} \int_{-\infty}^{\infty} |y_1| \exp\left[-\frac{1}{2} \frac{(1+y_2^2)y_1^2}{(1+y_2)^2} \right] dy_1.$$

Let

$$u = \frac{1}{2} \frac{(1+y_2^2)}{(1+y_2)^2} y_1^2;$$

then

$$du = \frac{(1+y_2^2)}{(1+y_2)^2} y_1\, dy_1,$$

and so

$$f_{Y_2}(y_2) = \frac{1}{2\pi} \frac{1}{(1+y_2)^2} \frac{(1+y_2)^2}{1+y_2^2} 2 \int_0^{\infty} e^{-u}\, du = \frac{1}{\pi} \cdot \frac{1}{1+y_2^2},$$

a Cauchy density. In other words, the ratio of two independent standard normal random variables has a Cauchy distribution.

To generate an r.v. from a Cauchy distribution we generate Z_1 and Z_2 from $N(0, 1)$ and take their ratio.

Example 2 Let X_i have a gamma distribution

$$f_{X_i}(x_i) = \begin{cases} \dfrac{x_i^{n_i - 1} e^{-x_i}}{\Gamma(\alpha)}, & x_i \geq 0, \quad n_i > 0 \\ 0, & \text{otherwise} \end{cases}$$

with parameters n_i and 1 for $i = 1, 2$, and assume X_1 and X_2 are independent. Suppose now that the distribution of $Y_1 = X_1/(X_1 + X_2)$ is desired. We have only the one function $y_1 = g_1(x_1, x_2) = x_1/(x_1 + x_2)$, so we have to select the other to use the transformation technique. Since x_1 and x_2 occur in the exponent of their joint density as their sum, $x_1 + x_2$ is a good choice. Let $y_2 = x_1 + x_2$; then $x_1 = y_1 y_2$, $x_2 = y_2 - y_1 y_2$, and

$$J = \begin{vmatrix} y_2 & y_1 \\ -y_2 & 1 - y_1 \end{vmatrix} = y_2.$$

Hence

$$f_{Y_1,Y_2}(y_1,y_2) = y_2 \frac{1}{\Gamma(n_1)} \frac{1}{\Gamma(n_2)} (y_1 y_2)^{n_1-1} (y_2 - y_1 y_2)^{n_2-1} e^{-y_2} I_D(y_1,y_2)$$

$$= \frac{1}{\Gamma(n_1)\Gamma(n_2)} y_1^{n_1-1}(1-y_1)^{n_2-1}$$

$$\times y_2^{n_1+n_2-1} e^{-y_2} I_{(0,1)}(y_1) I_{(0,\infty)}(y_2)$$

$$= \left[\frac{1}{B(n_1,n_2)} y_1^{n_1-1}(1-y_1)^{n_2-1} I_{(0,1)}(y_1) \right]$$

$$\times \left[\frac{1}{\Gamma(n_1+n_2)} y_2^{n_1+n_2-1} e^{-y_2} I_{(0,\infty)}(y_2) \right].$$

It turns out that Y_1 and Y_2 are independent and that Y_1 has a beta distribution with parameters n_1 and n_2.

Thus to generate a random variate from beta distribution we generate two gamma variates X_1 and X_2, then calculate $X_1/(X_1+X_2)$.

3.5.3 Multinormal Distribution

A random vector $X = (X_1, \ldots, X_n)$ has a multinormal distribution if the p.d.f. is given by

$$f_X(x) = \frac{1}{(2\pi)^{n/2}|\Sigma|^{1/2}} \exp\left[-\tfrac{1}{2}(x-\mu)^T \Sigma^{-1}(x-\mu)\right] \tag{3.5.10}$$

and denoted by $N(\mu, \Sigma)$.

Here $\mu = (\mu_1, \ldots, \mu_n)$ is the mean vector, Σ is the covariance $(n \times n)$ matrix

$$\Sigma = \begin{Vmatrix} \sigma_{11} & \sigma_{12} & \cdots & \sigma_{1n} \\ \sigma_{21} & \sigma_{22} & \cdots & \sigma_{2n} \\ \vdots & \vdots & & \vdots \\ \sigma_{n1} & \sigma_{n2} & \cdots & \sigma_{nn} \end{Vmatrix}, \tag{3.5.11}$$

which is positive definite and symmetric, $|\Sigma|$ is the determinant of Σ, and Σ^{-1} is the inverse matrix of Σ.

Inasmuch as Σ is positive definite and symmetric, there exists a unique lower triangular matrix

$$C = \begin{Vmatrix} c_{11} & 0 & \cdots & 0 \\ c_{21} & c_{22} & \cdots & 0 \\ \vdots & \vdots & & \vdots \\ c_{n1} & c_{n2} & \cdots & c_{nn} \end{Vmatrix} \tag{3.5.12}$$

such that

$$\Sigma = CC^T. \tag{3.5.13}$$

Then the vector X can be represented as

$$X = CZ + \mu, \tag{3.5.14}$$

where $Z = (Z_1, \ldots, Z_n)$ is a normal vector with zero mean and covariance matrix equal to identity matrix, that is, all components Z_i, $i = 1, \ldots, n$, of Z are distributed according to the standard normal distribution $N(0, 1)$.

In order to obtain C from $\Sigma = CC^T$ the so-called "square root method" can be used, which provides a set of recursive formulas for computation of the elements of C.

It follows from (3.5.14) that

$$X_1 = c_{11} Z_1 + \mu_1. \tag{3.5.15}$$

Therefore $\operatorname{var} X_1 = \sigma_{11} = c_{11}^2$ and $c_{11} = \sigma_{11}^{1/2}$. Proceeding with (3.5.14) we obtain

$$X_2 = c_{21} Z_1 + c_{22} Z_2 + \mu_2 \tag{3.5.16}$$

and

$$\operatorname{var} X_2 = \sigma_{22} = \operatorname{var}(c_{21} Z_1 + c_{22} Z_2). \tag{3.5.17}$$

From (3.5.15) and (3.5.16)

$$E[(X_1 - \mu_1)(X_2 - \mu_2)] = \sigma_{12} = E[c_{11} Z_1 (c_{21} Z_1 + c_{22} Z_2)]. \tag{3.5.18}$$

From (3.5.17) and (3.5.18)

$$c_{21} = \frac{\sigma_{12}}{c_{11}} = \frac{\sigma_{12}}{\sigma_{11}^{1/2}} \tag{3.5.19}$$

$$c_{22} = \left(\sigma_{22} - \frac{\sigma_{21}^2}{\sigma_{11}} \right)^{1/2}. \tag{3.5.20}$$

Generally, c_{ij} can be found from the following recursive formula:

$$c_{ij} = \frac{\sigma_{ij} - \sum_{k=1}^{j-1} c_{ik} c_{jk}}{\left(\sigma_{jj} - \sum_{k=1}^{j-1} c_{jk}^2 \right)^{1/2}}, \tag{3.5.21}$$

where

$$\sum_{k=1}^{0} c_{ik} c_{jk} = 0, \qquad 1 \le j \le i \le n.$$

Algorithm MN-1 describes the necessary steps for generating a multinormal variate.

Algorithm MN-1

1

$$c_{ij} \leftarrow \frac{\sigma_{ij} - \sum_{k=1}^{j-1} c_{ik} c_{jk}}{\left(\sigma_{jj} - \sum_{k=1}^{j-1} c_{jk}^2\right)^{1/2}},$$

where

$$\sum_{k=1}^{0} c_{ik} c_{jk} = 0, \qquad 1 \le j \le i \le n.$$

2 Generate $Z = (Z_1, \ldots, Z_n)$ from $N(0, 1)$.

3 $X \leftarrow CZ + \mu$.

4 Deliver X.

3.6 GENERATING FROM CONTINUOUS DISTRIBUTIONS

This section describes generating procedures for various single-variate continuous distributions.

3.6.1 Exponential Distribution

An exponential variate X has p.d.f.

$$f_X(x) = \begin{cases} \dfrac{1}{\beta} e^{-x/\beta}, & 0 \le x \le \infty, \beta > 0 \\ 0, & \text{otherwise} \end{cases} \tag{3.6.1}$$

denoted by $\exp(\beta)$.

Procedure E-1

By inverse transform method

$$U = F_X(X) = 1 - e^{-x/\beta} \tag{3.6.2}$$

so that

$$X = -\beta \ln(1 - U). \tag{3.6.3}$$

Since $1 - U$ is distributed in the same way as U, we have

$$X = -\beta \ln U. \tag{3.6.4}$$

For sampling purposes we may assume $\beta = 1$: if V is sampled from the standard exponential distribution exp(1), then $X = \beta V$ is from exp(β).

Algorithm E-1

1 Generate U from $\mathcal{U}(0, 1)$.
2 $X \leftarrow -\beta \ln U$.
3 Deliver X.

Although this technique seems very simple, the computation of the natural logarithm on a digital computer includes a power series expansion (or some equivalent approximation technique) for each uniform variate generated.

Procedure E-2

We now prove a proposition that can be useful for generating from exponential distribution exp(1).

Proposition Let $U_1, \ldots, U_n, U_{n+1}, \ldots, U_{2n-1}$ be independent uniformly distributed random variables, and let $U_{(1)}, \ldots, U_{(n-1)}$ represent the order statistics corresponding to the random sample $U_{n+1}, \ldots, U_{2n-1}$. Assume $U_{(0)} = 0$ and $U_{(n)} = 1$; then the r.v.'s

$$Y_k = (U_{(k-1)} - U_{(k)}) \ln \prod_{i=1}^{n} U_i, \qquad k = 1, \ldots, n \tag{3.6.5}$$

are independent and distributed exp(1).

Proof Denote

$$X_k = U_{(k)} - U_{(k-1)}, \qquad k = 1, \ldots, n-1$$

and

$$X_n = -\ln \prod_{i=1}^{n} U_i.$$

It will be shown in Section 3.6.2 that X_n is from the Erlang distribution, that is,

$$f_{X_n}(x) = \frac{x^{n-1} e^{-x}}{(n-1)!}, \qquad x \geq 0. \tag{3.6.6}$$

It is also known (Feller [11]) that the vector $(X_1, \ldots, X_{n-1})$ is distributed

$$f_{X_1, \ldots, X_{n-1}}(x_1, \ldots, x_{n-1}) = (n-1)! \tag{3.6.7}$$

inside the simplex

$$\sum_{k=1}^{n-1} x_k \leq 1, \qquad x_k \geq 0, \qquad k = 1, \ldots, n-1.$$

Therefore

$$f_{X_1,\ldots,X_n}(x_1,\ldots,x_n) = f_{X_1,\ldots,X_{n-1}}(x_1,\ldots,x_{n-1})f_{X_n}(x_n)$$

$$= x_n^{n-1}e^{-x_n}, \quad \sum_{k=1}^{n} x_k \le 1, x_k \ge 0, k = 1,\ldots,n.$$

We introduce

$$y_k = x_k x_n, \qquad k = 1,\ldots,n-1$$

and

$$y_n = (1 - x_1 - \cdots - x_{n-1})x_n.$$

Then

$$x_k = \frac{y_k}{\sum_{i=1}^{n} y_i}, \qquad k = 1,\ldots,n-1$$

$$x_n = \sum_{i=1}^{n} y_i.$$

The Jacobian of the transformation is

$$J = \left(\sum_{i=1}^{n} y_i\right)^{-(n-1)}. \tag{3.6.8}$$

Hence

$$f_{Y_1,\ldots,Y_n}(y_1,\ldots,y_n) = \left(\sum_{i=1}^{n} y_1\right)^{n-1} \exp\left(-\sum_{i=1}^{n} y_i\right)\left(\sum_{i=1}^{n} y_i\right)^{-(n-1)}$$

$$= \prod_{i=1}^{n} e^{-y_i}, \qquad y_i \ge 0, i = 1,\ldots,n. \tag{3.6.9}$$

Q.E.D.

For $n = 3$ we have

$$Y_1 = -U_{(1)}\ln\left(\prod_{i=1}^{3} U_i\right)$$

$$Y_2 = (U_{(1)} - U_{(2)})\ln\left(\prod_{i=1}^{3} U_i\right)$$

$$Y_3 = (U_{(2)} - 1)\ln\left(\prod_{i=1}^{3} U_i\right).$$

Algorithm E-2 describes the necessary steps.

Algorithm E-2

1 Generate $2n-1$ uniformly distributed random variates $U_1, \ldots, U_n, U_{n+1}, \ldots, U_{2n-1}$.

2 Arrange the variates $U_{n+1}, \ldots, U_{2n-1}$ in order of increasing magnitudes, that is, define them to be the order statistics $U_{(1)}, \ldots, U_{(n-1)}$.

3 $Y_k \leftarrow (U_{(k-1)} - U_{(k)}) \ln(\prod_{i=1}^{n} U_i)$, $k = 1, \ldots, n$.

4 Deliver Y_k, $k = 1, \ldots, n$, as an r.v. from exp(1).

Comparing (3.6.5) with the inverse transform method

$$Y_k = -\ln U_k, \qquad k = 1, \ldots, n,$$

we find that the advantage of Algorithm E-2 is that it requires only one computation of $\ln \prod_{k=1}^{n} U_k$ for generating n exponential variates simultaneously. In the same time the inverse transform method requires n computations of $\ln U_k$ for each variate Y_k, $k = 1, \ldots, n$, separately. The disadvantage of Algorithm E-2 is that it needs $2n-1$ uniform variates rather than n uniform variates for the inverse transform method. Additionally, Algorithm E-2 requires the arrangement of the uniform variates $U_{n+1}, \ldots, U_{2n-1}$ to be order statistics $U_{(1)}, \ldots, U_{(n-1)}$ and then calculation of $U_{(k-1)} - U_{(k)}$, which is also time consuming.

Simulating both algorithms we find that Algorithm E-2 is faster than the standard inverse Algorithm E-1 for $n = 3$ to $n = 6$. The optimal n is 4.

There are many alternative procedures (Ahrens and Dieter [1], Marsaglia [19]) for generating from exp(β) without the benefit of a logarithmic transformation, procedures that are based on the composition method, acceptance-rejection method, and Forsythe method [15] (see also example 1, Section 3.4.4). The reader is also referred to Fishman's monograph [12]. Before leaving the exponential distribution we want to introduce von Neumann's ingenious method [34] for generating from exp(1), a method that was later extended by Forsythe [15] and Ahrens and Dieter [2] for generating various distributions.

Let $\{X_i : i = 0, \ldots\}$ be a sequence of i.i.d. r.v.'s from the standard triangular distribution

$$f_X(x) = \begin{cases} \dfrac{2x}{\gamma}, & 0 \leq x \leq \gamma \\ \dfrac{2(1-x)}{1-\gamma}, & \gamma \leq x \leq 1 \end{cases}$$

and define an r.v. N, taking positive integer values through $\{X_i\}$ by the inequalities

$$X_1 \le X_0, \qquad \sum_{j=1}^{2} X_j \le X_0, \ldots, \sum_{j=1}^{N-1} X_j \le X_0, \sum_{j=1}^{N} X_j > X_0.$$

We accept the sequence $\{X_i\}$ if N is odd; otherwise we reject it and repeat the process until N turns out odd. Let T be the number of sequences rejected before an odd N appears ($T = 0, 1, \ldots$) and let X_0 be the value of the first variable in the accepted sequence; then $Y = T + X_0$ is from exp(1). It is shown in ref. 34 that generation of one exponentional variate in such a way requires on the average $(1 + e)\,(1 - e^{-1}) \approx 6$ random numbers.

3.6.2 Gamma Distribution

A random variable X has a gamma distribution if its p.d.f. is defined as

$$f_X(x) = \begin{cases} \dfrac{x^{\alpha-1} e^{-x/\beta}}{\beta^{\alpha}\Gamma(\alpha)}, & 0 \le x \le \infty, \alpha > 0, \beta > 0 \\ 0, & \text{otherwise,} \end{cases}$$

and is denoted by $G(\alpha, \beta)$. Note that for $\alpha = 1$, $G(1, \beta)$ is exp(β).

Inasmuch as the c.d.f. does not exist in explicit form for gamma distribution, the inverse transform method cannot be applied. Therefore alternative methods of generating gamma variates must be considered.

Procedure G-1

One of the most important properties of gamma distribution is the *reproductive property*, which can be successfully used for gamma generation. Let X_i, $i = 1, \ldots, n$, be a sequence of independent random variables from $G(\alpha_i, \beta)$. Then $X = \sum_{i=1}^{n} X_i$ is from $G(\alpha, \beta)$ where $\alpha = \sum_{i=1}^{n} \alpha_i$.

If α is an integer, say, $\alpha = m$, a random variate from gamma distribution $G(m, \beta)$ can be obtained by summing m independent exponential random variates exp(β), that is,

$$X = \beta \sum_{i=1}^{m} (-\ln U_i) = -\beta \ln \prod_{i=1}^{m} U_i, \tag{3.6.10}$$

which is called Erlang distribution and denoted $Er(m, \beta)$. Algorithm G-1 describes generating r.v.'s from $Er(m, \beta)$.

Algorithm G-1

1 $X \leftarrow 0$.
2 Generate V from exp(1).
3 $X \leftarrow X + V$.

4 If $\alpha = 1$, $X \leftarrow \beta X$ and deliver X.
5 $\alpha \leftarrow \alpha - 1$.
6 Go to step 2.

It is not difficult to see that the mean computation (CPU) time for generation from Erlang distribution is an increasing linear function of α. However, if α is nonintegral, (3.6.10) is not applicable and some difficulties arise while generating gamma variates.

For some time no exact method was known and approximate techniques were used. The most common method was the so-called probability switch method [24].

Let $m = [\alpha]$ be the integral part of α and let $\delta = \alpha - m$. With probability δ, generate a random variate from $G(m+1, \beta)$. With probability $1-\delta$, generate a random variate from $G(m, \beta)$. This mixture of gamma variates with integral shape parameters will approximate the desired gamma distribution. This technique will only work when $\alpha \geq 1$.

In the particular case when $\delta = \frac{1}{2}$ gamma variables can be generated exactly by adding half the square of a standard normal variate to the variate generated in (3.6.10).

Procedure G-2

Jöhnk [16] suggested a technique that exactly generates* variates from $G(\delta, \beta)$, where $0 < \delta < 1$.

Theorem 3.6.1 Let W and V be independent variates from beta distribution $Be(\delta, 1-\delta)$ (see Section 3.6.3) and exp(1), respectively. Then $X = \beta V W$ is a variate with $G(\delta, \beta)$.

Proof Let $u = v$ and let $x = \beta w v$. Then $w = x/\beta u$, and $v = u$. The Jacobian of this transformation is

$$J = \begin{vmatrix} \dfrac{-x}{u^2\beta} & \dfrac{1}{u\beta} \\ 1 & 0 \end{vmatrix} = \frac{-1}{u\beta}. \tag{3.6.11}$$

The joint distribution of (u, x) is therefore given by

$$f_{U,X}(u,x) = \begin{cases} \dfrac{\beta^{-\delta}}{\Gamma(\delta)\Gamma(1-\delta)} x^{\delta-1}\left(u - \dfrac{x}{\beta}\right)^{-\delta} e^{-u}; & 0 \leq x, u \leq \infty \\ 0, & \text{otherwise}. \end{cases} \tag{3.6.12}$$

*It is understood that when we say a method "exactly generates" random variables on a computer, that the exactness is limited by the computer used and by the randomness of the underlying pseudorandom number generator.

The marginal distribution for X is

$$f_X(x) = \int_{x/\beta}^{\infty} f_{U,X}(u,x)\,du = \begin{cases} \dfrac{\beta^{-\delta}}{\Gamma(\delta)} x^{\delta-1} e^{-x/\beta}, & x \geq 0 \\ 0, & x < 0, \end{cases}$$

which is $G(\delta, \beta)$. Q.E.D.

Algorithm G-2

1 Generate two variates W and V from $Be(\delta, 1-\delta)$ and $\exp(1)$, respectively.

2 Compute $X = \beta V W$ that is from $G(\delta, \beta)$.

3. Deliver X.

To generate a variate from $G(\alpha, \beta)$ we generate an r.v. Y from $Er(m, 1)$, then compute $X = \beta(Y + VW)$, which is from $G(\alpha, \beta)$. Here $\alpha = \delta + m$.

Recently, a number of procedures for sampling from $G(\alpha, \beta)$, based on the acceptance-rejection method, were suggested by Ahrens and Dieter [3], Cheng [9], Fishman [13], Tadikamalla [30, 31] and Wallace [35]. Let us consider some of them.

Procedure G-3

Wallace [35] suggested a procedure for generating from $G(\alpha, 1)$ with $\alpha > 1$ based on both the acceptance-rejection and probability switch methods.

Let

$$f_X(x) = Ch(x)g(x),$$

where $h(x)$ is a mixture of two Erlang distributions $Er(m, 1)$ and $Er(m+1, 1)$ equal to

$$h(x) = P\frac{x^{m-1}e^{-x}}{(m-1)!} + (1-P)\frac{x^m e^{-x}}{m!}, \qquad x \geq 0, \tag{3.6.13}$$

$$C = \frac{(m-1)!\,m^{\delta}}{\Gamma(\alpha)}, \tag{3.6.14}$$

and

$$g(x) = \left(\frac{x}{m}\right)^{\delta}\left[1 + \left(\frac{x}{m} - 1\right)(1-P)\right]^{-1}. \tag{3.6.15}$$

It can be found from (3.4.10) that the optimal P is equal to $1-\delta$, where $\delta = \alpha - [\alpha]$. It follows from (3.6.14) that the mean number of trials C is a monotone decreasing function of m for a fixed δ and

$$\lim_{m\to\infty} \frac{(m-1)!\,m^{\delta}}{\Gamma(m+\delta)} = 1,$$

that is, asymptotically the execution time does not depend on δ and achieves optimal efficiency $C=1$. Algorithm G-3 describes Wallace's procedure.

Algorithm G-3

1 Compute $\delta=\alpha-m$, where $m=[\alpha]$.
2 Generate $U_1,\ldots,U_m$ from $\mathcal{U}(0,1)$.
3 With probability $1-\delta$ compute

$$V=-\ln\prod_{i=1}^{m}U_i.$$

4 With probability δ compute

$$V=-\ln\prod_{i=1}^{m+1}U_i.$$

5 Generate another uniform variate U from $\mathcal{U}(0,1)$.
6 If $U\leq(V/m)^{\delta}/[1+((V(/m)-1)\delta]$, deliver V as an r.v. from $G(\alpha,1)$.
7 Go to step 2.

The following three procedures are reproduced with little change from Ref. 12.

Procedure G-4

Fishman [13] describes another procedure for generating from $G(\alpha,1)$, $\alpha\geq1$:

$$g(x)=x^{\alpha-1}\frac{\exp[-x(1-1/\alpha)]}{\alpha^{\alpha-1}\exp(1-\alpha)} \tag{3.6.16}$$

$$h(x)=\frac{1}{\alpha}e^{-x/\alpha} \tag{3.6.17}$$

$$C=\frac{\alpha^{\alpha}e^{1-\alpha}}{\Gamma(\alpha)},\qquad \alpha\geq1,\ x\geq0. \tag{3.6.18}$$

The probability of success on a trial is

$$\frac{1}{C}=\frac{\Gamma(\alpha)}{\alpha^{\alpha}e^{1-\alpha}}. \tag{3.6.19}$$

For large α the mean number of trials is

$$C\approx e\left(\frac{\alpha}{2\pi}\right)^{1/2}. \tag{3.6.20}$$

It is not difficult to see that the condition $U\leq g(Y)$, where the r.v. Y is from exp$(1/\alpha)$, can be written as $V_2 \geq (\alpha - 1)(V_1 - \ln V_1 - 1)$ and V_1 and V_2 are independent r.v.'s from exp(1).

Algorithm G-4

1 $A \leftarrow \alpha - 1$.
2 Generate V_1 and V_2 from exp(1).
3 If $V_2 < A(V_1 - \ln V_1 - 1)$, go to step 2.
4 Deliver V_1 as a variate from $G(\alpha, 1)$.

Procedure G-5

This procedure is due to Cheng [9] and describes gamma generation $G(\alpha, 1)$ for $\alpha > 1$ with execution time asymptotically independent of α.

In Cheng's procedure

$$h(x) = \begin{cases} \lambda \mu x^{\lambda-1}(\mu + x^{\lambda})^{-2}, & x \geq 0 \\ 0, & \text{otherwise} \end{cases} \tag{3.6.21}$$

$$C = \frac{4\alpha^{\alpha}}{\Gamma(\alpha) e^{\alpha} \lambda} \tag{3.6.22}$$

$$g(x) = x^{\alpha-\lambda}(\mu + x^{\lambda})^2 \frac{e^{\alpha - x}}{4\alpha^{\alpha+\lambda}}, \tag{3.6.23}$$

where

$$\mu = \alpha^{\lambda}, \qquad \lambda = (2\alpha - 1)^{1/2}.$$

The execution time C is a monotonically decreasing function of α such that, for $\alpha = 1$, $C = 1.47$, and for $\alpha = 2$, $C = 1.25$; asymptotically

$$\lim_{\alpha \to \infty} C = \frac{2}{\sqrt{\pi}} \approx 1.13. \tag{3.6.24}$$

Let $b = \alpha - \ln 4$ and $d = \alpha + 1/\lambda$. Then Cheng's algorithm can be written as follows.

Algorithm G-5

1 Sample U_1 and U_2 from $\mathcal{U}(0, 1)$.
2 $V \leftarrow \lambda \ln[U_1/(1 - U_2)]$.
3 $X \leftarrow \alpha e^{V}$.
4 If $b + d - X \geq \ln(U_1^2 U_2)$, deliver X.
5 Go to step 1.

Procedure G-6

Ahrens and Dieter [3] suggested an alternative procedure for generating from $G(\alpha, \beta)$ with $\alpha > 1$ and execution time independent of α asymptotically and equal to $\lim_{\alpha\to\infty} C = \sqrt{\pi}$. Their procedure makes use of the truncated Cauchy distribution.

Let

$$h(x) = \left[\frac{h_X(x)}{1 - H_X(0)}\right], \qquad x \geq 0 \tag{3.6.25}$$

and

$$g(x) = \left[1 + \left(\frac{x-\gamma}{\beta}\right)^2\right]\left(\frac{x}{\gamma}\right)^{\gamma} e^{-x+\gamma}, \qquad x \geq 0, \tag{3.6.26}$$

where

$$C = \pi\beta\gamma^{\gamma}\frac{[1 - H_X(0)]}{\Gamma(\alpha)e^{\gamma}} \tag{3.6.27}$$

and

$$h_X(x) = \frac{\beta}{\pi\left[\beta^2 + (x-\gamma)^2\right]} \tag{3.6.28}$$

$$H_X(x) = \tfrac{1}{2} + \pi^{-1}\tan^{-1}\left(\frac{x-\gamma}{\beta}\right), \qquad -\infty < x < \infty \tag{3.6.29}$$

are the p.d.f. and c.d.f. of the Cauchy distribution, respectively, with parameters $\gamma = \alpha - 1$, and $\beta = (2\alpha - 1)^{1/2}$.

It follows from (3.6.25) and (3.6.28) that $h(x)$ is the truncated Cauchy distribution with parameters γ and β.

To apply the acceptance condition $U \leq g(Y)$, we have to generate an r.v. Y from the truncated Cauchy distribution $h(y)$. The c.d.f. of Y is

$$H(y) = \frac{H_Y(y) - H_Y(0)}{1 - H_Y(0)}. \tag{3.6.30}$$

where $H_Y(y)$ is given in (3.6.29).

Substituting (3.6.29) in (3.6.30) and using the inverse transform formula $Y = H^{-1}(U)$, we obtain

$$Y = \beta \tan \pi\left\{U\left[1 - H_Y(0)\right] + H_Y(0) - \tfrac{1}{2}\right\} + \gamma \tag{3.6.31}$$

where by (3.6.29)

$$1 - H_Y(0) = \frac{1}{2} - \frac{\tan^{-1}(\gamma/\beta)}{\pi}. \tag{3.6.32}$$

It is readily seen that the condition $U \leq g(Y)$ is equivalent to

$$-V = \ln U \leq \ln g(Y) = \ln\left[1 + \frac{(Y' - \gamma)^2}{\beta^2}\right] + \gamma \ln \frac{Y'}{\gamma} - Y' + \gamma, \tag{3.6.33}$$

where V is from exp(1). $Y' = \gamma + \beta \tan \pi(U - \frac{1}{2})$ and can be found from $U = H_{Y'}(y)$.

Algorithm G-6

1 $\gamma \leftarrow \alpha - 1$.
2 Generate U from $\mathcal{U}(0, 1)$.
3 $Y' \leftarrow \gamma + \beta \tan \pi(U - \frac{1}{2})$.
4 Generate V from exp(1).
5 If $-V \leq \ln[1 + (Y' - \gamma)^2/\beta^2] + \gamma \ln(Y'/\gamma) - Y' + \gamma$, deliver Y'.
6 Go to step 2.

The following two procedures for generating from $G(\alpha, 1)$ are due to Tadikamalla [30, 31].

Procedure G-7

In this procedure [30] $h(x)$ is from $Er(m, \beta)$, that is,

$$h(x) = \frac{x^{(m-1)}e^{-x/\beta}}{\beta^m (m-1)!}, \qquad \beta > 0, m > 0, x \geq 0. \tag{3.6.34}$$

Then it is readily shown that

$$g(x) = \frac{x^\delta \exp[-x(1 - 1/\beta)]}{[\delta\beta/(\beta - 1)]^\delta e^{-\delta}}, \qquad x \geq 0 \tag{3.6.35}$$

$$C = \frac{\beta^m (m-1)!}{\Gamma(\alpha)} \left(\frac{\delta\beta}{\beta - 1}\right)^2 e^{-\delta} \tag{3.6.36}$$

where $\delta = \alpha - m$ and $m = [\alpha]$.

The value of α that maximizes the efficiency can be found from (3.4.10) and is equal to α/m.

Tadikamalla showed by simulation that his procedure is faster than Fishman's Procedure G-4 for $3 \leq \alpha \leq 19$ and is comparable for other values of α. For $1 \leq \alpha < 2$ both methods coincide. This is not surprising. The reason for the great efficiency of this procedure is that Erlang distribution $Er(m, \beta)$, with $n = [\alpha]$, approximates the gamma distribution $G(\alpha, \beta)$ better than the exponential distribution $\exp(\alpha)$ (see Procedure G-4) does.

In addition, Tadikamalla's procedure is better than Ahrens and Dieter's Procedure G-6 for $\alpha \leq 8$.

Algorithm G-7

1 Compute $\delta \leftarrow \alpha - m$, where $m = [\alpha]$.
2 Generate m independent random variates $U_1, \ldots, U_m$ from $\mathcal{U}(0, 1)$.
3 Compute $Y = -\beta \ln \prod_{i=1}^{m} U_i$.
4 Generate another uniform variate U from $\mathcal{U}(0, 1)$.
5 If

$$U \leq \frac{Y^{\delta} \exp[-Y(1 - 1/\beta)]}{\left[\dfrac{\delta\beta}{(\beta - 1)}\right]^{\delta} e^{-\delta}},$$

deliver Y.
6 Go to step 2.

Procedure G-8

In this procedure [31] $h(x)$ is from the Laplace (double exponential) distribution with location parameter $\alpha - 1$ and scale parameter θ, that is,

$$h(x) = \frac{\exp\{-|x - (\alpha - 1)|/\theta\}}{2\theta}, \qquad -\infty < x < \infty, \theta > 0. \tag{3.6.37}$$

Then it is readily shown that

$$g(x) = \left\{\frac{(\theta - 1)x}{\theta(\alpha - 1)}\right\}^{\alpha - 1} \exp\left\{-x + \frac{[|x - (\alpha - 1)| - (\alpha - 1)(\theta + 1)]}{\theta}\right\} \tag{3.6.38}$$

and

$$C = \frac{2\theta^{\alpha}}{\Gamma(\alpha)} \left\{\frac{\alpha - 1}{\theta - 1}\right\}^{\alpha - 1} \exp\left\{\frac{(\alpha - 1)(1 + \theta)}{\theta}\right\}. \tag{3.6.39}$$

Algorithm G-8

1 Generate a random variate Y from the Laplace distribution with location parameter $\alpha - 1$ and scale parameter θ.
2 If $Y < 0$, go to step 1.
3 Generate a uniform random variate from $\mathcal{U}(0, 1)$.
4 If $U \leq g(Y)$ (see (3.6.38)), deliver X.
5 Go to step 1.

Table 3.6.1 The Relative Efficiencies ($1/C$), and the Average Number of Random Numbers Required (N) for Certain Algorithms

	Fishman		Tadikamalla 1		Tadikamalla 2	
α	$1/C$	N	$1/C$	N	$1/C$	N
1.5	0.7953	2.5	0.7953	2.5	0.8642	2.3
2.5	0.6029	3.3	0.8871	3.4	0.7872	2.5
3.5	0.5047	4.0	0.9222	4.3	0.7565	2.6
5.5	0.3992	5.0	0.9520	6.3	0.7304	2.7
8.5	0.3194	6.3	0.9695	9.3	0.7174	2.8
10.5	0.2868	7.0	0.9755	11.3	0.7144	2.8
15.5	0.2355	8.5	0.9836	16.3	0.7132	2.8
20.5	0.2045	9.8	0.9876	21.3	0.7149	2.8
30.5	0.1674	11.9	0.9917	31.3	0.7195	2.8
100.5	0.0920	21.7	1.0000	101.0	0.7355	2.7

Tadikamalla [31] compared the relative efficiency and CPU timing of his procedures with Fishman's [13] and Ahrens and Dieter's procedures [3].

Table 3.6.1 gives the relative efficiencies and the number of uniform random numbers required for these procedures for some selected values of α. The efficiencies of Ahrens and Dieter's method are not given in Table 3.6.1 because these have to be calculated numerically and the details are not available in Ref. 3. For increasing values of α the efficiency of Fishman's algorithm decreases and the efficiencies of Tadikamalla's first algorithm (G-7) and of Ahrens and Dieter's algorithm increase. The efficiency of Tadikamalla's second algorithm (G-8) decreases as α increases up to a certain value and then it increases again. Also note that the number of uniforms required for Tadikamalla's second algorithm (G-8) remains fairly constant.

Table 3.6.2 gives the CPU timings for these four methods on an IBM 370/165 computer, for selected values of α. These timings are based on generating 10,000 variates and using the subroutine TIMER available on the IBM computer.

The following observations can be made about the methods compared above.

1 Fishman's procedure is the simplest of all the procedures and the CPU time per trial is constant for any α. As α increases, the number of trials required for one gamma variate increases (efficiency decreases), and thus the CPU time per variate increases with α.

Table 3.6.2 Average CPU Times (in Microseconds) to Generate One Gamma Variate on the IBM 370/165 Computer

α	Fishman	Tadikamalla 1	Ahrens and Dieter	Tadikamalla 2
1.5	127	137	N/A	138
2.5	175	176	N/A	152
3.5	213	184	225	157
5.5	260	225	218	162
8.5	334	307	210	166
10.5	380	354	209	167
15.5	473	470	203	167
20.5	559	596	194	166
30.5	693	850	190	165
50.5	—	—	181	164
100.5	—	—	171	162

2 Tadikamalla's first procedure (G-7), is also simple, and in this case the number of trials per gamma variate decreases as α increases. However, the CPU time per trial increases with α (more uniforms are required per trial). The average CPU time per variate for this procedure increases with α. Tadikamalla's procedure is faster than Fishman's procedure for $3 \leq \alpha \leq 19$ and the same as Fishman's procedure for $1 \leq \alpha < 2$.

3 Tadikamalla's second procedure, (G-8), is faster than Fishman's and Tadikamalla's first procedure (G-7) for $\alpha > 2$ and is faster than Ahrens and Dieter's for all α. The average CPU time required per variate for Tadikamalla's second procedure remains fairly constant for medium and large values of α.

3.6.3 Beta Distribution

An r.v. X has a beta distribution if the p.d.f. is

$$f_X(x) = \frac{\Gamma(\alpha+\beta)}{\Gamma(\alpha)\Gamma(\beta)} x^{\alpha-1}(1-x)^{\beta-1}, \qquad \alpha > 0, \beta > 0, 0 \leq x \leq 1 \quad (3.6.40)$$

and is denoted by $Be(\alpha, \beta)$. There are several ways of generating from $Be(\alpha, \beta)$.

Procedure Be-1

This procedure is based on the result from Section 3.5.2 (example 2) that says: if Y_1 and Y_2 are independent r.v.'s from $G(\alpha, 1)$ and $G(\beta, 1)$, respectively, then

$$X = \frac{Y_1}{Y_1 + Y_2} \tag{3.6.41}$$

is from $Be(\alpha, \beta)$.

The corresponding algorithm is as follows.

Algorithm Be-1

1 Generate Y_1 from $G(\alpha, 1)$.
2 Generate Y_2 form $G(\beta, 1)$.
3 $X \leftarrow Y_1/(Y_1 + Y_2)$.
4 Deliver X.

Procedure Be-2

Another approach when α and β are integers is based on the theory of order statistics. Let $U_1, \ldots, U_{\alpha+\beta-1}$ be random variates from $\mathscr{U}(0, 1)$. Then the αth order statistic $U_{(\alpha)}$ is from $Be(\alpha, \beta)$. The algorithm is extremely simple.

Algorithm Be-2

1 Generate $(\alpha + \beta - 1)$ uniform random variates $U_1, \ldots, U_{\alpha+\beta-1}$ from $\mathscr{U}(0, 1)$.
2 Find $U_{(\alpha)}$, which is from $Be(\alpha, \beta)$.

It can be shown that the total number of comparisons needed to find $U_{(\alpha)}$ is equal to $(\alpha/2)\ (\alpha + 2\beta - 1)$, that is, this procedure is not efficient for large α and β.

Many procedures for sampling from $Be(\alpha, \beta)$ with nonintegral α and β have been proposed recently (see Ahrens and Dieter [3], Cheng [8], Jöhnk [16], and Michailov [22]). We consider only a few of them.

Procedure Be-3

The simplest procedure for generating from $Be(\alpha, \beta)$ with arbitrary nonintegral α and β uses the mode

$$f_X(x^*) = M = \frac{\Gamma(\alpha + \beta)}{\Gamma(\alpha)\Gamma(\beta)} \left(\frac{\alpha - 1}{\alpha + \beta - 2} \right)^{\alpha - 1} \left(\frac{\beta - 1}{\alpha + \beta - 2} \right)^{\beta - 1}, \tag{3.6.42}$$

which corresponds to $x^* = (\alpha - 1)/(\alpha + \beta - 2)$.

The following algorithm, Be-3, is based on the acceptance-rejection Algorithm AR-2.

Algorithm Be-3

1

$$M \leftarrow f_X\left(\frac{\alpha-1}{\alpha+\beta-2}\right) = \frac{\Gamma(\alpha+\beta)}{\Gamma(\alpha)\Gamma(\beta)}\left(\frac{\alpha-1}{\alpha+\beta-2}\right)^{\alpha-1}\left(\frac{\beta-1}{\beta+\alpha-2}\right)^{\beta-1}.$$

2 Generate U_1 and U_2 from $\mathcal{U}(0, 1)$.

3 If $MU_2 \le [\Gamma(\alpha+\beta)/\Gamma(\alpha)\Gamma(\beta)]U_1^{\alpha-1}(1-U_1)^{\beta-1}$, deliver U_1 as a variate from $Be(\alpha, \beta)$.

4 Go to step 2.

Procedure Be-4

This procedure is due to Jöhnk [16] and is based on the following theorem.

Theorem 3.6.2 Let U_1 and U_2 be two uniform variates from $\mathcal{U}(0, 1)$ and let $Y_1 = U_1^{1/\alpha}$ and $Y_2 = U_2^{1/\beta}$. If $Y_1 + Y_2 \le 1$, then

$$X = \frac{Y_1}{Y_1 + Y_2} \tag{3.6.43}$$

is from $B(\alpha, \beta)$.

Proof It is obvious that

$$f_{Y_1}(y_1) = \alpha y_1^{\alpha-1}, \qquad 0 \le y_1 \le 1 \tag{3.6.44}$$

$$f_{Y_2}(y_2) = \beta y_2^{\beta-1}, \qquad 0 \le y_2 \le 1 \tag{3.6.45}$$

and

$$f_{Y_1Y_2}(y_1, y_2) = \alpha\beta y_1^{\alpha-1} y_2^{\beta-1}. \tag{3.6.46}$$

Let $X = Y_1/(Y_1 + Y_2)$ and $W = Y_1 + Y_2$. The Jacobian

$$J = \begin{vmatrix} \dfrac{\partial y_1}{\partial x} & \dfrac{\partial y_1}{\partial w} \\ \dfrac{\partial y_2}{\partial x} & \dfrac{\partial y_2}{\partial w} \end{vmatrix} = \begin{vmatrix} w & x \\ -w & 1-x \end{vmatrix} = w \tag{3.6.47}$$

and

$$f_{X,W}(x, w) = \alpha\beta x^{\alpha-1}(1-x)^{\beta-1} w^{\alpha+\beta-1}, \qquad 0 \le x \le 1, 0 \le w \le 2. \tag{3.6.48}$$

By Bayes' formula

$$f_X(x|0 \le W \le 1) = \frac{f_{X,W}(x, 0 \le W \le 1)}{\Pr(0 \le W \le 1)} \tag{3.6.49}$$

$$f_{X,W}(x, 0 \le W \le 1) = \int_0^1 f_{X,W}(x, w)\, dw = \frac{\alpha\beta}{\alpha+\beta} x^{\alpha-1}(1-x)^{\beta-1},$$

$$0 \le x \le 1 \tag{3.6.50}$$

$$\Pr(0 \le W \le 1) = \int_0^1 f_{X,W}(x, 0 \le W \le 1)\, dx \tag{3.6.51}$$

$$= \frac{\alpha\beta}{\alpha+\beta} \frac{\Gamma(\alpha)\Gamma(\beta)}{\Gamma(\alpha+\beta)}.$$

Substituting (3.6.51) and (3.6.50) into (3.6.49), we obtain

$$f_X(x|0 \le W \le 1) = \frac{\Gamma(\alpha+\beta)}{\Gamma(\alpha)\Gamma(\beta)} x^{\alpha-1}(1-x)^{\beta-1}, \qquad 0 \le x \le 1$$

Q.E.D.

The efficiency of the method is equal to

$$\frac{1}{C} = P(Y_1 + Y_2 \le 1) = \Pr(0 \le W \le 1) = \frac{\alpha\beta}{\alpha+\beta} \frac{\Gamma(\alpha)\Gamma(\beta)}{\Gamma(\alpha+\beta)}. \tag{3.6.52}$$

For integer α and β

$$C = \frac{(\alpha+\beta)!}{\alpha!\beta!}. \tag{3.6.53}$$

Table 3.6.3 represents the mean number of trials C as a function of α and β. Asymptotically,

$$\lim_{\substack{\alpha\to\infty \\ \beta>0}} C = \lim_{\substack{\beta\to\infty \\ \alpha>0}} C = \lim_{\substack{\alpha\to\infty \\ \beta\to\infty}} C = \infty.$$

Thus for large α or β Jöhnk's procedure is not efficient.

Table 3.6.3 The Mean Number of Trials as a Function of α and β

	β		
α	1	3	5
1	2	4	6
3	4	20	56
5	6	56	252

Algorithm Be-4

1 $j \leftarrow 1$.
2 Generate U_j and U_{j+1} from $\mathcal{U}(0, 1)$.
3 $Y_1 \leftarrow U_j^{1/\alpha}$.
4 $Y_2 \leftarrow U_{j+1}^{1/\beta}$.
5 If $Y_1 + Y_2 \geq 1$, go to step 2.
6 $j \leftarrow j + 2$.
7 Deliver $X = Y_1/(Y_1 + Y_2)$.

Procedure Be-5

This procedure is based on the results of examples 6 and 7 from Section 3.4.3. As follows from (3.4.20) and (3.4.24), the efficiencies of the acceptance-rejection method AR-3 are, respectively,

$$\frac{1}{C} = \frac{\beta\Gamma(\alpha)\Gamma(\beta)}{\Gamma(\alpha+\beta)} \tag{3.5.54}$$

$$\frac{1}{C} = \frac{\alpha\Gamma(\alpha)\Gamma(\beta)}{\Gamma(\alpha+\beta)} \tag{3.5.55}$$

in examples 6 and 7.

For integer α and β we have, respectively,

$$\frac{1}{C} = \frac{(\alpha-1)!\beta!}{(\alpha+\beta-1)!} \tag{3.6.56}$$

$$\frac{1}{C} = \frac{\alpha!(\beta-1)!}{(\alpha+\beta-1)!}. \tag{3.6.57}$$

In both cases (3.6.56) and (3.6.57) the efficiencies are a little higher than in Jöhnk's procedure Be-4 (see (3.6.53)). It is interesting to note that for $\beta > \alpha$ it is more efficient to represent $f_X(x)$ in the form of (3.4.18) through (3.4.20) and for $\alpha > \beta$ it is more efficient to represent $f_X(x)$ in the form of (3.4.22) through (3.4.24).

Procedure Be-6

In this procedure $h(x)$ is $Be(m, n)$, that is,

$$h(x) = \frac{(m+n-1)!}{(m-1)!(n-1)!} x^{m-1}(1-x)^{n-1}, \qquad 0 \leq x \leq 1 \tag{3.6.58}$$

where $m = [\alpha]$ and $n = [\beta]$. Then

$$Cg(x) = \frac{B(m,n)}{B(\alpha,\beta)} x^{\delta_1}(1-x)^{\delta_2}, \qquad 0 \leq x \leq 1$$

where $\delta_1 = \alpha - m$, $\delta_2 = \beta - n$, and $B(r, s) = \Gamma(r)\Gamma(s)/\Gamma(r+s)$. It is quite easy to prove that the function $y = x^{\delta_1}(1-x)^{\delta_2}$ is concave on [0, 1] and achieves its unique maximum

$$y^* = \frac{\delta_1^{\delta_1}\delta_2^{\delta_2}}{(\delta_1+\delta_2)^{\delta_1+\delta_2}} \text{ at the point } x^* = \frac{\delta_1}{\delta_1+\delta_2}.$$

Now we set

$$g(x) = \frac{x^{\delta_1}(1-x)^{\delta_2}}{y^*} = x^{\delta_1}(1-x)^{\delta_1}\frac{(\delta_1+\delta_2)^{\delta_1+\delta_2}}{\delta_1^{\delta_1}\delta_2^{\delta_2}} \tag{3.6.59}$$

and

$$C = \frac{B(m,n)}{B(\alpha,\beta)}y^* = \frac{B(m,n)}{B(\alpha,\beta)}\frac{\delta_1^{\delta_1}\delta_2^{\delta_2}}{(\delta_1+\delta_2)^{\delta_1+\delta_2}}. \tag{3.6.60}$$

The efficiency of the procedure is

$$\frac{1}{C} = \frac{B(\alpha,\beta)}{B(m,n)}\frac{(\delta_1+\delta_2)^{\delta_1+\delta_2}}{\delta_1^{\delta_1}\delta_2^{\delta_2}}.$$

It is easy to see that

$$1 \geq \frac{1}{C} \geq \frac{(\delta_1+\delta_2)^{\delta_1+\delta_2}}{(m+n+1)(m+n)\delta_1^{\delta_1}\delta_2^{\delta_2}}. \tag{3.6.61}$$

Comparing (3.6.61) with (3.6.56) and (3.6.57), we can also readily prove that Procedure Be-6 is more efficient than Procedure Be-5 for $\alpha \geq 2$, $\beta \geq 2$.

Algorithm Be-6

1 Generate U from $\mathcal{U}(0, 1)$.
2 Generate Y from $Be(m, n)$.
3 If

$$U \leq Y^{\delta_1}(1-Y)^{\delta_2}\frac{(\delta_1+\delta_2)^{\delta_1+\delta_2}}{\delta_1^{\delta_1}\delta_2^{\delta_2}},$$

deliver Y.
4 Go to step 1.

Remark If $\delta_1 = 0$, then $g(x) = (1-x)^{\delta_2}$, $y^* = 1$, and $C = B(m,n)/B(m,\beta)$. If $\delta_2 = 0$, then $g(x) = x^{\delta_1}$, $y^* = 1$, and $C = B(m,n)/B(\alpha,n)$. If $\delta_1 = \delta_2 = 0$, then $C = 1$.

3.6.4 Normal Distribution

A random variable X has a normal distribution if the p.d.f. is

$$f_X(x) = \frac{1}{\sigma\sqrt{2\pi}} \exp\left[-\frac{(x-\mu)^2}{2\sigma^2} \right], \qquad -\infty < x < \infty. \quad (3.6.62)$$

and is denoted $N(\mu, \sigma^2)$. Here μ is the mean and σ^2 is the variance.

Since $X = \mu + \sigma Z$, where Z is the standard normal variable denoted by $N(0, 1)$, we consider only generation from $N(0, 1)$. As we mentioned in Section 3.2, the inverse transform method cannot be applied to the normal distribution and some alternative procedures have to be employed. We consider some of them. More about generation from normal distribution can be found in Fishman [12].

Procedure N-1

This approach is due to Box and Muller [6]. Let us prove that, if U_1 and U_2 are independent random variates from $\mathcal{U}(0, 1)$, then the variates

$$Z_1 = (-2 \ln U_1)^{1/2} \cos 2\pi U_2 \quad (3.6.63)$$

$$Z_2 = (-2 \ln U_1)^{1/2} \sin 2\pi U_2$$

are independent standard normal deviates. To see this let us rewrite the system (3.6.63) as

$$Z_1 = (2V)^{1/2} \cos 2\pi U \quad (3.6.64)$$

$$Z_2 = (2V)^{1/2} \sin 2\pi U,$$

where V is from exp(1) and $U_2 = U$. It follows from (3.6.64) that

$$Z_1^2 + Z_2^2 = 2V \qquad \text{and} \qquad \frac{Z_2}{Z_1} = \tan 2\pi U.$$

The Jacobian of the transformation

$$J = \begin{vmatrix} \dfrac{\partial u}{\partial z_1} & \dfrac{\partial u}{\partial z_2} \\ \dfrac{\partial v}{\partial z_1} & \dfrac{\partial v}{\partial z_2} \end{vmatrix} = \begin{vmatrix} \dfrac{-z_2 \cos^2 2\pi u}{2\pi z_1^2} & \dfrac{\cos^2 2\pi u}{2\pi z_1} \\ z_1 & z_2 \end{vmatrix}$$

$$= \begin{vmatrix} \dfrac{-z_2}{4\pi v} & \dfrac{z_1}{4\pi v} \\ z_1 & z_2 \end{vmatrix} = -\frac{1}{4\pi v}\left(z_2^2 + z_1^2\right) = -\frac{1}{2\pi}$$

and

$$f_{Z_1Z_2}(z_1, z_2) = f_{Z,V}(u, v)|J| = \frac{1}{2\pi}\exp\left(-\frac{z_1^2 + z_2^2}{2}\right). \tag{3.6.65}$$

The last formula represents the joint p.d.f. of two independent standard normal deviates.

Algorithm N-1

1 Generate two independent random variates U_1 and U_2 from $\mathcal{U}(0, 1)$.
2 Compute Z_1 and Z_2 simultaneously by substituting U_1 and U_2 in the system of equations (3.6.63).

Procedure N-2

This procedure is based on the acceptance-rejection method. Let the r.v. X be distributed

$$f_X(x) = \sqrt{\frac{2}{\pi}}\, e^{-x^2/2}, \qquad x \geq 0. \tag{3.6.66}$$

Since the standard normal distribution is symmetrical about zero, we can assign a random sign to the r.v. generated from (3.6.66) and obtain an r.v. from $N(0, 1)$.

To generate an r.v. from (3.6.66) write $f_X(x)$ as

$$f_X(x) = Ch(x)g(x)$$

where

$$h(x) = e^{-x} \tag{3.6.67}$$

$$C = \sqrt{\frac{2e}{\pi}} \tag{3.6.68}$$

$$g(x) = \exp\left[-\frac{(x-1)^2}{2}\right]. \tag{3.6.69}$$

The efficiency of the method is equal to $\sqrt{\pi/2e} \simeq 0.76$.

The acceptance condition

$$U \leq g(Y) \text{ is } U \leq \exp\left[-(Y-1)^2/2\right], \tag{3.6.70}$$

which is equivalent to

$$-\ln U \geq \frac{(Y-1)^2}{2}, \tag{3.6.71}$$

where Y is from $\exp(1)$.

Since $-\ln U$ is also from exp(1), the last inequality can be written

$$V_2 \geq \frac{(V_1 - 1)^2}{2}, \tag{3.6.72}$$

where both $V_2 = -\ln U$ and $V_1 = Y$ are from exp(1).

Algorithm N-2

1 Generate V_1 and V_2 from exp(1).
2 If $V_2 < (V_1 - 1)^2/2$, go to step 1.
3 Generate U from $\mathcal{U}(0, 1)$.
4 If $U \geq 0.5$, deliver $Z = -V_1$.
5 Deliver $Z = V_1$.

Remark In order to obtain Algorithm N-2 we can represent $f_X(x)$ as

$$f_X(x) = Ch_{Y_1}(x)(1 - H_{Y_2}(x)),$$

where

$$h_{Y_1}(x) = h(x) = e^{-x}$$

$$H_{Y_2}(T(x)) = 1 - e^{-T(x)}$$

$$T(x) = \tfrac{1}{2}(x-1)^2,$$

and then apply Algorithm AR-3′.

Procedure N-3

In this procedure we make use of the logistic distribution [32]

$$h(x, \theta) = \frac{e^{-x/\theta}}{\theta[1 + e^{-x/\theta}]^2}, \qquad -\infty < x < \infty. \tag{3.6.73}$$

It is shown numerically in Ref. 32 that $\theta^* = 0.626657$,

$$\frac{1}{C} = 0.9196 \tag{3.6.74}$$

and

$$g(x) = 0.25\left[1 + \exp(-1.5957x)^2 \exp\left(\frac{-x^2}{2} + 1.5957x\right)\right]. \tag{3.6.75}$$

Algorithm N-3 is as follows.

Algorithm N-3

1 Generate U_1 and U_2 from $\mathcal{U}(0, 1)$.
2 $Y \leftarrow -0.626657 \ln(1/U - 1)$.
3 If $U \leq g(Y)$, deliver Y.
4 Go to step 1.

Procedure N-4

This procedure is based on the relationship between the normal distribution with chi-squared distribution and a vector uniformly distributed on the n-dimensional unit sphere.

Let $Z_1, \ldots, Z_n$ be i.i.d. r.v.'s distributed $N(0, 1)$ and let $X = (\sum_{i=1}^{n} Z_i^2)^{1/2}$; then it can be shown by the multivariate transformation method that the vector

$$Y = (Y_1, \ldots, Y_n) = \left(\frac{Z_1}{X}, \ldots, \frac{Z_n}{X} \right) \tag{3.6.76}$$

is distributed uniformly on the n-dimensional unit sphere.*

Now taking into account that $X^2 = \sum_{i=1}^{n} Z_i^2$ has the chi-squared distribution with n degrees of freedom (see Section 3.6.8), the algorithm for generating from $N(0, I)$, where I is a unit matrix of size n, is as follows.

Algorithm N-4

1 Generate a random vector $Y = (Y_1, \ldots, Y_n)$ uniformly distributed on the n-dimensional unit sphere.

2 Generate a chi-square distributed random variate χ^2 with n degrees of freedom.

3 $Z_k = XY_k$, $k = 1, \ldots, n$.

4 Deliver $Z = (Z_1, \ldots, Z_n)$.

Since the efficiency of the algorithm for generating $Y = (Y_1, \ldots, Y_n)$ (see example 5, Section 3.4.2) decreases when n increases, it would be interesting to find the optimal n in order to minimize the CPU time while sampling from $N(0, I)$.

Procedure N-5

This procedure relies on the central limit theorem, which says that if X_i, $i = 1, \ldots, n$, are i.i.d. r.v.'s with $E(X_i) = \mu$ and $\mathrm{var}(X_i) = \sigma^2$, then

$$Z = \frac{\sum_{i=1}^{n} X_i - n\mu}{n^{1/2}\sigma} \tag{3.6.77}$$

converges asymptotically with n to $N(0, 1)$. Consider the particular case

*An alternative algorithm for generating a vector uniformly distributed on the n-dimensional unit sphere is given in example 5, Section 4.3.2.

when all X_i, $i = 1, \ldots, n$, are from $\mathcal{U}(0, 1)$. We find that

$$\mu = \tfrac{1}{2}$$

$$\sigma = \frac{1}{\sqrt{12}}$$

$$Z = \frac{\sum_{i=1}^{n} U_i - n/2}{\sqrt{\frac{n}{12}}}. \tag{3.6.78}$$

A good approximation can already be obtained for $n = 12$. In this case

$$Z = \sum_{i=1}^{12} U_i - 6. \tag{3.6.79}$$

Algorithm N-5 is straightforward.

Algorithm N-5

1 Generate 12 uniformly distributed random variates $U_1, \ldots, U_{12}$ from $\mathcal{U}(0, 1)$.
2 $Z \leftarrow \sum_{i=1}^{12} U_i - 6$.
3 Deliver Z.

Procedure N-6

Another approximation technique for generating from $N(0, 1)$ is given in Tocher [33]; it makes use of the following approximation:

$$e^{-x^2/2} \simeq \frac{2e^{-kx}}{(1 + e^{-kx})^2} \tag{3.6.80}$$

for $x > 0$ and $k = \sqrt{8/\pi}$.

The c.d.f. for the approximation is

$$F_X(x) = \frac{1}{1 - e^{-kx}} - 1.$$

The inverse transformation is

$$X = \frac{1}{k} \ln \frac{1 + U}{1 - U}. \tag{3.6.81}$$

Attaching a random sign κ to this variate we obtain the desired variate

$$Z = \kappa X.$$

Algorithm N-6

1 Generate U_1 and U_2 from $\mathcal{U}(0, 1)$.
2 $X \leftarrow \sqrt{\pi/8}\ \ln[(1 + U_1)/1 - U_1)]$.
3 If $U_2 \leq 0.5$, deliver $Z = -X$.
4 Deliver $Z = X$.

3.6.5 Lognormal Distribution

Let X be from $N(\mu, \sigma^2)$. Then $Y = e^X$ has the lognormal distribution with p.d.f.

$$f_Y(y) = \begin{cases} \dfrac{1}{\sqrt{2\pi}\,\sigma y} \exp\left[-\dfrac{(\ln y - \mu)^2}{2\sigma^2}\right], & 0 \leq y \leq \infty \\ 0, & \text{otherwise.} \end{cases} \tag{3.6.82}$$

Algorithm LN-1

1 Generate Z from $N(0, 1)$.
2 $X \leftarrow \mu + \sigma Z$.
3 $Y \leftarrow e^X$.
4 Deliver Y.

3.6.6 Cauchy Distribution

An r.v. X has a Cauchy distribution denoted by $C(\alpha, \beta)$ if the p.d.f. is equal to

$$f_X(x) = \frac{\beta}{\pi\left[\beta^2 + (x - \alpha)^2\right]}, \qquad \alpha > 0, \beta > 0, -\infty < x < \infty \tag{3.6.83}$$

and the c.d.f. is equal to

$$F_X(X) = \tfrac{1}{2} + \pi^{-1} \tan^{-1}\left(\frac{X - \alpha}{\beta}\right). \tag{3.6.84}$$

Applying the inverse transform method, we obtain

$$X = F_X^{-1}(U) = \alpha + \beta \tan\left[\pi\left(U - \tfrac{1}{2}\right)\right] = \alpha - \frac{\beta}{\tan(\pi U)}. \tag{3.6.85}$$

Algorithm C-1 describes the necessary steps.

Algorithm C-1

1 Generate U from $U(0, 1)$.
2 $X \leftarrow \alpha - \beta/\tan(\pi U)$.
3 Deliver X.

The next algorithm is based on the following two properties:

(a) If Z_1 and Z_2 are independent variates from $N(0, 1)$ then $Y = Z_1/Z_2$ is from $C(0, 1)$.

(b) If X is from $C(0, 1)$ then $Y = \beta X + \alpha$ is from $C(\alpha, \beta)$. The last property can be obtained directly from the transformation formula (3.5.9)

$$f_Y(y) = f_X(x)\left|\frac{dx}{dy}\right| I_\kappa(x)$$

Algorithm C-2

1 Generate Z_1 and Z_2 from $N(0, 1)$.
2 $X \leftarrow \beta Z_1/Z_2 + \alpha$.
3 Deliver X.

The third algorithm is based on the following property [18]:

(c) If Y_1 and Y_2 are independent r.v.'s both from $\mathcal{U}(-\frac{1}{2}, \frac{1}{2})$ and $Y_1^2 + Y_2^2 \le \frac{1}{4}$ then $X = Y_1/Y_2$ is from $C(0, 1)$.

Algorithm C-3

Generate U_1 and U_2 from $\mathcal{U}(0, 1)$.
2 $Y_1 \leftarrow U_1 - \frac{1}{2}$ and $Y_2 \leftarrow U_2 - \frac{1}{2}$.
3 If $Y_1^2 + Y_2^2 > \frac{1}{4}$ go to 1.
4 $X \leftarrow \beta Y_1/Y_2 + \alpha$.
5 Deliver X.

The efficiency of the algorithm is

$$P\left(Y_1^2 + Y_2^2 \le \tfrac{1}{4}\right) = \tfrac{\pi}{4},$$

so the algorithm is relatively efficient.

3.6.7 Weibul Distribution

An r.v. has a Weibul distribution if the p.d.f. is equal to

$$f_X(x) = \begin{cases} \dfrac{\alpha}{\beta^\alpha} x^{\alpha-1} e^{-(x/\beta)^\alpha}, & 0 \le x < \infty, \alpha > 0, \beta > 0 \\ 0, & \text{otherwise} \end{cases} \tag{3.6.86}$$

and is denoted by $W(\alpha, \beta)$.

To generate X by the inverse transformation method note that

$$U = F_X(x) = 1 - e^{-(x/\beta)^\alpha} \tag{3.6.87}$$

so

$$X = \beta(-\ln(1 - U))^{1/\alpha}. \tag{3.6.88}$$

Since $1 - U$ is also from $\mathcal{U}(0, 1)$, we have

$$X = \beta(-\ln U)^{1/\alpha} \tag{3.6.89}$$

or

$$\left(\frac{X}{\beta}\right)^{\alpha} = -\ln U. \tag{3.6.90}$$

Taking into account that $-\ln(U)$ is from $\exp(1)$, the algorithm for generating an r.v. from a Weibul distribution can be written as follows.

Algorithm W-1

1 Generate V from $\exp(1)$.
2 $X \leftarrow \beta V^{1/\alpha}$.
3 Deliver X.

3.6.8 Chi-Square Distribution

Let $Z_1, \ldots, Z_k$ be from $N(0, 1)$. Then

$$Y = \sum_{i=1}^{k} Z_i^2 \tag{3.6.91}$$

has the chi-square distribution with k degrees of freedom and is denoted $\chi^2(k)$.

Formula (3.6.91) says, "the sum of the squares of independent standard normal random variables has a chi-square distribution with degrees of freedom equal to the number of terms in the sum". One approach for generating a chi-square variate from $\chi^2(k)$ is to generate k standard normal random variables and then apply (3.6.91).

Another approach makes use of the fact that $\chi^2(k)$ is a particular case of a gamma density with gamma parameters α and β equal, respectively, to $k/2$ and 2.

Consider two cases.

CASE 1 If k is even, then Y can be computed as

$$Y = -2\ln\left(\prod_{i=1}^{k/2} U_i\right). \tag{3.6.92}$$

Formula (3.6.92) requires $k/2$ uniform variates compared to k in (3.6.91). It also requires one logarithmic transformation, compared to k logarithmic and k cosine or sine transformations for generating Z_i from $N(0, 1)$, $i = 1, \ldots, k$ (see (3.6.63) and (3.6.64)).

CASE 2 If k is odd, then

$$Y = -2 \ln \left(\prod_{i=1}^{k/2-1/2} U_i \right) + Z^2, \tag{3.6.93}$$

where Z is from $N(0, 1)$ and U_i is from $\mathcal{U}(0, 1)$.

For $k > 30$ the normal approximation for chi-square variates can be used based on the following formula [24]:

$$Z = \sqrt{2Y} - \sqrt{2k-1}\,.$$

Solving for Y, the chi-square variate, we obtain

$$Y = \frac{(Z + \sqrt{2k-1}\,)^2}{2}. \tag{3.6.94}$$

Remark Let Y_1, Y_2, and Y_3 be chi-square random variables with degrees of freedom $2(\alpha + \beta)$, 2α, and 2β, respectively; then

$$W = \frac{Y_1}{Y_2 + Y_3}$$

has a beta density with parameters α and β. Applying formula (3.6.92), we get

$$W = \frac{\ln (U_1 U_2 \cdots U_{(\alpha+\beta)})}{\ln (U_{\alpha+\beta+1} \cdots U_{2(\alpha+\beta)})}.$$

3.6.9 Student's *t* Distribution

Let Z have a standard normal distribution, let Y have a chi-square distribution with k degrees of freedom, and let Z and Y be independent; then

$$X = \frac{Z}{\sqrt{Y/k}} \tag{3.6.95}$$

has a Student's t distribution with k degrees of freedom. To generate X we simply generate Z as described in Section 3.6.4 and Y as described in Section 3.6.8 and apply (3.6.95). For $k \geq 30$ the normal approximation can be used.

3.6.10 *F* Distribution

Let Y_1 be a chi-square random variable with k_1 degrees of freedom; let Y_2 be a chi-square random variable with k_2 degrees of freedom, and let Y_1

and Y_2 be independent. Then the random variable

$$X = \frac{Y_1/k_1}{Y_2/k_2} \tag{3.6.96}$$

is distributed as an F distribution with k_1 and k_2 degrees of freedom. To generate an F variate we first produce two chi-square variates and then use (3.6.96).

Remark 1. If X has an F distribution with k and k_2 degrees of freedom, then $1/X$ has an F distribution with k_2 and k_1 degrees of freedom.

Remark 2. If X is an F-distributed random variable with k_1 and k_2 degrees of freedom, then

$$W = \frac{k_1 X/k_2}{1 + k_1 X/k_2} \tag{3.6.97}$$

has a beta density with parameters $\alpha = k_1/2$ and $\beta = k_2/2$.

3.7 GENERATING FROM DISCRETE DISTRIBUTIONS

In this section we describe several procedures for generating stochastic variates from most of the well known discrete distributions. We start with the inverse transform method, which is generally easily implemented and is widely used.

Let X be a discrete r.v. with probability mass function (p.m.f.)

$$\Pr(X = x_k) = P_k, \qquad k = 0, 1, \ldots \tag{3.7.1}$$

and with c.d.f.

$$g_k = \Pr(X \leq X_k) = \sum_{i=0}^{k} P_i. \tag{3.7.2}$$

Then

$$\Pr(g_{k-1} < U \leq g_k) = \int_{g_{k-1}}^{g_k} du = g_k - g_{k-1} = P_k, \quad g_{-1} = 0, \tag{3.7.3}$$

where U is from $\mathcal{U}(0, 1)$. Thus

$$X = \min\{x : g_{k-1} < U \leq g_k\}. \tag{3.7.4}$$

Algorithm IT-2, which is called the inverse transform algorithm, describes generating discrete r.v.'s. This algorithm is based on logical comparison of U with g_k's and is as follows.

Algorithm IT-2

1 $C \leftarrow P_0$.
2 $B \leftarrow C$.
3 $K \leftarrow 0$.
4 Generate U from $\mathcal{U}(0,1)$.
5 If $U \leq B$ $(U \leq g_k)$, deliver $X = x_k$.
6 $K \leftarrow K+1$.
7 $C \leftarrow A_{k+1}C$ $(P_{k+1} = A_{k+1}P_k)$.
8 $B \leftarrow B + C$ $(g_{k+1} = g_k + P_{k+1})$.
9 Go to step 5.

Here P_0 and $A_{k+1} = P_{k+1}/P_k$ are distributed dependent. The recurrent formulas

$$P_{k+1} = A_{k+1}P_k \tag{3.7.5}$$

$$g_{k+1} = g_k + P_{k+1} \tag{3.7.6}$$

in steps 7 and 8 are straightforward for calculation.

Most discrete r.v.'s are integers nonnegative valued, that is, $x_k = k$, $k = 0, 1, \ldots$. Later, we consider only these r.v.'s. It is easy to see that the mean number of trials

$$C = 1 + \sum_{k=1}^{\infty} x_k P_k = \sum_{k=1}^{\infty} kP_k = 1 + E(X) \tag{3.7.7}$$

is equal to the expected value plus one additional trial.

Table 3.7.1 represents the values of P_0, A_{k+1}, and C for most well known discrete distributions.

In order to generate an r.v. from a specified discrete distribution, we take the corresponding values P_0 and A_{k+1} from Table 3.7.1 and then run Algorithm IT-2.

In many cases we can improve the efficiency of the inverse transform method IT-2 by starting the search of X at $k = m$, m being an interior point (for example, mode, median, etc.), rather than at $k = 0$. We assume that tables of P_k and g_k are available.

The procedure is as follows. If $U \geq g_m$, then

$$g_{m+i} = g_{m+i-1} + P_{m+i} \tag{3.7.8}$$

$$P_{m+i} = P_{m+i-1}A'_{m+i}, \qquad i = 1, 2, \ldots. \tag{3.7.9}$$

If $U < g_m$, then

$$g_{m-i} = g_{m-i+1} - P_{m-i+1} \tag{3.7.10}$$

$$P_{m-i} = P_{m-i+1}A''_{m-i}, \qquad i = 1, 2, \ldots, m, \tag{3.7.11}$$

Table 3.7.1 Discrete Distributions

Distribution	Notation	P_0	$A_{k+1} = \frac{P_{k+1}}{P_k}$	Mean Number of Trials C
Binomial				
$P_x = \binom{n}{x} p^x (1-p)^{n-x}$ $x = 0, 1, \ldots, n, 0 < p < 1$	$B(n, p)$	$(1-p)^n$	$\frac{(n-k)p}{(k+1)(1-p)}$	$1 + np$
Poisson				
$P_x = \frac{e^{-\lambda}\lambda^x}{x!}$ $x = 0, 1, \ldots; \lambda > 0$	$P(\lambda)$	$e^{-\lambda}$	$\frac{\lambda}{k+1}$	$1 + \lambda$
Geometric				
$P_x = p(1-p)^x$ $x = 0, 1, \ldots; p > 0$	$Ge(p)$	p	$1 - p$	$1 + \frac{1-p}{p}$
Negative binomial				
$P_x = \binom{x+r-1}{x} p^r (1-p)^x$ $x = 0, 1, \ldots; p < 1$	$NB(r, p)$	p^r	$\frac{(r+k)(1-p)}{k+1}$	$1 + \frac{r(1-p)}{p}$
Hypergeometric				
$P_x = \frac{\binom{n_1}{x}\binom{n-n_1}{m-x}}{\binom{n}{m}}$ $\max(0, n_1 + m - n)$ $\leq x \leq \min(n_1, m)$	$H(n, m, n_1)$	$\frac{(n-n_1)!(n-m)!}{(n-2n_1)!n!}$	$\frac{(n_1-k)(m-k)}{(k+1)(n-n_1-m+1+k)}$	$1 + \frac{n_1 m}{n}$

where A'_{m+i} and A''_{m-i} are distribution dependent and their values are available to compute.

Algorithm IT-3 describes the necessary steps.

Algorithm IT-3

1. $D \leftarrow g_m$.
2. $E \leftarrow P_m$.
3. Generate U from $\mathcal{U}(0, 1)$.
4. $K \leftarrow m$.
5. If $U > g_k$, go to step 12.
6. $D \leftarrow D - E \quad (g_{k-1} = g_k - P_k)$.
7. If $U > D$, deliver $X = K$; go to step 1.
8. $K \leftarrow K - 1$.
9. If $K = 0$, deliver $X = K$; go to step 1.
10. $E \leftarrow EA''_{k-1} \quad (P_{k-1} = A''_{k-1}P_k)$.
11. Go to step 6.
12. $K \leftarrow K + 1$.
13. $E \leftarrow EA'_{k+1} \quad (P_{k+1} = A'_{k+1}P_k)$.
14. $D \leftarrow D + E$.
15. If $U \leq D$, deliver $X = K$.
16. Go to step 12.

Table 3.7.2 represents the values of P_0, m(mode), A'_{k+1} and A''_{k+1} for most well known discrete distributions.

It is easy to see that for an integer m the number of trials (number of logical comparisons of U with g'_x s) is the following r.v.:

$$\eta = \begin{cases} 2 + (m - X), & \text{if } x = 0, 1, \ldots, m \\ 1 + (X - m), & \text{if } x = m + 1, m + 2, \ldots. \end{cases} \tag{3.7.12}$$

The mean number of trials is

$$\begin{aligned} C = E(\eta) &= \sum_{x=0}^{m} [2 + (m - x)] P_x - \sum_{x=m+1}^{\infty} [1 + (x - m)] P_x \\ &= \sum_{x=0}^{m} P_x + \sum_{x=0}^{m} P_x + \sum_{x=m+1}^{\infty} P_x + m \sum_{x=0}^{m} P_x - m \sum_{x=m+1}^{\infty} P_x - \sum_{x=0}^{m} xP_x \\ &\quad + \sum_{x=m+1}^{\infty} xP_x = g_m + 1 + mg_m - m(1 - g_m) - \sum_{x=0}^{m} xP_x + E(X) \\ &\quad - \sum_{x=0}^{m} xP_x = 1 + E(X) - \gamma(m), \end{aligned} \tag{3.7.13}$$

Table 3.7.2 Discrete Unimodal Distributions

Distribution	Notation	P_0	Modal Value m	$A'_{k+1} = \frac{p_{k+1}}{p_k}$	$A''_{k-1} = \frac{p_{k-1}}{p_k}$
Binomial					
$P_x = \binom{n}{x} p^x (1-p)^{n-x}$	$B(n,p)$	$(1-p)^n$	$[(n+1)p]$	$\frac{n-k}{k+1} \cdot \frac{p}{1-p}$	$\frac{k}{n-k+1} \cdot \frac{1-p}{p}$
$x = 0, 1, \ldots, n, p > 0$					
Poisson					
$P_x = \frac{e^{-\lambda}\lambda^x}{x!}$	$P(\lambda)$	$e^{-\lambda}$	$[\lambda]$	$\frac{\lambda}{k+1}$	$\frac{k}{\lambda}$
$x = 0, 1, \ldots, \lambda > 0$					
Negative binomial					
$P_x = \binom{x+r-1}{x} p^r (1-p)^x$	$NB(r,p)$	p^r	$[\frac{(r-p)(1-p)}{p}]$	$\frac{(r+k)(1-p)}{k+1}$	$\frac{k}{(r+k-1)(1-p)}$
$x = 0, 1, \ldots, 0 < p < 1$					
Hypergeometric					
$P_x = \frac{\binom{n_1}{x}\binom{n-n_1}{m-x}}{\binom{n}{m}}$	$H(n,m,n_1)$	$\frac{(n-n_1)!(n-m)!}{(n-2n_1)!n!}$	$[(m+1)(n_1+1)(n+2)]$	$\frac{(n_1-k)(m-k)}{(k+1)(n-n_1-m+1+k)}$	$\frac{k(n-n_1-m+k)}{(n_1-k+1)(m-k+1)}$
$\max(0, n_1+m-n)$ $\le x \le \min(n_1, m)$					

where

$$\gamma(m)=2\sum_{x=0}^{m} xP_x - g_m + m - 2mg_m. \tag{3.7.14}$$

It follows from (3.7.7) and (3.7.13) that Algorithm IT-3 is more efficient than Algorithm IT-2 for m such that $\gamma(m)>0$. However, $\gamma(m)$ is not necessarily positive for each m.

The following example illustrates this point.

Example 1 Assume that the r.v. X has the following p.m.f.:

$$P_x=\begin{cases}\frac{1}{4}, & x=0\\ \frac{1}{2}, & x=1\\ \frac{1}{4}, & x=2\\ 0, & \text{otherwise.}\end{cases}$$

Let $m=1$; then $\gamma(1)=2\Sigma_{x=0}^{1}xP_x-g_m+1-2g_m=1-\frac{3}{4}+1-\frac{3}{2}=-0.25<0$, and therefore Algorithm IT-2 is more efficient than Algorithm IT-3.

Nevertheless, in many cases it is possible to choose the starting point m in such a way that $\gamma(m)>0$, and therefore it is possible for IT-3 to be more efficient than IT-2.

Lemma 3.7.1 If there exist $m>0$ such that

$$P_0 \le \sum_{x=1}^{m}(2x-1)P_x, \qquad \text{for } g_m \le \tfrac{1}{2}, \tag{3.7.15}$$

then $\gamma(m)>0$.

Proof Condition $P_0\le\Sigma_{x=1}^{m}(2x-1)P_x$ is equivalent to

$$2\sum_{x=0}^{m}xP_x-g_m>0, \tag{3.7.16}$$

and, correspondingly, condition $g_m\le\frac{1}{2}$, $m>0$, is equivalent to

$$m-2mg_m>0. \tag{3.7.17}$$

Both (3.7.16) and (3.7.17) yield $\gamma(m)>0$. Q.E.D.

Note 1 We can see that Lemma 3.7.1 is valid if $P_0\le\Sigma_{x=1}^{m}P_x$.

This condition is not restrictable and holds for practically all discrete distributions.

Lemma 3.7.2 $\gamma(m)$ achieves its maximum at points m_0 or m_0+1 where $m_0 = \max\{m : g_m \le \frac{1}{2}\}$, depending, correspondingly, on whether $g_{m_0} + g_{m_0+1} \le 1$ or $g_{m_0} + g_{m_0+1} > 1$.

Proof It is straightforward to obtain from (3.7.14) that

$$\Delta\gamma(m) = \gamma(m+1) - \gamma(m) = 1 - g_m - g_{m+1}. \tag{3.7.18}$$

For $m < m_0$ we have $g_m + g_{m+1} \le 1$, and therefore $\Delta\gamma(m) \ge 0$; for $m > m_0$ we have, correspondingly, $g_m + g_{m+1} > 1$ and $\Delta\gamma(m) < 0$. Therefore $\gamma(m)$ is a unimodal function with the maximum at points m_0 or $m_0 + 1$, depending on whether $g_{m_0} + g_{m_0+1} \le 1$ or $g_{m_0} + g_{m_0+1} > 1$. Q.E.D.

Note 2 In other words, Lemma 3.7.2 says that $\gamma(m)$ achieves its maximum at the median or at a point neighboring the median on the left.

As a corollary from these two lemmas we obtain the following theorem.

Theorem 3.7.1 The optimal starting point in Algorithm IT-3 is either the median $m_0 = \max\{m : g_m \le \frac{1}{2}\}$, if $P_0 \le \sum_{x=1}^{m_0+1}(2x-1)P_x$ and $g_{m_0} + g_{m_0+1} \le 1$, or $m_0 + 1$, if $P_0 \le \sum_{x=1}^{m_0+1}(2x-1)P_x$ and $g_{m_0} + g_{m_0+1} > 1$.

Note 3 Theorem 3.7.1 is valid not only for integer nonnegative valued r.v.'s, but for any discrete r.v. with values $x_0, x_1, \ldots$, since Algorithm IT-3 is determined not by the sequence $x_0, x_1, \ldots$, but by its indices $0, 1, \ldots$.

In the rest of this chapter we consider some alternative procedures for generating discrete r.v.'s. Generally, procedures for generating discrete variates are simpler than procedures for generating continuous variates, and we describe them only briefly.

3.7.1 Binomial Distribution

An r.v. X has a binomial distribution if the p.m.f. is equal to

$$P_x = \binom{n}{x} p^x (1-p)^{n-x}, \qquad x = 0, \ldots, n \tag{3.7.19}$$

and is denoted by $B(n,p)$. Here $0 < p < 1$ is the probability of success in a single trial, and n is the number of trials.

To apply the inverse transform method IT-2 we must check the following condition after step 5: if $K = n - 1$, terminate the procedure with $X = K = n$.

It is also worthwhile to note that, if Y is from $B(n,p)$, then $n-Y$ is from $B(n,1-p)$. Hence for purposes of efficiency we generate X from $B(n,p)$ according to

$$X=\begin{cases} Y\sim B(n,p) \text{ if } p\le\frac{1}{2} \\ Y\sim B(n,1-p) \text{ if } p>\frac{1}{2}. \end{cases} \tag{3.7.20}$$

For larger n the inverse-transform procedure becomes time consuming, and we can consider the normal distribution as an approximation to the binomial.

As n increases the distribution of

$$Z=\frac{X-np+\frac{1}{2}}{\left[np(1-p)\right]^{1/2}} \tag{3.7.21}$$

approaches $N(0,1)$.

To obtain a binomial variate we generate Z from $N(0,1)$, solve (3.7.21) with respect to X, and round to nonnegative integer, that is,

$$X=\max\left\{0,\left[-0.5+np+Z(np(1-p))^{1/2}\right]\right\}, \tag{3.7.22}$$

where $[\alpha]$ denotes the integer part of α.

We should consider replacing the binomial with the approximate normal when $np>10$ for $p\ge\frac{1}{2}$ and $n(1-p)>10$ for $p<\frac{1}{2}$.

It is shown [22] that, if m is the mode, then for large n the mean number of trials in Algorithm IT-3 is equal to

$$C=1.5+\sqrt{\frac{2}{\pi}}\cdot\sqrt{np(1-p)}\,. \tag{3.7.23}$$

Comparing both Algorithms IT-2 and IT-3 (compare (3.7.7) with (3.7.23)), we can see that for large n the mean number of trials is proportional to np and $\sqrt{np(1-p)}$, respectively.

So for large n Algorithm IT-3 is essentially more efficient than Algorithm IT-2.

The acceptance-rejection method can also successfully be implemented for generating from $B(n,p)$ (see Ahrens and Dieter [4] and Marsaglia [20]). Description of algorithms for this and their efficiency can be found in Fishman's monograph [12].

3.7.2 Poisson Distribution

An r.v. X has a Poisson distribution if the p.m.f. is equal to

$$P_x=\frac{\lambda^x e^{-\lambda}}{x!},\qquad x=0,1,\ldots;\lambda>0 \tag{3.7.24}$$

and is denoted by $P(\lambda)$.

It is well known (Feller [11]) that, if the time intervals between events are from $\exp(1/\lambda)$, the number of events occurring in an unit interval of time is from $P(\lambda)$.

Mathematically, it can be written

$$\sum_{i=0}^{X} T_i \leq 1 \leq \sum_{i=0}^{X+1} T_i, \tag{3.7.25}$$

where T_i, $i = 0, 1, \ldots, X + 1$, are from $\exp(1/\lambda)$.

Since $T_i = -(1/\lambda) \ln U_i$, the last formula can be written as

$$-\sum_{i=0}^{X} \ln U_i \leq \lambda \leq -\sum_{i=0}^{X+1} \ln U_i, \qquad X = 0, 1, \ldots \tag{3.7.26}$$

or

$$\prod_{i=0}^{X} U_i \geq e^{-\lambda} \geq \prod_{i=0}^{X+1} U_i, \qquad X = 0, 1, \ldots. \tag{3.7.27}$$

The following algorithm is written with respect to (3.7.25):

1 $A \leftarrow 1 \quad (g_k = 1)$.
2 $K \leftarrow 0$.
3 Generate U_k from $\mathcal{U}(0, 1)$.
4 $A \leftarrow U_k A \quad (g_{k+1} = g_k U_k)$.
5 If $A < e^{-\lambda}$, deliver $X = K$.
6 $K \leftarrow K + 1$.
7 Go to step 3.

For large $\lambda (\lambda > 10)$ we can approximate the Poisson distribution by normal distribution. As λ increases, the distribution of

$$Z = \frac{X - \lambda + \frac{1}{2}}{\lambda^{1/2}} \tag{3.7.28}$$

approaches $N(0, 1)$.

To obtain a Poisson variate we generate Z from $N(0, 1)$, then by analogy with (3.7.22) we obtain

$$X = \max\left(0, \left[\lambda + Z^{1/2} - 0.5\right]\right), \tag{3.7.29}$$

where $[\alpha]$ is the integer part of α.

It is shown in Ref. 22 that, if m is the mode, then for large n the mean execution time in Algorithm IT-3 is similar to (3.7.23) and is equal to

$$C = 1.5 + \sqrt{\frac{2}{\pi}}\, \lambda^{1/2}. \tag{3.7.30}$$

The mean number of trials in both Algorithms IT-2 and IT-3 are proportional, respectively, to λ and $\lambda^{1/2}$, and therefore Algorithm IT-3 is again essentially more efficient than Algorithm IT-2.

3.7.3 Geometric Distribution

An r.v. has the geometric distribution if the p.m.f. is equal to

$$P_X = p(1-p)^x, \qquad x = 0, 1, \ldots, \quad 0 < p < 1 \tag{3.7.31}$$

and is denoted by $Ge(p)$. Geometric distribution describes the number of trials to the first success in a serial of Bernoulli trials.

The following procedure describes generating from $Ge(p)$ and is based on the relationship between exponential and geometric distribution. Let Y be from $\exp(\beta)$; then

$$\Pr(x < Y \leq x+1) = \frac{1}{\beta}\int_x^{x+1} e^{-y/\beta}\, dy = e^{-x/\beta}(1 - e^{-1/\beta}), \tag{3.7.32}$$

which is $Ge(p = 1 - e^{-1/\beta})$.

For $\beta = -1/\ln(1-p)$ (3.7.32) is identical to (3.7.31). Therefore

$$X = \left[\frac{\ln U}{\ln(1-p)}\right] = \left[-\frac{V}{\ln(1-p)}\right], \tag{3.7.33}$$

where $V = -\ln(U)$ is a standard exponential variate, that is, X is from $Ge(p)$. Hence to generate an r.v. from $Ge(p)$ we generate an r.v. from the exponential distribution with $\beta = -1/\ln(1-p)$ and round the value to an integer.

CPU time for this procedure is constant, whereas the CPU time for the inverse transform method is proportional to $1/p$. However, because this procedure requires generation from the exponential distribution and rounding, it is more efficient than Algorithm IT-2 only for $p < 0.25$.

3.7.4 Negative Binomial Distribution

The p.m.f. for the negative binomial distribution is

$$P_x = \binom{x+r-1}{x} p^r (1-p)^x, \qquad x = 0, 1, \ldots; p > 0 \tag{3.7.34}$$

and is denoted by $NB(r, p)$. When r is an integer the distribution is called Pascal distribution, which describes the number of successes occurring before the rth failure in a series of Bernoulli trials. This implies that geometric distribution is a special case of Pascal distribution with $r = 1$.

The following algorithm describes generating from Pascal distribution with parameters r and p denoted $PS(r,p)$.

1 $X \leftarrow 0$.
2 $Y \leftarrow 0$.
3 Generate U_{X+Y} from $\mathcal{U}(0, 1)$.
4 If $U_{X+Y} > p$, go to step 8.
5 $Y \leftarrow Y + 1$.
6 If $Y = r$, deliver X.
7 Go to step 3.
8 $X \leftarrow X + 1$.
9 Go to step 3.

An alternative procedure is based on the *reproductive property* of the negative binomial distribution analogous to that for the gamma distribution. Let X_i, $i = 1, \ldots, n$, denote a sequence of i.i.d. r.v.'s from $NB(r_i, p)$. Then $X = \sum_{i=1}^{n} X_i$ is from $PS(r,p)$, where $r = \sum_{i=1}^{n} r_i$.

Suppose that $r_i = 1$, $i = 1, \ldots, r$, which means that X_i, $i = 1, \ldots, r$, are from $Ge(p)$; then $X = \sum_{i=1}^{r} X_i$ is from $NB(r,p)$.

The algorithm is straightforward and contains the following steps:

1 Generate $X_1, \ldots, X_r$ from $Ge(p)$.
2 $X \leftarrow \sum_{i=1}^{r} X_i$.
3 Deliver X.

This procedure is more efficient than the inverse transform method IT-2 for $p > 0.75$.

Another possible method for generating an r.v. from $NB(r,p)$ makes use of the following relationship (see Johnson and Kotz [18, p. 127]):

$$\Pr(X \leq k) = \Pr(Y \geq r), \tag{3.7.35}$$

where X is from $NB(r,p)$ and Y is from $B(p, r + k)$. The reader is asked to describe an algorithm based on (3.7.35), assuming that r.v. Y from $B(p, r + k)$ is given.

The next procedure is based on the relationship between negative binomial distribution with gamma and Poisson distributions.

Suppose we have a mixture of Poisson distributions, such that the parameter λ of the Poisson distributions

$$P(X = x \mid \lambda) = \frac{e^{-\lambda}\lambda^x}{x!}, \qquad x = 0, 1, \ldots$$

varies according to $G(\alpha, \beta)$, that is,

$$f_\Lambda(\lambda) = \frac{1}{\beta^\alpha \Gamma(\alpha)} \lambda^{\alpha-1} e^{-\lambda/\beta}, \qquad \lambda \geq 0, \alpha > 0, \beta > 0. \tag{3.7.36}$$

Then

$$P(X=x) = \int_0^\infty P(X=x|\lambda) f_\Lambda(\lambda)\, d\lambda \tag{3.7.37}$$

$$= [\beta^\alpha \Gamma(\alpha)]^{-1} \int_0^\infty \frac{\lambda^{\alpha-1} e^{-\lambda/\beta} \lambda^x e^{-\lambda}}{x!} d\lambda$$

$$= [\beta^\alpha \Gamma(\alpha)]^{-1} \int_0^\infty \lambda^{\alpha+x-1} \exp\left[-\lambda(\beta^{-1}+1)\right] d\lambda$$

$$= \binom{\alpha+x-1}{x} \left(\frac{\beta}{\beta+1}\right)^x \left(\frac{1}{\beta+1}\right)^\alpha.$$

So X is from $NB(\alpha, 1/(\beta+1))$.

It is obvious that, when λ is from $G(r,(1-p)/p)$, (3.7.37) is identical to (3.7.34). The algorithm is as follows:

1 Generate an r.v. λ from $G(r, p/(1-p))$.
2 Generate X from $P(\lambda)$.
3 Deliver X.

It is not difficult to see that an alternative algorithm for generating an r.v. from $NB(r,p)$ is the following:

1 Generate λ from $G(r, 1)$.
2 Generate X from $P(\lambda p/(1-p))$
3 Deliver X.

3.7.5 Hypergeometric Distribution

An r.v. X has a hypergeometric distribution if the p.m.f. is equal to

$$P_x = \frac{\binom{n_1}{x}\binom{n-n_1}{m-x}}{\binom{n}{m}}, \qquad \max(0, n_1+m-n) \leq x \leq \min(n_1, m) \tag{3.7.38}$$

and is denoted $H(n, m, n_1)$. Hypergeometric distribution describes sampling without replacement from finite population. It has three parameters, n, m, and n_1, which have the following meanings: n, the size of the total population in two classes, m, the size of the sample ($m < n$) that is taken from the total population n without replacement, and n_1, the size of the population in the first class ($n - n_1$ is the size of population in the second class).

Generation from $H(n, m, n_1)$ involves simulating a sampling experiment without replacement, which is merely a Bernoulli trials method of generating from $B(n,p)$ with n and p altering (varying) depending, respectively, on the total number of elements that have been previously drawn from the total population and the number of the first class elements that have been drawn.

The original value $n = n_0$ is reduced according to the formula

$$n_i = n_{i-1} - 1, \qquad i = 1, \ldots, m \tag{3.7.39}$$

when an element in a sample of m is drawn.

Similarly, the value $p = p_0 = n_1/n$, when the ith element in a sample of n elements is drawn, becomes

$$p_i = \frac{n_{i-1}p_{i-1} - \delta}{n_{i-1} - 1}, \qquad i = 1, \ldots, m, \tag{3.7.40}$$

where $\delta = 1$ when the sample elements $(i = 1)$ belong to the first class, and $\delta = 0$ when the sample elements $(i - 1)$ belong to the second class.

EXERCISES

1 Describe an algorithm for generating from Laplace (double exponential distribution)

$$f_X(x) = \tfrac{1}{2}\beta^{-1}\exp\left[\frac{-1|x-\theta|}{\beta}\right], \qquad \beta > 0, \; -\infty < x < \infty.$$

using the inverse transform method

2 Apply the inverse transform method for generating from extreme value distribution

$$f_X(x) = \frac{1}{\sigma}\exp\left[-\frac{1}{\sigma}(x-\mu) - \exp\left(\frac{-(x-\mu)}{\sigma}\right)\right], \qquad -\infty < x < \infty.$$

3 Describe an algorithm for generating from logistic distribution

$$f_X(x) = \frac{\exp[-(x-\alpha)/\beta]}{\beta[1 + \exp[-(x-\alpha)/\beta]]^2}, \qquad -\infty < x < \infty, \beta > 0, \alpha > 0.$$

4 Consider the triangular random variable with the density function

$$f_X(x) = \begin{cases} 0, & \text{if } x < 2a \text{ or } x \geq 2b \\ \dfrac{x - 2a}{(b-a)^2}, & \text{if } 2a \leq x < a + b \\ \dfrac{(2b - x)}{(b-a)^2}, & \text{if } a + b \leq x < 2b, \end{cases}$$

and the distribution function

$$F_X(x) = \begin{cases} 0, & \text{if } x < 2a \\ \dfrac{(x-2a)^2}{2(b-a)^2}, & \text{if } 2a \le x < a+b \\ 1 - \dfrac{(2b-x)^2}{2(b-a)^2} & \text{if } a+b \le x < 2b \\ 1, & \text{if } x \ge 2b \end{cases}$$

This random variable can be considered as a sum of two independent random variables uniformly distributed between a and b. Show that, applying the inverse method, we obtain

$$X = \begin{cases} 2a + (b-a)\sqrt{2U} & \text{if } 0 \le U < 0.5 \\ 2b + (a-b)\sqrt{2(1-U)}, & \text{if } 0.5 \le U < 1. \end{cases}$$

5 Let

$$f_X(x) = \begin{cases} c_i x, & x_{i-1} \le x \le x_i, \quad i = 1, \ldots, n \\ 0, & \text{Otherwise} \end{cases}$$

$$x_0 = a, \qquad x_n = b, \qquad c_i \ge 0, \qquad a \ge 0.$$

Using the inverse transform method, prove that

$$X = \left[x_{i-1}^2 + \frac{2(U - F_{i-1})}{c_i} \right]^{1/2}$$

where $F_i = \sum_{j=1}^{i} \int_{x_{j-1}}^{x_j} c_j x \, dx$. Describe an algorithm for generating from $f_X(x)$.

6 Let $X_1, \ldots, X_n$ be i.i.d. r.v.'s from $\exp(\lambda)$.

(a) Show that $Y_1 = \min(X_1, \ldots, X_n)$ is distributed $\exp(n\lambda)$.
(b) Describe an algorithm for generating from Y_1.

7 Let $U_1, \ldots, U_{\alpha+\beta-1}$ be from $\mathcal{U}(0,1)$. Prove that the αth order statistic $U_{(\alpha)}$ is from $Be(\alpha, \beta)$.

8 The joint density of the r.v.'s X and Y is of the form $f(u^2 + v^2)$ for all u and v. Show that their ratio X/Y has a Cauchy density.

9 Describe two alternative algorithms, correspondingly, for examples 4 and 5 of Section 3.4 by making use c' Theorem 3.4.2.

10 Describe algorithms for generating from the following p.d.f.'s:

(a) $f_{X,Y}(x,y) = ce^{-(x+y)}$, $x \ge 0$, $y \ge 0$.
(b) $f_{XY}(xy) = cxe^{-xy}$, $0 \le x \le 2$, $y \ge 0$.
(c) For generating from $N(\mu, \Sigma)$ where $\mu = (\mu_1, \mu_2, \mu_3) = (1, 2, 3)$, and

$$\Sigma = \begin{vmatrix} 1 & 1 & 0 \\ 1 & 2 & 0 \\ 0 & 0 & 3 \end{vmatrix}.$$

11 Let Y_1 and Y_2 be i.i.d. r.v.'s from $\mathcal{U}(-\frac{1}{2}, \frac{1}{2})$. Prove that, if

$$Y_1^2 + Y_2^2 \leq \tfrac{1}{4},$$

then Y_1/Y_2 is from $C(0, 1)$.

12 Let $V_1, \ldots, V_n$ be i.i.d. r.v.'s from $\exp(1)$ and let $X = \sum_{i=1}^n V_i$. Prove that the vector

$$Y = (Y_1, \ldots, Y_n) = \left(\frac{V_1}{X}, \ldots, \frac{V_n}{X}\right)$$

is distributed uniformly on the simplex $\sum_{i=1}^n y_i = 1$, $0 < y_i < 1$, $i = 1, \ldots, n$.

13 Let $Z_1, \ldots, Z_n$ be i.i.d. r.v.'s distributed $N(0, 1)$ and let $X = (\sum_{i=1}^n Z_i^2)^{1/2}$. Prove that the vector

$$Y = (Y_1, \ldots, Y_n) = \left(\frac{Z_1}{X}, \ldots, \frac{Z_n}{X}\right)$$

is distributed uniformly on the sphere $\sum y_i^2 = 1$.

14 Let $U_{(1)}, \ldots, U_{(n)}$ be order statistics from $\mathcal{U}(0, 1)$. Prove that the vector $X = (X_1, \ldots, X_n)$,

$$X_1 = U_{(1)}, X_2 = U_{(2)} - U_{(1)}, \ldots, X_n = U_{(n)} - U_{(n-1)},$$

is distributed uniformly inside the simplex $\sum_{i=1}^n x_i \leq 1$, $x_i > 0$.

15 Consider the p.d.f.

$$f_X(x) = \sqrt{\frac{2}{\pi}}\, e^{-x^2/2}, \qquad x \geq 0.$$

Let

$$h(x, \beta) = \beta e^{-x/\beta}, \; x > 0, \; \beta > 0$$

$$g(x) = \exp\left(\frac{x}{\beta} - \frac{x^2}{2}\right)$$

$$C = \beta\sqrt{\frac{2}{\pi}}\,.$$

Using (3.4.10), prove that the maximum efficiency is achieved when $\beta = 1$.

16 Describe an algorithm for generating from $Be(\alpha, \beta)$, making use of the inequality

$$x^{\alpha-1}(1-x)^{\beta-1} \leq x^{\alpha-1} + (1-x)^{\beta-1}$$

and assuming

$$h(x) = \frac{\alpha\beta}{\alpha+\beta}\left[x^{\alpha-1} + (1-x)^{\beta-1}\right], \quad 0 \leq x \leq 1, \beta > 0$$

$$g(x) = \left[x^{\alpha-1} + (1-x)^{\beta-1}\right]^{-1} x^{\alpha-1}(1-x)^{\beta-1}$$

$$C = \frac{(\alpha+\beta)}{\alpha\beta}\,\frac{\Gamma(\alpha+\beta)}{\Gamma(\alpha)\Gamma(\beta)}\,.$$

Compare the efficiency of this procedure with the efficiency of Jöhnk's procedure, Be-4.

17 Describe an algorithm for generating from $G(\alpha, 1)$ by the acceptance-rejection method AR-1, assuming

$$h(x) = p\,Er(\beta, m) + (1-p)\,Er(\beta, m+1),$$

that is, $h(x)$ is a mixture of two Erlang distributions, where $m = [\alpha]$ and $\beta = \alpha/m$.

18 Prove that Procedure Be-6 is more efficient than Procedure Be-5 for $\alpha \geq 2$, $\beta \geq 2$.

19 Describe an acceptance-rejection algorithm for generating an r.v. from $N(0, 1)$, representing $f_X(x) = Cg(x)h(x, \beta)$ and assuming that

$$h(x, \beta) = (\pi\beta)^{-1}\left[1 + \left(\frac{x}{\beta}\right)^2\right]^{-1}, \qquad -\infty < x < \infty.$$

Verify that the optimal $\beta = 1$, the efficiency $1/C = e^{1/2}/\sqrt{2\pi} = 0.6578$, and

$$g(x) = 0.8243(1 + x^2)e^{-x^2/2}.$$

From Tadikamalla and Johnson [32].

20 Describe an algorithm for generating from truncated Erlang distribution

$$f_X(x) = c\frac{x^{m-1}e^{-x/\beta}}{\beta^m(m-1)!}, \qquad x \geq 1, \beta > 0, m = 1, 2, \ldots,$$

and find c.

21 Prove that, if $f_X(x)$ can be represented as $f_X(x) = Ch_{Y_1}(x)[1 - H_{Y_2}(T(x))]$, then Algorithm AR-3 can be rewritten as AR-3′.

22 The p.m.f. for the uniform discrete distribution is

$$P_x = \frac{1}{b-a+1}, \qquad x = a, a+1, \ldots, b,$$

where b and a are integers and $b > a$. Prove that $X = [a + (b - a + 1)U]$ has the desired distribution, and describe an algorithm for generating an r.v. from P_x. Here $[\alpha]$ is the integer part of α.

23 Let Y be from Bernoulli distribution, that is,

$$P_Y = p^y(1-p)^{1-y}, \qquad y = 0, 1, 0 < p < 1.$$

Prove that, if $Y_1, \ldots, Y_n$ are i.i.d. r.v.'s from Bernoulli distribution, then $X = \sum_{i=1}^n Y_i$ is from $B(n,p)$. Describe an algorithm for generating an r.v. from $B(n,p)$, using the above result. For purposes of efficiency use the fact that if X is from $B(n,p)$, then $n - X$ is from $B(n, 1-p)$.

24 Prove (3.7.25), that is, if the time intervals between events are from $\exp(1/\lambda)$, then the number of events occurring in a unit interval of time is from $P(\lambda)$.

25 Prove that $y = x^{\delta_1}(1-x)^{\delta_2}$ is a concave function on [0, 1] and has a maximum equal to

$$y^* = \frac{\delta_1^{\delta_1}\delta_2^{\delta_2}}{(\delta_1+\delta_2)^{\delta_1+\delta_2}} \quad \text{at the point} \quad x^* = \frac{\delta_1}{\delta_1+\delta_2}.$$

26 Let X and X_1 be i.i.d. r.v.'s and let $Y = \alpha X + (1-\alpha)X_1$, where $0 \leq |\alpha| \leq 1$. Prove that the correlation coefficient

$$\rho_{X,Y} = \frac{\alpha}{\sqrt{\alpha^2 + (1-\alpha)^2}}.$$

Describe an algorithm for generating a pair of r.v.'s (X, Y) for which $\rho_{XY} = \beta$.

27 Prove Theorems 3.4.2 and 3.4.3.

28 By analogy with Theorem 3.4.2 formulate a theorem that is a multidimensional version of Algorithm AR-1, and prove it.

29 Let $X = (X_1, \ldots, X_n)$ be i.i.d. r.v.'s uniformly distributed inside an n-dimensional unit sphere. Prove that the vector $Y = CS$ is uniformly distributed inside the ellipsoid

$$Y^T \Sigma Y \leq K^2,$$

where Σ is a symmetric and positively defined $(n \times n)$ matrix and C is the lower triangular matrix (3.5.13), such that $\Sigma = C^T C$. *Hint:* Use the fact that the vector $W = (W_1, \ldots, W_n) = KX$ is uniformly distributed inside the n-dimensional sphere

$$W^T W = W_1^2 + W_2^2 + \cdots + W_n^2 \leq K^2$$

with radius K.

REFERENCES

1 Ahrens, J. H. and U. Dieter, Computer methods for sampling from the exponential and normal distributions, *Comm. Assoc. Comp. Mach.*, **15**, 1972, 873–882.

2 Ahrens, J. H. and U. Dieter, Extensions of Forsythe's method for random sampling from the normal distribution, *Math. Comp.*, **27**, 1973, 927–937.

3 Ahrens, J. H. and U. Dieter, Computer methods for sampling from gamma, beta, poisson and binomial distributions, *Computing*, **12**, 1974, 223–246.

4 Ahrens, J. H. and U. Dieter, *Non-Uniform Random Numbers*, Institut für Mathematische Statistik, Technische Hochschule in Graz, Austria, 1974.

5 Anderson, T. W., *An Introduction to Multivariate Statistical Analysis*, Wiley, New York, 1958.

6 Box, G. E. P. and M. E. Muller, A note on the generation of random normal deviates, *Ann. Math. Stat.*, **29**, 1958, 610–611.

7 Butler, J. W., Machine sampling from given probability distributions, in *Symposium on Monte Carlo Methods*, edited by M. A. Meyer, Wiley, New York, 1956.

8 Cheng, R. C. H., Generating Beta variates with non-integral shape parameters, Comm. Assoc. Comp. Mach., **21**, 1978, 317–322.

9 Cheng, R. C. H., The generation of gamma variables, *Appl. Stat.*, **26**, 1977, 71–75.

10 Ermakov, J. M., *Monte Carlo Method and Related Questions*, Nauka, Moscow, 1976 (in Russian).

11 Feller, W., *An Introduction to Probability Theory and Its Applications*, Wiley, New York, 1950.

12 Fishman, G. S., *Principles of Discrete Event Simulation*, Wiley, New York, 1978.

13 Fishman, G. S., Sampling from the gamma distribution on a computer, *Comm. Assoc. Comp. Mach.*, **19**, 1976, 407–409.

14 Fishman, G. S., Sampling from the Poisson distribution on a computer, *Computing*, **17**, 1976, 147–156.

15 Forsythe, G. E., Von Neumann's comparison method for random sampling and from the normal and other distributions," *Math. Comp.*, **26**, 1972, 817–826.

16 Jöhnk, M. D., Erzeugung von Betraverteilten and Gammaverteilten Zuffalszahlen, *Metrika*, **8**, 1964, 5–15.

17 Johnson, N. L. and S. Kotz, *Discrete Distributions*, Houghton-Mifflin, 1969.

18 Johnson, N. L. and S. Kotz, *Continuous Univariate Distributions*, Vols. 1 and 2, Houghton-Mifflin, 1970.

19 Marsaglia, G., Generating exponential random variables, *Ann. Math. Stat.*, **32**, 1961, 899–900.

20 Marsaglia, G., Generating discrete random variables in a computer, *Comm. Assoc. Comp. Mach.*, **6**, 1963, 37–38.

21 Marsaglia, G., M. D. MacLaren, and T. A. Bray, A fast procedure for generating normal random variables, *Comm. Assoc. Comp. Mach.*, **7**, 1964.

22 Michailov, S. A., *Some Problems in the Theory of the Monte Carlo Methods*, Nauka, Novosibirsk, U.S.S.R., 1974 (in Russian).

23 Mood, A. M., F. A. Graybill, and D. C. Boes, *Introduction to the Theory of Statistics*, 3rd ed., McGraw-Hill, New York, 1974, 4–10.

24 Naylor, T. H. et al., *Computer Simulation Techniques*, Wiley, New York, 1966.

25 Neuts, M., *Probability*, Allyn and Bacon, 1972.

26 Philips, D. T. and C. Beightler, Procedures for generating gamma variates with non-integer parameter sets, *J. Stat. Comp. Simulation*, 1972, 197–208.

27 Relles, D., A simple algorithm for generating binomial random variables, *J. Amer. Stat. Assoc.*, **67**, 1972, 612–613.

28 Robinson, D. W. and P. A. W. Lewis, Generating gamma and Cauchy random variables: An extension to the Naval Postgraduate School random number package, 1975.

29 Sobol, J. M., *Computational Methods of Monte Carlo*, Nauka, Moscow, 1973 (in Russian).

30 Tadikamalla, P. R., Computer generation of gamma random variables, *Comm. Assoc. Comp. Mach.*, **21**, 1978, 419–422.

31 Tadikamalla, P. R., Computer generation of gamma random variables, II, *Comm. Assoc. Comp. Mach.*, **21**, 1978, 925–928.

32 Tadikamalla, P. R., and M. E. Johnson, Simple rejection methods for sampling from the normal distribution, in *Proceedings of the First International Conference of Mathematical Modeling*, X. J. Avula, Ed., St. Louis, Missouri, 1977, 573–577.

33 Tocher, K. D., *The Art of Simulation*, Van Nostrand, Princeton, New Jersey, 1963.

34 von Neumann, J., Various techniques used in connection with random digits, *U.S. Nat. Bur. Stand. Appl. Math. Ser.*, No. 12, pp. 36–38, 1951.

35 Wallace, N. D., Computer generation of gamma random variates with non-integral shape parameters, *Comm. Assoc. Comp. Mach.*, **17**, 1974, 691–695.

36 Walker, A. J., An efficient method for generating discrete random variables with general distributions, *Trans. Math. Software*, **3**, No. 3, September 1977, 253–257.

37 Whittaker, J., Generating gamma and beta random variables with non-integrable shape parameters, *Appl. Stat.*, **23**, 1974, 210–214.

38 Wilde, D. J., *Optimum Seeking Methods*, Prentice-Hall, Englewood Cliffs, New Jersey, 1964.

39 Yakowitz, S. J., *Computation Probability and Simulation*, Addison-Wesely, Reading, Massachusetts, 1977.

CHAPTER 4

Monte Carlo Integration and Variance Reduction Techniques

4.1 INTRODUCTION

The importance of good numerical integration schemes is evident. There are many deterministic quadrature formulas for computation of ordinary integrals with well behaved integrands. The Monte Carlo method is not competitive in this case.

But if the function fails to be regular (i.e., to have continuous derivatives of moderate order), numerical analytic techniques, such as the trapezoidal and Simpson's rules become less attractive. Especially in the case of multidimensional integrals, application of such rules (formulas) runs into severe difficulties. It is often more convenient to compute such integrals by a Monte Carlo method, which, although less accurate than conventional quadrature formulas, is much simpler to use.

It is shown that each *integral* can be represented as an *expected value* (*parameter*) and the problem of estimating an integral by the Monte Carlo Method is equivalent to the problem of estimating an unknown parameter. For convenience we use the expression "estimating the integral" rather than "estimating the unknown parameter." In Section 4.3.12 we consider several practical examples of estimating such parameters (integrals).

4.2 MONTE CARLO INTEGRATION

In this section we consider two simple techniques for computing one-dimensional integrals,

$$I = \int_a^b g(x)\,dx, \tag{4.2.1}$$

by a Monte Carlo method. The first technique is called "the hit or miss Monte Carlo method," and is based on the geometrical interpretation of an integral as an area; the second technique is called "the sample-mean Monte Carlo method," and is based on the representation of an integral as a mean value.

4.2.1 The Hit or Miss Monte Carlo Method

Consider the problem of calculating the one-dimensional integral (4.2.1) where, for simplicity, we assume that the integrand $g(x)$ is bounded

$$0 \le g(x) \le c, \qquad a \le x \le b.$$

Let Ω denote the rectangle (Fig. 4.2.1)

$$\Omega = \{(x,y) : a \le x \le b, 0 \le y \le c\}.$$

Let (X, Y) be a random vector uniformly distributed over the rectangle Ω with probability density function (p.d.f.)

$$f_{XY}(x,y) = \begin{cases} \dfrac{1}{c(b-a)}, & \text{if } (x,y) \in \Omega \\ 0, & \text{otherwise.} \end{cases} \tag{4.2.2}$$

What is the probability p that the random vector (X, Y) falls within the area under the curve $g(x)$? Denoting $S = \{(x,y) : y \le g(x)\}$ and observing that the area under the curve $g(x)$ is

$$\text{area under } g(x) = \text{area } S = \int_a^b g(x)\,dx,$$

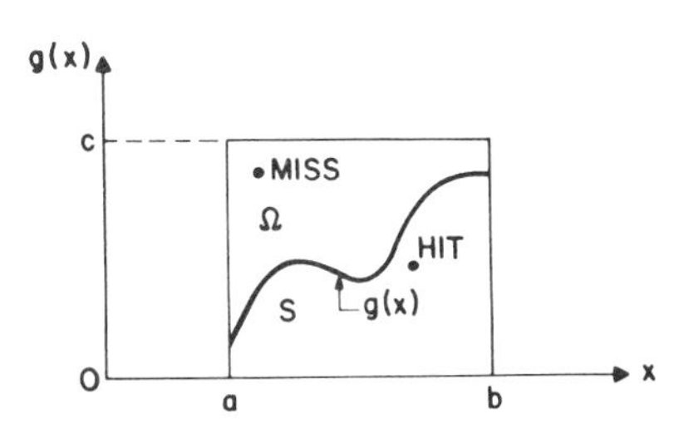

Fig. 4.2.1 Graphical representation of the hit or miss Monte Carlo method.

we obtain

$$p = \frac{\text{area } S}{\text{area } \Omega} = \frac{\int_a^b g(x)\,dx}{c(b-a)} = \frac{I}{c(b-a)}. \tag{4.2.3}$$

Let us assume that N independent random vectors (X_1, Y_1), $(X_2, Y_2), \ldots, (X_N, Y_N)$ are generated. The parameter p can be estimated by

$$\hat{p} = \frac{N_H}{N}, \tag{4.2.4}$$

where N_H is the number of occasions on which $g(X_i) \geq Y_i$, $i = 1, 2, \ldots, N$, that is, the number of "hits," and $N - N_H$ is the number of "misses"; we score a miss if $g(X_i) < Y_i$, $i = 1, \ldots, N$, as depicted in Fig. 4.2.1.

It follows from (4.2.3), and (4.2.4) that the integral I can be estimated by

$$I \approx \theta_1 = c(b-a)\frac{N_H}{N}. \tag{4.2.5}$$

In other words, to estimate the integral I we take a sample N from the distribution (4.2.2), count the number N_H of hits (below the curve $g(x)$), and apply (4.2.5).

Since each of the N trials constitutes a Bernoulli trial with probability p of a hit, then

$$E(\theta_1) = c(b-a)E\left(\frac{N_H}{N}\right) = c(b-a)\frac{E(N_H)}{N} = pc(b-a) = I, \tag{4.2.6}$$

that is, θ_1 is an unbiased estimator of I.

The variance of $\hat{p}$ is

$$\operatorname{var}\hat{p} = \operatorname{var}\left(\frac{N_H}{N}\right) = \frac{1}{N^2}\operatorname{var}(N_H) = \frac{1}{N}p(1-p), \tag{4.2.7}$$

which, together with (4.2.3), gives

$$\operatorname{var}\hat{p} = \frac{1}{N}\frac{I}{[c(b-a)]^2}[c(b-a) - I]. \tag{4.2.8}$$

Thus

$$\begin{aligned}\operatorname{var}\theta_1 &= [c(b-a)]^2 \operatorname{var}\hat{p} = [c(b-a)]^2 \frac{1}{N}p(1-p) \\ &= \frac{I}{N}[c(b-a) - I]\end{aligned} \tag{4.2.9}$$

and the standard deviation

$$\sigma_{\theta_1} = [\operatorname{var}\theta_1]^{1/2} = N^{-1/2}\{I[c(b-a) - I]\}^{1/2}.$$

Note that the precision of the estimator θ_1, which is measured by the inverse of standard deviation, is of order $N^{-1/2}$.

How many trials do we have to perform, according to the hit or miss Monte Carlo method, if we require

$$P[|\theta_1 - I| < \varepsilon] \geq \alpha? \tag{4.2.10}$$

Chebyshev's inequality,

$$P[|\theta_1 - I| < \varepsilon] \geq 1 - \frac{\operatorname{var} \theta_1}{\varepsilon^2}, \tag{4.2.11}$$

together with (4.2.10), gives

$$\alpha \leq 1 - \frac{\operatorname{var} \theta_1}{\varepsilon^2}. \tag{4.2.12}$$

Substituting (4.2.9) in (4.2.12), we obtain

$$\alpha \leq 1 - \frac{p(1-p)[c(b-a)]^2}{N\varepsilon^2}. \tag{4.2.13}$$

Solving (4.2.13) with respect to N, we have

$$N \geq \frac{(1-p)p[c(b-a)]^2}{(1-\alpha)\varepsilon^2}, \tag{4.2.14}$$

which is the required number of trials for (4.2.10) to hold.

When N is sufficiently large we can apply the central limit theorem, which says that for N sufficiently large the random variable (r.v.)

$$\hat{\theta}_1 = \frac{\theta_1 - I}{\sigma_{\theta_1}} \tag{4.2.15}$$

is distributed approximately according to the standard normal distribution, that is,

$$P(\theta_1 \leq x) \simeq \phi(x), \tag{4.2.16}$$

where

$$\phi(x) = \frac{1}{\sqrt{2}\,\pi} \int_{-\infty}^{x} e^{-t^2/2}\, dt. \tag{4.2.17}$$

We can easily verify that the confidence interval with level $1 - 2\alpha$ for I is

$$\theta_1 \pm z_\alpha \frac{[\hat{p}(1-\hat{p})]^{1/2}(b-a)c}{N^{1/2}}, \tag{4.2.18}$$

where

$$z_\alpha = \phi^{-1}(\alpha). \tag{4.2.19}$$

Hammersley and Handscomb [10] write:

> Historically, hit or miss methods were once the ones most usually propounded in explanation of Monte Carlo techniques; they were of course, the easiest methods to understand (particularly if explained in the kind of graphical language involving a curve in a rectangle).

Hit or Miss Monte Carlo Method Algorithm

1 Generate a sequence $\{U_j\}_{j=1}^{2N}$ of $2N$ random numbers.

2 Arrange the random numbers into N pairs (U_1, U_1'), $(U_2, U_2'), \ldots, (U_N, U_N')$ in any fashion such that each random number U_1 is used exactly once.

3 Compute

$$X_i = a + U_i(b-a) \qquad \text{and} \qquad g(X_i), \qquad i = 1, 2, \ldots, N.$$

4 Count the number of cases N_H for which $g(X_i) > cU_i'$.

5 Estimate the integral I by

$$\theta_1 = c(b-a)\frac{N_H}{N}.$$

4.2.2 The Sample-Mean Monte Carlo Method

Another way of computing the integral

$$I = \int_a^b g(x)\,dx$$

is to represent it as an expected value of some random variable. Indeed, let us rewrite the integral as

$$I = \int_a^b \frac{g(x)}{f_X(x)} f_X(x)\,dx, \tag{4.2.20}$$

assuming that $f_X(x)$ is any p.d.f. such that $f_X(x) > 0$ when $g(x) \neq 0$.

Then

$$I = E\left[\frac{g(X)}{f_X(X)}\right], \tag{4.2.21}$$

where the random variable X is distributed according to $f_X(x)$.

Let us assume for simplicity

$$f_X(x) = \begin{cases} \dfrac{1}{b-a}, & \text{if } a < x < b, \\ 0, & \text{otherwise;} \end{cases} \tag{4.2.22}$$

then

$$E[g(X)] = \frac{I}{b-a} \tag{4.2.23}$$

and

$$I = (b-a)E[g(X)]. \tag{4.2.24}$$

An unbiased estimator of I is its sample mean

$$\theta_2 = (b-a)\frac{1}{N}\sum_{i=1}^{N} g(X_i). \tag{4.2.25}$$

The variance of θ_2 is equal to $E(\theta_2^2) - [E(\theta_2)]^2$, so that

$$\operatorname{var}\theta_2 = \operatorname{var}\left[\frac{1}{N}(b-a)\sum_{i=1}^{N} g(X_i)\right] = \frac{1}{N}\left[(b-a)^2\int_a^b g^2(x)\frac{1}{b-a}\,dx - I^2\right]$$

$$= \frac{1}{N}\left[(b-a)\int_a^b g^2(x)\,dx - I^2\right]. \tag{4.2.26}$$

Sample-Mean Monte Carlo Algorithm

1 Generate a sequence $\{U_i\}_{i=1}^N$ of N random numbers.
2 Compute $X_i = a + U_i(b-a)$, $i = 1, \ldots, N$.
3 Compute $g(X_i)$, $i = 1, \ldots, N$.
4 Compute the sample mean θ_2 according to (4.2.25), which estimates I.

4.2.3 Efficiency of Monte Carlo Method

Suppose two Monte Carlo methods exist for estimating the integral I. Let θ_1 and θ_2 be two estimates produced by these methods such that

$$E(\theta_1) = E(\theta_2) = I. \tag{4.2.27}$$

We denote by t_1 and t_2 the units of computing time required for evaluating the random variables θ_1 and θ_2, respectively. Let the variance associated with the first method be $\operatorname{var}\theta_1$ and that associated with the second method be $\operatorname{var}\theta_2$. Then we say that the first method is more efficient than the second method if

$$\varepsilon = \frac{t_1 \operatorname{var}\theta_1}{t_2 \operatorname{var}\theta_2} < 1. \tag{4.2.28}$$

Let us compare now the efficiency of the hit or miss Monte Carlo method with that of the sample-mean Monte Carlo method.

Proposition 4.2.1 $\operatorname{var}\theta_2 \le \operatorname{var}\theta_1$.

Proof Subtracting (4.2.26) from (4.2.9), we obtain

$$\operatorname{var}\theta_1 - \operatorname{var}\theta_2 = \frac{1}{N}(b-a)\left[cI - \int_a^b g^2(x)\,dx\right]. \tag{4.2.29}$$

Note that

$$g(x) \le c; \tag{4.2.30}$$

therefore

$$cI - \int_a^b g^2(x)\,dx \ge 0$$

and further

$$\operatorname{var}\theta_1 - \operatorname{var}\theta_2 \ge 0. \qquad \text{Q. E. D.}$$

Assuming that the computing times t_1 and t_2 for θ_1 and θ_2 are approximately equal, we conclude that the sample-mean method is more efficient than the hit or miss method.

If $\operatorname{var}\theta_1$ and $\operatorname{var}\theta_2$ are unknown, we can replace them by their estimators

$$S^2 = \frac{1}{N-1}\left[\sum_{i=1}^{N} g(X_i)(b-a) - \theta\right]^2 \tag{4.2.31}$$

and then estimate by

$$\varepsilon = \frac{t_1 S_1^2}{t_2 S_2^2}. \tag{4.2.32}$$

It is interesting to note that, estimating the integral by θ_1 and θ_2, we do not need to know the function $g(x)$ explicitly. We need only evaluate $g(x)$ at any point x.

4.2.4 Integration in the Presence of Noise

Suppose now that $g(x)$ is measured with some error, that is, we observe $\tilde{g}(x_i) = g(x_i) + \varepsilon_i$, $i = 1, 2, \ldots, N$, instead of g, where ε_i are independent identically distributed (i.i.d.) random variables with

$$E(\varepsilon) = 0, \qquad \operatorname{var}(\varepsilon) = \sigma^2 \tag{4.2.33}$$

and

$$|\varepsilon| < k < \infty. \tag{4.2.34}$$

Let (X, Y) be a random vector distributed

$$f_{X,Y}(x,y) = \begin{cases} \dfrac{1}{c_1(b-a)}, & a \le x \le b, 0 \le y \le c_1 \\ 0, & \text{otherwise,} \end{cases}$$

where

$$c_1 \ge g(x) + k.$$

Then, by analogy with θ_1 for the hit or miss method, we obtain

$$\tilde{\theta}_1 = c_1(b-a)\frac{N_H}{N}, \tag{4.2.35}$$

where N_H is the number of hits, that is $\hat{g}(X_i) \ge Y_i$, $i = 1, \ldots, N$. By analogy with θ_2 for the sample-mean Monte Carlo method with

$$f_X(x) = \begin{cases} \dfrac{1}{b-a}, & a \le x \le b \\ 0, & \text{otherwise,} \end{cases}$$

we obtain

$$\tilde{\theta}_2 = \frac{1}{N}(b-a)\sum_{i=1}^{N} \hat{g}(X_i). \tag{4.2.36}$$

We can show that both r.v.'s $\tilde{\theta}_1$ and $\tilde{\theta}_2$ are unbiased and converge almost surely (a.s.) and in mean square to I and that the sample-mean method is again more efficient than the hit or miss method.

4.3 VARIANCE REDUCTION TECHNIQUES

Variance reduction can be viewed as a means to use known information about the problem. In fact, if nothing is known about the problem, variance reduction cannot be achieved. At the other extreme, that is, complete knowledge, the variance is equal to zero and there is no need for simulation. Variance reduction cannot be obtained from nothing; it is merely a way of not wasting information. One way to gain this information is through a direct crude simulation of the process. Results from this simulation can then be used to define variance reduction techniques that will refine and improve the efficiency of a second simulation. Therefore the more that is known about the problem, the more effective the variance reduction techniques that can be employed. Hence it is always important to clearly define what is known about the problem. Knowledge of a process to be simulated can be qualitative, quantitative, or both.

4.3.1 Importance Sampling

Let us consider the problem of estimating the multiple integral*

$$I = \int g(x)\,dx, \qquad x \in D \subset R^n. \tag{4.3.1}$$

We suppose that $g \in L^2(x)$ (in other words, that $\int g^2(x)\,dx$ exists and therefore that I exists).

The basic idea of this technique [14] consists of concentrating the distribution of the sample points in the parts of the region D that are of most "importance" instead of spreading them out evenly. By analogy with (4.2.20) and (4.2.21) we can represent the integral (4.3.1) as

$$I = \int \frac{g(x)}{f_X(x)} f_X(x)\,dx = E\left[\frac{g(X)}{f_X(X)}\right]. \tag{4.3.2}$$

Here X is any random vector with p.d.f. $f_X(x)$, such that $f_X(x) > 0$ for each $x \in D \subset R^n$. The function $f_X(x)$ is called the importance sampling distribution. It is obvious from (4.3.2) that $\zeta = g(X)/f_X(X)$ is an unbiased estimator of I, with the variance

$$\operatorname{var} \zeta = \int \frac{g^2(x)}{f_X(x)}\,dx - I^2. \tag{4.3.3}$$

In order to estimate the integral we take a sample $X_1, \ldots, X_N$ from p.d.f. $f_X(x)$ and substitute its values in the sample-mean formula

$$\theta_3 = \frac{1}{N} \sum_{i=1}^{N} \frac{g(X_i)}{f_X(X_i)}. \tag{4.3.4}$$

We now show how to choose the distribution of the r.v. X in order to minimize the variance of ζ, which is the same as to minimize the variance of θ_3.

Theorem 4.3.1 The minimum of $\operatorname{var} \zeta$ is equal to

$$\operatorname{var} \zeta_0 = \left(\int |g(x)|\,dx\right)^2 - I^2 \tag{4.3.5}$$

*Formula (4.3.1) is a Lebesque integral and it is assumed that the domain of integration is bounded (has finite measure). Readers not familiar with Lebesque integrals may assume it to be a Riemann integral.

and occurs when the r.v. X is distributed with p.d.f.

$$f_X(x) = \frac{|g(x)|}{\int |g(x)|\, dx}. \tag{4.3.6}$$

Proof Formula (4.3.5) follows directly if we substitute (4.3.6) into (4.3.3). In order to prove that $\operatorname{var} \zeta_0 \leq \operatorname{var} \zeta$ it is enough to prove that

$$\left(\int |g(x)|\, dx\right)^2 \leq \int \frac{g^2(x)}{f_X(x)}\, dx, \tag{4.3.7}$$

which can be obtained from Cauchy-Schwarz inequality.

Indeed,

$$\begin{aligned}\left(\int |g(x)|\, dx\right)^2 &= \left(\int \frac{|g(x)|}{[f_X(x)]^{1/2}} [f_X(x)]^{1/2}\, dx\right)^2 \\ &\leq \int \frac{g^2(x)}{f_X(x)}\, dx \int f_X(x)\, dx = \int \frac{g^2(x)}{f_X(x)}\, dx.\end{aligned} \tag{4.3.8}$$

Q.E.D.

Corollary If $g(x) > 0$, then the optimal p.d.f. is

$$f_X(x) = \frac{g(x)}{I} \tag{4.3.9}$$

and $\operatorname{var} \zeta = 0$.

This method is unfortunately useless, since the optimal density contains the integral $\int |(g(x))|\, dx$, which is practically equivalent to computing I. In the case where $g(x)$ has constant sign it is precisely equivalent to calculating I. But if we already know I, we do not need Monte Carlo methods to estimate it.

Not all is lost, however. The variance can be essentially reduced if $f_X(x)$ is chosen in order to have a shape similar to that of $|g(x)|$. When choosing $f_X(x)$ in such a way we have to take into consideration the difficulties of sampling from such a p.d.f., especially if $|g(x)|$ is not a well behaved function. In estimating the integral, we can save CPU time if the sample $X_1, \ldots, X_N$ will be taken in the subregion $D' = \{x : g(x) \neq 0\}$ of D. This is the same as defining

$$f_X(x) > 0, \text{ if } g(x) \neq 0 \quad \text{and} \quad f_X(x) = 0, \text{ if } g(x) = 0. \tag{4.3.10}$$

Consider the problem of choosing the parameters of the distribution $f_X(x)$ in an optimal way. We assume that the p.d.f. $f_X(x)$ is determined up to the vector of parameters α, that is, $f_X(x) = f_X(x, \alpha)$. For instance, if $f_X(x)$ represents one-dimensional normal distribution, that is, $X \sim N(\mu, \sigma^2)$, then the unknown parameters can be the expected value μ and the variance σ^2. We want to choose the vector of parameters α to minimize the variance of θ_3, that is,

$$\min_{\alpha} \operatorname{var}\left[\theta_3 = \frac{1}{N}\sum_{i=1}^{N} \frac{g(X_i)}{f_X(x_i, \alpha)}\right] = \frac{1}{N}\min_{\alpha}\left[\int \frac{g^2(x)}{f_X(x, \alpha)}\, dx - I^2\right]. \tag{4.3.11}$$

The last problem is equivalent to

$$\min_{\alpha} \int \frac{g^2(x)}{f_X(x, \alpha)}\, dx. \tag{4.3.12}$$

The function

$$\int \frac{g^2(x)}{f_X(x, \alpha)}\, dx \tag{4.3.13}$$

can be multiextremal and generally it is difficult to find the optimal α. Some techniques for global optimization are discussed in Chapter 7.

4.3.2 Correlated Sampling

Correlated sampling is one of the most powerful variance reduction techniques.

Frequently, the primary objective of a simulation study is to determine the effect of a small change in the system. The sample-mean Monte Carlo method would make two independent runs, with and without the change in the system being simulated, and subtract the results obtained. Unfortunately, the difference being calculated is often small compared to the separate results, while the variance of the difference will be the sum of the variances in the two runs, which is usually significant. If, instead of being independent, the two simulations use the same random numbers, the results can be highly positively correlated, which provides a reduction in the variance. Another way of viewing correlated sampling through random numbers control is to realize that the use of the same random numbers generates identical histories in those parts of the two systems that are the same. Thus the aim of correlated sampling is to produce a high positive correlation between two similar processes so that the variance of the difference is considerably smaller than it would be if the two processes were statistically independent.

Unfortunately, there is no general procedure that can be implemented in correlated sampling. However, in the following two situations correlated sampling can be successfully employed.

1 The value of a small change in a system is to be calculated.
2 The difference in a parameter in two or more similar cases is of more interest than its absolute value.

Let us assume that we desire to estimate

$$\Delta I = I_1 - I_2, \tag{4.3.14}$$

where*

$$I_1 = \int g_1(x) f_1(x)\, dx, \qquad x \in D_1 \subset R^n \tag{4.3.15}$$

and

$$I_2 = \int g_2(x) f_2(x)\, dx, \qquad x \in D_2 \subset R^n. \tag{4.3.16}$$

Then the procedure for correlated sampling is as follows:

1 Generate $X_1, \ldots, X_N$ from $f_1(x)$ and $Y_1, \ldots, Y_N$ from $f_2(x)$.
2 Estimate ΔI using

$$\Delta\theta = \frac{1}{N}\sum_{i=1}^{N} g_1(X_i) - \frac{1}{N}\sum_{i=1}^{N} g_2(Y_i) = \frac{1}{N}\sum_{i=1}^{N} \Delta_i, \tag{4.3.17}$$

where

$$\Delta_i = g_1(X_1) - g_2(Y_1).$$

The variance of $\Delta\theta$ is

$$\sigma^2 = \sigma_1^2 + \sigma_2^2 - 2\operatorname{cov}\left(\hat{\theta}_1, \hat{\theta}_2\right), \tag{4.3.18}$$

where

$$\hat{\theta}_1 = \frac{1}{N}\sum_{i=1}^{N} g_1(X_i) \tag{4.3.19}$$

$$\hat{\theta}_2 = \frac{1}{N}\sum_{i=1}^{N} g_2(Y_i) \tag{4.3.20}$$

$$\sigma_1^2 = E\left(\hat{\theta}_1 - I_1\right)^2 \tag{4.3.21}$$

$$\sigma_2^2 = E\left(\hat{\theta}_2 - I_2\right)^2 \tag{4.3.22}$$

*Introducing $g(x) = \phi(x)/f_X(x)$, where $f_X(x)$ is a p.d.f., integral $I = \int \phi(x)\, dx$ can be written as $I = \int g(x) f(x)\, dx$. An unbiased estimator of the last integral is

$$\eta_4 = g(X) \tag{4.3.13}$$

and the integral can be estimated by

$$\theta_4 = \frac{1}{N}\sum_{i=1}^{N} g(X_i). \tag{4.3.14}$$

and

$$\operatorname{cov}(\hat{\theta}_1, \hat{\theta}_2) = E\left[(\hat{\theta}_1 - I_1)(\hat{\theta}_2 - I_2)\right]. \tag{4.3.23}$$

Now if $\hat{\theta}_1$ and $\hat{\theta}_2$ are statistically independent, then

$$\operatorname{cov}(\hat{\theta}_1, \hat{\theta}_2) = 0 \tag{4.3.24}$$

and

$$\sigma^2 = \sigma_1^2 + \sigma_2^2. \tag{4.3.25}$$

However, if the random variables X and Y are positively correlated and if $g_1(x)$ is similar to $g_2(x)$ in shape, then the random variables $\hat{\theta}_1$ and $\hat{\theta}_2$ will also be positively correlated, that is, $\operatorname{cov}(\hat{\theta}_1, \hat{\theta}_2) > 0$, and the variance of $\Delta\theta$ may be greatly reduced.

Thus the key to reducing the variance of $\Delta\theta$ is to insure positive correlation between the estimates $\hat{I}_1$ and $\hat{I}_2$. This can be achieved in several ways. The easiest way is to obtain correlated samples through random number control. Specifically, this can be accomplished by using the same (common) sequence of random numbers $U_1, \ldots, U_N$ in both simulations, that is, the sequences $X_1, \ldots, X_N$ and $Y_1, \ldots, Y_N$ are generated using $X_i = F_1^{-1}(U_i)$ and $Y_i = F_2^{-1}(U_i)$, respectively. Clearly, if f_X is similar to f_Y, the r.v.'s X_i and Y_i will be highly positively correlated since they both used the same random numbers.

It is difficult to be specific as to how random number control should be applied generally. As a rule, however, to achieve maximum correlation common random numbers should be used whenever the similarities in problem structure will permit this. Such an example is given in Section 6.7.2, while comparing some output parameters of regenerative processes.

4.3.3 Control Variates

The use of control variates is another technique for reducing the variance. In this technique, instead of estimating a parameter directly, the difference between the problem of interest and some analytical model is considered.

Application of control variates is very general [10, 12, 13]. Most of them concern queues and queueing networks (see Sections 4.3.13 and 6.7). Our nomenclature follows Lavenberg and Welch's paper [13].

A random variate C is a control variate for Y if it is correlated with Y and if its expectation μ_C is known. The control variate C is used to construct an estimator for μ that has a smaller variance than the estimator Y. For any β

$$Y(\beta) = Y - \beta(C - \mu_C) \tag{4.3.26}$$

is an unbiased estimator of μ. Now

$$\mathrm{var}[Y(\beta)] = \mathrm{var}[Y] - 2\beta\,\mathrm{cov}[Y, C] + \beta^2\,\mathrm{var}[C]. \quad (4.3.27)$$

Hence if

$$2\beta\,\mathrm{cov}[Y, C] > \beta^2\,\mathrm{var}[C],$$

variance reduction is achieved. The value of β that minimizes $\mathrm{var}[Y(\beta)]$ is easily to be found as

$$\beta^* = \frac{\mathrm{cov}[Y, C]}{\mathrm{var}[C]}$$

and the minimum variance is equal to

$$\mathrm{var}[Y(\beta^*)] = (1 - \rho_{YC}^2)\,\mathrm{var}[Y], \quad (4.3.28)$$

where ρ_{YC} is the correlation coefficient between Y and C. Hence the more C is correlated with Y, the greater the reduction in variance.

Another type of control variate is one for which the mean $E(C)$ is unknown but is equal to μ, that is, $E(C) = E(Y) = \mu$. Any linear combination

$$Y(\beta) = \beta Y + (1 - \beta)C$$

is again an unbiased estimator of μ, and if Y and C are correlated, variance reduction will be achieved.

We now extend the above results to the case of more than one control variate. Let $\mathbf{C} = (C_1, \ldots, C_Q)$ be a vector of Q control variates, let $\boldsymbol{\mu}_C$ be the known mean vector corresponding to $\mathbf{C}$, that is, $\boldsymbol{\mu}_C = (\mu_1, \ldots, \mu_Q)$, where $\mu_q = E[C_q]$, and let $\boldsymbol{\beta}$ be any vector. Then

$$Y(\boldsymbol{\beta}) = Y - \boldsymbol{\beta}^t(\mathbf{C} - \boldsymbol{\mu}_C) \quad (4.3.29)$$

is an unbiased estimator of μ. Here t is the transpose operator. The vector $\boldsymbol{\beta}^*$ that minimizes $\mathrm{var}[Y(\boldsymbol{\beta})]$ (see [13]) is

$$\boldsymbol{\beta}^* = \boldsymbol{\sigma}_{YC}\Sigma_C^{-1}, \quad (4.3.30)$$

where Σ_C is the covariance matrix of $\mathbf{C}$ and $\boldsymbol{\sigma}_{YC}$ is a Q-dimensional vector whose components are the covariances between Y and C_q's. The resulting minimum variance is

$$\mathrm{var}[Y(\boldsymbol{\beta}^*)] = (1 - R_{YC}^2)\,\mathrm{var}[Y], \quad (4.3.31)$$

where

$$R_{YC}^2 = \frac{\boldsymbol{\sigma}_{YC}^t \Sigma_C^{-1} \boldsymbol{\sigma}_{YC}}{\mathrm{var}[Y]}. \quad (4.3.32)$$

As before the larger the multiple correlation coefficient R^2_{YC} between $\mathbf{C}$ and Y, the greater the variance reduction.

Again, if $Y_1, \ldots, Y_{Q+1}$ are $Q+1$ different unbiased estimators of unknown μ, then

$$\sum_{i=1}^{Q+1} \beta_i Y_i \tag{4.3.33}$$

where $\sum_{i=1}^{Q+1} \beta_i = 1$ is also an unbiased estimator of μ.

For practical application of control variables there are two key problems. First, control variables must be found that are highly correlated with the estimators of interest. Second, since the vector $\boldsymbol{\sigma}_{YC}$ and the matrix Σ_C are in general unknown, the optimum coefficient vector $\boldsymbol{\beta}^*$ is unknown and must be estimated. Further, its estimation must be incorporated into effective statistical procedures, and we now turn our attention to these questions.

Let Y_k, $k = 1, \ldots, K$, be a sample from $f_Y(y)$. An unbiased estimator of μ is

$$\bar{Y} = \left(\frac{1}{K}\right) \sum_{k=1}^{K} Y_k.$$

The variance of $\bar{Y}$ is equal to

$$\sigma^2(\bar{Y}) = \frac{\sigma^2(Y)}{K}$$

and is estimated by

$$\hat{\sigma}^2(\bar{Y}) = \frac{\hat{\sigma}^2(Y)}{K} = \left(\frac{1}{K(K-1)}\right) \sum_{k=1}^{K} (Y_k - \bar{Y})^2. \tag{4.3.34}$$

The random variable

$$\frac{(\bar{Y} - \mu)}{\hat{\sigma}(\bar{Y})}$$

has approximately a t-distribution with $K-1$ degrees of freedom. The confidence interval can be found from

$$\text{prob}\left\{\bar{Y} - t_{K-1}\left(1 - \frac{\alpha}{2}\right)\hat{\sigma}(\bar{Y}) \le \mu \le \bar{Y} + t_{K-1}\left(1 - \frac{\alpha}{2}\right)\hat{\sigma}(\bar{Y})\right\} \approx 1 - \alpha. \tag{4.3.35}$$

Let $\mathbf{C}_k$ be the value of $\mathbf{C}$ for the kth run. Then if the optimum coefficient vector $\boldsymbol{\beta}^*$ were known, we would use the estimator

$$Y_k(\boldsymbol{\beta}^*) = Y_k - \boldsymbol{\beta}^{*\prime}(\mathbf{C}_k - \boldsymbol{\mu}_C) \tag{4.3.36}$$

for the kth replication. The estimator based on K runs would be

$$\bar{Y}(\boldsymbol{\beta}^*) = \left(\frac{1}{K}\right) \sum_{k=1}^{K} Y_k(\boldsymbol{\beta}^*),$$

and a confidence interval could be obtained by replacing $\bar{Y}$ and Y_k with $\bar{Y}(\boldsymbol{\beta}^*)$ and $Y_k(\boldsymbol{\beta}^*)$, respectively, in (4.3.34) and (4.3.35). In this case ($\boldsymbol{\beta}^*$ known)

$$\frac{\sigma^2(\bar{Y}(\boldsymbol{\beta}^*))}{\sigma^2(\bar{Y})} = 1 - R_{YC}^2, \tag{4.3.37}$$

and the variance reduction given by (4.3.31) would be obtained. Furthermore, the ratio of the mean confidence interval widths would be approximately proportional to the ratio of the standard deviations, and hence confidence interval width would be reduced by approximately $(1 - R_{YC}^2)^{1/2}$.

However, in practice $\boldsymbol{\beta}^*$ is unknown and hence must be estimated. We estimate it by the sample equivalent of (4.3.30), that is, by

$$\hat{\boldsymbol{\beta}}^* = \hat{\boldsymbol{\sigma}}_{YC} \hat{\Sigma}_C^{-1}, \tag{4.3.38}$$

where $\hat{\boldsymbol{\sigma}}_{YC}$ and $\hat{\Sigma}_C$ are the sample covariance vector and sample covariance matrix whose elements are given by

$$(\hat{\sigma}_{YC})_q = \left(\frac{1}{K-1}\right) \sum_{k=1}^{K} (Y_k - \bar{Y})(C_{qk} - \bar{C}_q)$$

and

$$(\hat{\Sigma}_C)_{qr} = \left(\frac{1}{K-1}\right) \sum_{k=1}^{K} (C_{qk} - \bar{C}_q)(C_{rk} - \bar{C}_r),$$

where C_{qk} is the qth element of $\mathbf{C}_k$ and $\bar{C}_q$ is the average of C_{qk}, $k = 1, \ldots, K$. Substituting $\hat{\boldsymbol{\beta}}^*$ for $\boldsymbol{\beta}^*$ in (4.3.3.6), we obtain

$$Y_k(\hat{\boldsymbol{\beta}}^*) = Y_k - \hat{\boldsymbol{\beta}}^{*\prime}(\mathbf{C}_k - \boldsymbol{\mu}_C)$$

and

$$\bar{Y}(\hat{\boldsymbol{\beta}}^*) = \left(\frac{1}{K}\right) \sum_{k=1}^{K} Y_k(\hat{\boldsymbol{\beta}}^*).$$

In general, $\bar{Y}(\hat{\boldsymbol{\beta}}^*)$ is a biased estimator of μ since $\hat{\boldsymbol{\beta}}^*$ and $\bar{\mathbf{C}}$ are dependent. Also, the $Y_k(\hat{\boldsymbol{\beta}}^*)$ are dependent, so we cannot directly use the t-statistic to obtain a confidence interval for μ. However, if we assume $\mathbf{Z} = \begin{pmatrix} \mathbf{Y} \\ \mathbf{C} \end{pmatrix}$ to have a multivariate normal distribution, then it is shown in

[13] that $\bar{Y}(\hat{\boldsymbol{\beta}}^*)$ is an unbiased estimator of μ and

$$\frac{\bar{Y}(\hat{\boldsymbol{\beta}}^*) - \mu}{\hat{\sigma}\left(\bar{Y}(\hat{\boldsymbol{\beta}}^*)\right)} \tag{4.3.39}$$

has a t-distribution with $K-Q-1$ degrees of freedom. Hence a confidence interval can be obtained from

$$\text{prob}\left\{ \bar{Y}(\hat{\boldsymbol{\beta}}^*) - t_{K-Q-1}\left(1-\frac{\alpha}{2}\right)\hat{\sigma}\left(\bar{Y}(\hat{\boldsymbol{\beta}}^*)\right) \right.$$
$$\left. \leq \mu \leq \bar{Y}(\hat{\boldsymbol{\beta}}^*) + t_{K-Q-1}\left(1-\frac{\alpha}{2}\right)\hat{\sigma}\left(\bar{Y}(\hat{\boldsymbol{\beta}}^*)\right) \right\} = 1-\alpha. \tag{4.3.40}$$

Further, the ratio $\sigma^2(\bar{\mathbf{Z}})(\bar{Y}(\hat{\boldsymbol{\beta}}^*))/\sigma^2(\bar{\mathbf{Y}})$ is given [13] by

$$\frac{\sigma^2\left(\bar{Y}(\hat{\boldsymbol{\beta}}^*)\right)}{\sigma^2(\bar{Y})} = \left(\frac{K-2}{K-Q-2}\right)\left(1-R^2_{YC}\right). \tag{4.3.41}$$

We can see from (4.3.41) that there exists a trade-off between $(K-2)/(K-Q-2)$ and $1-R^2_{YC}$. At one extreme, if K is not large with respect to Q, the factor $(K-2)/(K-Q-2)$ can nullify the potential variance reduction. At the other extreme we expect the factor $1-R^2_{YC}$ to be a decreasing function with respect to Q. It was indicated in [13] that for finite K the number of control variates Q has to be relatively small. It would be interesting to find the optimal Q as a function of K by making some assumptions about R_{YC}.

The major cost involved in the application of control variables is the effort required to develop a reasonable set of control variates. This requires understanding the model in sufficient detail to define possible control variables and estimators of interest.

There are only a few published reports describing the application of control variables for practical problems. However, judging from them we hope that variance reduction in the range 0.25 to 0.75 could be realized in practical situations.

Now we consider how the control variates can be used in estimating the integral

$$I = E[g(X)] = \int g(x) f_X(x)\, dx. \tag{4.3.42}$$

Let $g_0(x)$ be a function that approximates $g(x)$ well and let the expectation $E[g_0(x)]$ be known. The function $g_0(x)$ is a control variate for $g(x)$. Denoting $Y = g(x)$, $C = g_0(x)$, and $\mu_C = \int g_0(x) f_X(x)\, dx$, we have for any β

$$Y(\beta) = Y - \beta(C - \mu_C),$$

which is an unbiased estimator of the integral I.

Taking a sample $X_1, \ldots, X_N$ from $f_X(x)$, we can estimate the integral I by

$$\theta_5 = \frac{1}{N} \sum_{i=1}^{N} \left[g(X_i) - \beta^* g_0(X_i) \right] + \beta^* \mu_C,$$

where β^* is the optimal β, which minimizes var$[Y(\beta)]$. The efficiency of this technique depends on how well $g_0(x)$ approximates $g(x)$. But it is sometimes difficult to find a $g_0(x)$ that approximates $g(x)$ well enough and such that $E[g_0(x)]$ is known.

In many cases no approximation is known for $g(x)$. This can be overcome by simulating some values of X (making a pilot run) and plotting the results.

The extension to the case of Q control variates (see (4.3.29)) in calculating the integral I is as follows. Let $\boldsymbol{\phi}(X) = [\phi_1(X), \ldots, \phi_Q(X)]$ be a vector of control variates, with known mean vector $\boldsymbol{\mu}_\phi$, that is, $\mu_q = E[\phi_q(X)]$. Then for any vector $\boldsymbol{\beta}$

$$Y(\boldsymbol{\beta}) = g(X) - \beta(\boldsymbol{\phi}(X) - \boldsymbol{\mu}_\phi) \tag{4.3.43}$$

is an unbiased estimator of μ. Denoting $Y = g(X)$, $\boldsymbol{\phi}(X) = \mathbf{C}$, $\boldsymbol{\mu}_\phi = \boldsymbol{\mu}_C$, we obtain formula (4.3.29).

4.3.4 Stratified Sampling

This technique is well known in statistics [3]. For stratified sampling we break the region D into m disjoint subregions D_i, $i = 1, 2, \ldots, m$, that is, $D = \cup_{i=1}^{m} D_i$, $D_k \cap D_j = \varnothing$, $k \neq j$ where $\varnothing$ is an empty set. Then define

$$I_i = \int_{D_i} g(x) f_X(x)\, dx, \tag{4.3.44}$$

which can be estimated separately by the Monte Carlo method (for instance by the sample-mean Monte Carlo).

The idea of this technique is similar to the idea of importance sampling: we also take more observations (samples) in the parts of the region D that are more "important," but the effect of reducing the variance is achieved by concentrating more samples in more important subsets D_i, rather than by choosing the optimal p.d.f.

Let us define

$$P_i = \int_{D_i} f_X(x)\, dx. \tag{4.3.45}$$

It is obvious that $\sum_{i=1}^{m} P_i = 1$ and

$$I = \int_D g(x) f_X(x)\, dx = \sum_{i=1}^{m} \int_{D_i} g(x) f_X(x)\, dx = \sum_{i=1}^{m} I_i. \tag{4.3.46}$$

Introducing

$$g_i(x) = \begin{cases} g(x), & \text{if } x \in D_i \\ 0, & \text{otherwise,} \end{cases} \tag{4.3.47}$$

we can rewrite integral I_i as

$$I_i = \int_{D_i} P_i g(x) \frac{f_X(x)}{P_i}\, dx = P_i \int_D g_i(x) \frac{f_X(x)}{P_i}\, dx = P_i E[g_i(X)], \tag{4.3.48}$$

where

$$\int_{D_i} \frac{f_X(x)}{P_i}\, dx = 1.$$

Inasmuch as I_i is expressed as an expected value, the sample-mean estimator for I_i can be written as

$$Y_i = P_i g(X_i), \tag{4.3.49}$$

where the r.v. X_i is distributed according to $f_X(x)/P_i$ on D_i.

The integral I_i can be estimated by

$$\tau_i = \frac{P_i}{N_i} \sum_{k_i=1}^{N_i} g(X_{k_i}), \qquad k_i = 1, \ldots, N_i,\ i = 1, \ldots, m \tag{4.3.50}$$

and the integral I by

$$\theta_6 = \sum_{i=1}^{m} \tau_i = \sum_{i=1}^{m} \frac{P_i}{N_i} \sum_{k_i=1}^{N_i} g(X_{k_i}). \tag{4.3.51}$$

We may quickly verify that

$$\operatorname{var} \theta_6 = \sum_{i=1}^{m} \frac{P_i^2}{N_i} \operatorname{var} g(X_i) = \sum_{i=1}^{m} \frac{P_i^2 \sigma_i^2}{N_i}, \tag{4.3.52}$$

where

$$\sigma_i^2 = \operatorname{var} g(X_i) = \frac{1}{P_i} \int_{D_i} g^2(x) f_X(x)\, dx - \frac{I_i^2}{P_i^2}.$$

If stratification is well carried out, the variance of θ_6 may be less than the variance of the sample-mean method θ_4 with $\sum_{i=1}^{m} N_i = N$.

Once the subsets $D_1, \ldots, D_m$ are selected, the next requirement is to define the number of samples to assign to each interval. More specifically, let N_i be the number of samples assigned to the subset D_i where

$$\sum_{i=1}^{m} N_i = N. \tag{4.3.53}$$

The following theorem tells us how to stratify in an optimal way.

Theorem 4.3.2 For given partitioning $D = \cup_{i=1}^{m} D_i$

$$\min\left(\operatorname{var}\theta_6 = \sum_{i=1}^{m} \frac{P_i^2}{N_i}\sigma_i^2\right) \tag{4.3.54}$$

subject to

$$\sum_{i=1}^{m} N_i = N$$

occurs when

$$N_i = N\frac{P_i\sigma_i}{\sum_{j=1}^{m} P_j\sigma_j} \tag{4.3.55}$$

and is equal to

$$\frac{1}{N}\left[\sum_{i=1}^{m} P_i\sigma_i\right]^2. \tag{4.3.56}$$

The proof of the theorem is left to the reader.

Thus when the stratification regions are prescribed the minimum variance of θ_6 occurs when the N_i are proportional to $P_i\sigma_i$.

This theorem, as well as Theorem 4.3.1, has no important direct application because the values of σ_i are usually unknown.

One practical suggestion is to make a small "pilot" run to obtain rough estimates for σ_i. Such estimates would be of help in determining the optimal N_i, with the appropriate trade-off between the cost of sampling and the degree of precision desired.

Let us choose $N_i = P_i N$ (we assume that P_i can be calculated analytically).

Proposition 4.3.1 $\operatorname{var}\theta_6 \le \operatorname{var}\theta_4$, that is, if the sample size N_i in each subset D_i is proportional to P_i (i.e., if $N_i = NP_i$), then the variance of the stratified sampling method will be less or equal to the variance of the sample-mean method.

Proof Substituting $N_i = NP_i$ in (4.3.52), we obtain

$$\operatorname{var} \theta_6 = \frac{1}{N} \sum_{i=1}^{m} P_i \operatorname{var} g(X_i). \tag{4.3.57}$$

From the Cauchy-Schwarz inequality we have

$$I^2 = \left(\sum_{i=1}^{m} I_1 \right)^2 = \left[\sum_{i=1}^{m} \frac{I_i}{\sqrt{P_i}} \sqrt{P_i} \right]^2 \tag{4.3.58}$$

$$\leq \sum_{i=1}^{m} \frac{I_i^2}{P_i} \sum_{i=1}^{m} P_i = \sum_{i=1}^{m} \frac{I_i^2}{P_i}.$$

Multiplying (4.3.52) by P_i and summing over i from 1 to m, we obtain

$$\sum_{i=1}^{m} P_i \operatorname{var} g(X_i) = \int_D g^2(x) f_X(x)\, dx - \sum_{i=1}^{m} \frac{I_i^2}{P_i}, \tag{4.3.59}$$

which together with (4.3.58) can be written as

$$\sum_{i=1}^{m} P_i \operatorname{var} g(X_i) \leq \int_D g^2(x) f_X(x)\, dx - I^2 = N \operatorname{var} \theta_4. \tag{4.3.60}$$

Comparing (4.3.57) and (4.3.60), we immediately receive the proof of this proposition. Q.E.D.

In other words, proposition 4.3.1 states: There is no function $g(x) \in L_2(D_i, f)$ such that the stratified sampling method would be worse than the sample-mean method while choosing $N_i = P_i N$. Of course, if the last assumption is not true, the stratified sampling method may be worse than the sample-mean method. In exercise 6 such an example is presented.

It can be proven that the efficiency of stratified sampling in comparison with the sample-mean method is approximately m^2. In the particular case when $P_i = 1/m$ and $N_i = N/m$, we obtain the so-called *systematic sampling* method [8].

The procedure for systematic sampling is as follows:

1 Divide the range [0, 1] of the cumulative distribution into m intervals each of width $1/m$.

2 Generate $\{U_{k_i}, k_i = 1, \ldots, N/m; i = 1, \ldots, m\}$ from $\mathcal{U}(0, 1)$.

3 $Y_{k_i} \leftarrow (i - 1 + U_{k_i})/m$; $k_i = 1, \ldots, N/m$; $i = 1, \ldots, m$.

4 $X_{k_i} \leftarrow F^{-1}(Y_{k_i})$.

The estimator for the integral I is

$$\theta_6 = \frac{1}{N} \sum_{i=1}^{m} \sum_{k=1}^{N/m} \left[g(X_{k_i}) \right]$$

and the sample variance is

$$S^2 = \frac{N}{N-1}\left(\frac{1}{N}\sum_{k=1}^{N}\theta_k - \theta_6\right)^2,$$

where $\theta_k = (1/m)\Sigma_{i=1}^{m} g(X_{k_i})$.

4.3.5 Antithetic Variates

This technique is due to Hammersley and Morton [11]. In this technique we seek two unbiased estimators Y' and Y'' for some unknown parameter I (in our case I is the unknown integral), having strong negative correlation. Note that $\frac{1}{2}(Y' + Y'')$ will be an unbiased estimator of I with variance

$$\operatorname{var}\left[\tfrac{1}{2}(Y' + Y'')\right] = \tfrac{1}{4}\operatorname{var} Y' + \tfrac{1}{4}\operatorname{var} Y'' + \tfrac{1}{2}\operatorname{cov}(Y', Y''), \quad (4.3.61)$$

and it follows from the last equation that, if the covariance $\operatorname{cov}(Y', Y'')$ is strongly negative, the method of antithetic variates can be effective in reducing the variance.

As an example, consider the integral

$$I = \int_0^1 g(x)\,dx,$$

which is equal to

$$I = \frac{1}{2}\int_0^1 [g(x) + g(1-x)]\,dx. \quad (4.3.62)$$

The estimator of I is then

$$Y = \tfrac{1}{2}(Y' + Y'') = \tfrac{1}{2}[g(U) + g(1-U)]. \quad (4.3.63)$$

Y is an unbiased estimator of I, because both $Y' = g(U)$ and $Y'' = g(1-U)$ are unbiased estimators of I. To estimate I we can take a sample of size N from the uniform distribution and find

$$\theta_7 = \frac{1}{2N}\sum_{i=1}^{N}[g(U_i) + g(1-U_i)]. \quad (4.3.64)$$

The time required for one computation by (4.3.64) is twice that required by the sample-mean method. Therefore the estimator (4.3.64) will be more efficient than the estimator θ_2 (4.2.25) with $a = 0$ and $b = 1$ only if

$$\operatorname{var}\theta_7 \le \tfrac{1}{2}\operatorname{var}\theta_2.$$

Proposition 4.3.2 If $g(x)$ is a continuous monotonically nonincreasing (nondecreasing) function with continuous first derivatives, then

$$\operatorname{var}\theta_7 \le \tfrac{1}{2}\operatorname{var}\theta_2. \quad (4.3.65)$$

Proof Let us assume without loss of generality that $N = 1$. It follows from (4.3.61) that

$$\operatorname{var}\theta_7 = \frac{1}{4}\int_0^1 g^2(x)\,dx + \frac{1}{4}\int_0^1 g^2(1-x)\,dx + \frac{1}{2}\int_0^1 g(x)g(1-x)\,dx - I^2 = \frac{1}{2}\int_0^1 g^2(x)\,dx + \frac{1}{2}\int_0^1 g(x)g(1-x)\,dx - I^2. \tag{4.3.66}$$

Therefore

$$2\operatorname{var}\theta_7 - \operatorname{var}\theta_2 = \int_0^1 g(x)g(1-x)\,dx - I^2.$$

The theorem will be proved if we prove

$$\int_0^1 g(x)g(1-x)\,dx \le I^2. \tag{4.3.67}$$

Let us assume that $g(x)$ is a monotonically nondecreasing function with continuous first derivatives (the proof when $g(x)$ is nonincreasing is similar), such that $g(1) > g(0)$. Let us introduce another auxiliary function

$$\phi(x) = \int_0^x g(1-t)\,dt - xI \tag{4.3.68}$$

such that $\phi(0) = \phi(1) = 0$. The first derivative

$$\phi'(x) = g(1-x) - I \tag{4.3.69}$$

is also a monotone function and $\phi'(0) > 0$, $\phi'(1) < 0$. Therefore $\phi(x) \ge 0$, $x \in [0, 1]$, and obviously

$$\int_0^1 \phi(x)g'(x)\,dx \ge 0. \tag{4.3.70}$$

Integrating (4.3.70) by parts, we get

$$\int_0^1 \phi'(x)g(x)\,dx \le 0, \tag{4.3.71}$$

and substituting (4.3.69) into (4.3.71), we obtain (4.3.67). Q.E.D.

More generally, let

$$I = \int_{-\infty}^{\infty} g(x)f_X(x)\,dx, \qquad x \in R^1. \tag{4.3.72}$$

Then by analogy with (4.3.64) an unbiased estimator of I is

$$\theta_7 = \frac{1}{2N}\sum_{i=1}^{N}\left[g(X_i) + g(X_i')\right], \tag{4.3.73}$$

where

$$X_i = F^{-1}(U_i) \tag{4.3.74}$$

$$X_i' = F^{-1}(1 - U_i) \tag{4.3.75}$$

and $F_X(x)$ is the cumulative distribution function (c.d.f.) of X. The pairs X_i and X_i' are, of course, correlated since the same random numbers U_i, $i = 1, \ldots, N$, were used to generate both r.v.'s X_i and X_i'. Furthermore, these r.v.'s are negatively correlated and therefore θ_7 may have a smaller variance than θ_4.

Let us rewrite (4.3.51) for the case when the region $D = \{x : x \in [0, 1]\}$. We have

$$\theta_6 = \sum_{i=1}^{m} \frac{(\alpha_i - \alpha_{i-1})}{N_i} \sum_{j=1}^{N_i} g\left[\alpha_{i-1} + (\alpha_i - \alpha_{i-1})U_{ji}\right], \tag{4.3.76}$$

where $0 = \alpha_0 < \alpha_1 < \cdots < \alpha_m = 1$, $P_i = \alpha_i - \alpha_{i-1}$, and U_{ij} is a sample from $\mathcal{U}(0, 1)$. Letting $m = 2$, $N_i = N$, and denoting $\alpha_1 = \alpha$, we get for (4.3.76)

$$\theta_6 = \frac{1}{N} \sum_{j=1}^{N} \left\{\alpha g(\alpha U_{j1}) + (1 - \alpha) g\left[\alpha + (1 - \alpha) U_{j2}\right]\right\}. \tag{4.3.77}$$

Let us now make U_{ji} dependent. Assuming $U_{j1} = U_{j2} = U_j$, we obtain

$$\theta_7' = \frac{1}{N} \sum_{j=1}^{N} \left\{\alpha g(\alpha U_j) + (1 - \alpha) g\left[\alpha + (1 - \alpha) U_j\right]\right\} \tag{4.3.78}$$

or, alternatively, assuming $U_{j1} = 1 - U_{j2} = U_j$, we have

$$\theta_7'' = \frac{1}{N} \sum_{j=1}^{N} \left\{\alpha g(\alpha U_j) + (1 - \alpha) g\left[1 - (1 - \alpha) U_j\right]\right\} \tag{4.3.79}$$

It is easy to see that both θ_7' and θ_7'' are estimates of the antithetic variates type. If $\alpha = \frac{1}{2}$, then (4.3.79) reduces to (4.3.64).

Consider now a case with two strata for (4.3.72). Assume the domain of $f_X(x)$ is broken up by x_α into the ranges $-\infty < x < x_\alpha$ and $x_\alpha < x < \infty$. By analogy with (4.3.79) an unbiased estimator of I is

$$\theta_7 = \frac{1}{N} \sum_{i=1}^{N} \left[\alpha g(X_i) + (1 - \alpha) g(X_i')\right] \tag{4.3.80}$$

where

$$X_i = F^{-1}(\alpha U_i) \tag{4.3.81}$$

$$X_i' = F^{-1}\left[\alpha + (1 - \alpha) U_i\right]. \tag{4.3.82}$$

In the particular case when $\alpha = \frac{1}{2}$ (4.3.82) reduces to (4.3.73).

We can try to obtain an α that minimizes $\operatorname{var}\theta_7$ in (4.3.80). Generally, this problem is difficult to solve because θ_7 does need to be unimodal with respect to α. In Chapter 7 some techniques for multiextremal optimization are considered.

4.3.6 Partition of the Region.

In this technique [21] we break region D into two parts $D = D_1 \cup D_2$, representing the integral I as

$$I = \int_D g(x)\,dx = \int_{D_1} g(x)\,dx + \int_{D_2} g(x)\,dx. \tag{4.3.83}$$

Let us assume that the integral

$$I_1 = \int_{D_1} g(x)\,dx \tag{4.3.84}$$

can be calculated analytically, and let us define a truncated p.d.f.

$$h(x) = \begin{cases} \dfrac{f_X(x)}{1-P}, & \text{if } x \in D_2 \\ 0, & \text{otherwise} \end{cases} \tag{4.3.85}$$

where $P = \int_{D_1} f_X(x)\,dx$.

Formula (4.3.83) can be written as

$$\begin{aligned} I &= I_1 + \int_{D_2} g(x)\,dx \\ &= I_1 + \int_{D_2} \frac{g(x)}{h(x)} h(x)\,dx \\ &= I_1 + E\left[\frac{g(X)}{h(X)}\right] = I_1 + (1-P)E\left[\frac{g(X)}{f_X(X)}\right]. \end{aligned} \tag{4.3.86}$$

An unbiased estimator of I is then

$$Y = I_1 + (1-P)\frac{g(X)}{f_X(X)} \tag{4.3.87}$$

and the integral I can be estimated by

$$\theta_8 = I_1 + (1-P)\frac{1}{N}\sum_{i=1}^{N} \frac{g(X_i)}{f_X(X_i)}. \tag{4.3.88}$$

Proposition 4.3.3 $\operatorname{var}\theta_8 \le (1-P)\operatorname{var}\theta_3.$ (4.3.89)

Proof We have from (4.3.4) that

$$N\operatorname{var}\theta_3 = \int_D \frac{g^2(x)}{f_X(x)}\,dx - I^2$$

$$= \int_{D_1} \frac{g^2(x)}{f_X(x)}\,dx + \int_{D_2} \frac{g^2(x)}{f_X(x)}\,dx - I^2 \tag{4.3.90}$$

and, correspondingly, from (4.3.88) that

$$N\operatorname{var}\theta_8 = (1-P)^2 \int_{D_2} \frac{g^2(x)}{f_X^2(x)}\frac{f_X(x)}{(1-P)}\,dx - \left[(1-P)\int_{D_2} \frac{g(x)}{f_X(x)}\frac{f_X(x)}{(1-P)}\,dx\right]^2$$

$$= (1-P)\int_{D_2} \frac{g^2(x)}{f_X(x)}\,dx - \left(\int_{D_2} g(x)\,dx\right)^2. \tag{4.3.91}$$

Multiplying (4.3.90) by $(1-P)$ and subtracting (4.3.91), we obtain

$$N[(1-P)\operatorname{var}\theta_3 - \operatorname{var}\theta_8] = (1-P)\int_{D_1} \frac{g^2(x)}{f_X(x)}\,dx - (1-P)I^2 + \left(\int_{D_2} g(x)\,dx\right)^2 \tag{4.3.92}$$

$$= (1-P)\int_{D_1} \frac{g^2(x)}{f_X(x)}\,dx - (1-P)I^2 + (I - I_1)^2.$$

Now introducing

$$C^2 = \int_{D_1} \frac{g^2(x)}{f_X(x)}\,dx - \frac{I_1^2}{P} = \int_{D_1} \left(\frac{g(x)}{f_X(x)} - \frac{I_1}{P}\right)^2 f_X(x)\,dx, \tag{4.3.93}$$

we have

$$N[(1-P)\operatorname{var}\theta_3 - \operatorname{var}\theta_8] = (1-P)C^2 + \left(P^{1/2}I - P^{-1/2}I_1\right)^2 \ge 0,$$

and Proposition 4.3.3 is proved. Q.E.D.

As a result of the proposition, we find that this technique is at least $(1-P)^{-1}$ times more efficient than the sample-mean Monte Carlo method.

4.3.7 Reducing the Dimensionality

This approach is due to Buslenko [21] and is sometimes called *expected value*.

Let us assume that the integral

$$I = \int_D g(x) f_X(x)\, dx, \qquad x \in D \subset R^n \tag{4.3.94}$$

can be represented as

$$I = \int_{D_1} dy \int_{D_2} g(y, z) f(y, z)\, dz, \tag{4.3.95}$$

where

$$y = (x_1, \ldots, x_t) \in D_1 \subset R^t$$

and

$$z = (x_{t+1}, \ldots, x_n) \in D_2 \subset R^{n-t}.$$

Assume also that the integration with respect to z can be performed analytically, that is, the marginal p.d.f.

$$f_Y(y) = \int_{D_2} f_{Y,Z}(y, z)\, dz \tag{4.3.96}$$

and the conditional expectation

$$E_Z[g(Z|Y)] = \int_{D_2} g(y, z) f(z|y)\, dz \tag{4.3.97}$$

can be found analytically.

It is obvious that

$$I = \int_{D_1} E_Z[g(Z|Y) f_Y(y)]\, dy = E_Y E_Z[g(Z|Y)]. \tag{4.3.98}$$

An unbiased estimator of I is

$$\eta_9 = E_Z[g(Z|Y)], \tag{4.3.99}$$

and it can be estimated by

$$\theta_9 = \frac{1}{N} \sum_{i=1}^{N} E_Z[g(Z|Y_i)], \tag{4.3.100}$$

where Y_i, $i = 1, \ldots, N$ are distributed with p.d.f. $f_Y(y)$.

Proposition 4.3.4 If integration can be performed analytically with respect to some variables, then the variance will be reduced, that is,

$$\operatorname{var} \eta_9 \leq \operatorname{var} \eta_4, \tag{4.3.101}$$

where η_4 is the sample-mean estimator (see (4.3.13)).

Proof The proof is quite simple. Denote $V = g(Y, Z)$. Now using the well known formula [17]

$$\operatorname{var} V = \operatorname{var}_Y\{E_Z(V|Y)\} + E_Y\{\operatorname{var}_Z(V|Y)\} \tag{4.3.102}$$

and noticing that $\eta_4 = V$, $\eta_9 = E_Z[g(Z|Y)] = E_Z(V|Y)$, and $E_Y \operatorname{var}_Z(V|Y) \geq 0$, the result follows immediately. Q.E.D.

4.3.8 Conditional Monte Carlo

If the problem under consideration is very complex—the sample space is complicated, or the p.d.f. is difficult to generate from—then it may be possible to embed the given sample space in a much larger space in which the desired density function appears as a conditional probability. Simulation of the large problem can be much simpler than the original complex problem and, despite the added computation required to calculate the conditional probabilities, the gain in efficiency can be quite high.

This technique was developed by Trotter and Tukey [24]. Our nomenclature follows Hammersley and Handscomb's book [10].

Consider again the problem of estimating

$$I = \int_D g(x) f_X(x)\, dx = E[g(X)]. \tag{4.3.103}$$

Let D be embedded in a product space $\Omega = D \times R$. Each point of $\Omega = D \times R$ can be written in the form $z = (x, y)$, where $x \in D$ and $y \in R$. Let $h(z) = h(x, y)$ be an arbitrary density function, let $\phi(z) = \phi(x, y)$ be an arbitrary real function, both defined on Ω, and let

$$\psi(x) = \int_R \phi(x, y)\, dy. \tag{4.3.104}$$

We also assume that both $h(z)$ and $\psi(x)$ are never zero. We may regard x and y as the first and second coordinates of z so that x is a function of z, which maps the points Ω onto D.

Let dz denote the volume element swept out in Ω when x and y sweep out volume elements dx and dy in D and R, respectively. The Jacobian of the transformation $z = (x, y)$ is

$$\mathcal{J}(z) = \mathcal{J}(x, y) = \frac{dx\, dy}{dz}. \tag{4.3.105}$$

We define the weight function

$$w(z) = \frac{f_X(x)\phi(z)\mathcal{G}(z)}{\psi(x)h(z)}. \tag{4.3.106}$$

Then we have the following identity:

$$\begin{aligned} I &= \int_D g(x)f_X(x)\,dx = \int_D dx\frac{g(x)f_X(x)}{\psi(x)}\int_R \phi(x,y)\,dy \\ &= \int_{D\times R} \frac{g(x)f_X(x)\phi(z)}{\psi(x)h(z)}h(z)\,dx\,dy \\ &= \int_{D\times R} g(x)w(z)h(z)\frac{dx\,dy}{\mathcal{G}(z)} = \int_\Omega g(x)w(z)h(z)\,dz, \end{aligned} \tag{4.3.107}$$

from which we can see that

$$I = E[g(X)w(Z)], \tag{4.3.108}$$

where X is the first coordinate of the random vector Z sampled from Ω with p.d.f. $h(z)$.

The unbiased estimator of I is then of the form

$$\eta_{10} = g(X)w(Z). \tag{4.3.109}$$

Both functions ϕ and h, and also the region R, are at our disposal; we may choose them to simplify the sampling procedure and to minimize the variance of the estimator η_{10}.

We now consider a particular case. Let $h(z)$ be a given distribution on the product space $\Omega = D\times R$, and let $f_X(x) = f_X(x|y_0)$ be the conditional distribution of $h(z)$ given $y = y_0$. If we write $P(y)$ for the p.d.f. of Y when $Z = (X, Y)$ has p.d.f. $h(z)$, we have

$$h(z)\,dz = f_X(x|y)P(y)\,dx\,dy, \tag{4.3.110}$$

and comparison of (4.3.106) and (4.3.103) gives

$$I(z) = \frac{h(z)}{f_X(x|y)P(y)}. \tag{4.3.111}$$

In particular

$$\mathcal{G}(x,y_0) = \frac{h(x,y_0)}{f_X(x)P(y_0)}. \tag{4.3.112}$$

By eliminating $f_X(x)$ from (4.3.106) and (4.3.112), we get

$$w(z) = \frac{h(x,y_0)\mathcal{G}(x,y)\phi(x,y)}{h(x,y)\mathcal{G}(x,y_0)P(y_0)\psi(x)}. \tag{4.3.113}$$

This leads to the following rule. Suppose that $Z=(X, Y)$ is distributed on Ω with p.d.f. $h(z)=h(x,y)$; then

$$\eta_{10}=g(X)w(Z),$$

where $w(Z)$ is given by (4.3.113), is an unbiased estimator of the conditional expectation of $g(X)$ given that $Y=y_0$. Note that this rule requires neither sampling from the possibly awkward space D nor evaluation of the possibly complicated function f, and ϕ is available for variance reduction.

4.3.9 Random Quadrature Method

Ermakov [4] suggested a quite general method of Monte Carlo integration based on orthonormal functions. We need some preliminary results before describing this method.

Let $\phi_i(x)$, $i=0,1,\ldots,m$, be a system of orthonormal functions over the region D, that is,

$$\langle\phi_i,\phi_j\rangle=\int_D\phi_i(x)\phi_j(x)\,dx=\begin{cases}0, & i\neq j\\ 1, & i=j,\end{cases} \tag{4.3.114}$$

and let

$$g(x)\approx g_m(x)=\sum_{i=0}^{m}c_i\phi_i(x) \tag{4.3.115}$$

be an interpolation formula for a given function $g(x)$. The problem is to choose c_i, for a given set of points $x_i\in D$, in such a way that

$$g_m(x_i)=g(x_i), \qquad i=0,1,\ldots,m; \tag{4.3.116}$$

that is, at points x_i we require coincidence in both the original $g(x)$ and the approximated function $g_m(x)$. To find c_i we have to solve the following system of linear equations with respect to c_i:

$$\begin{aligned}c_0\phi_0(x_0)+c_1\phi_1(x_0)+\cdots+c_m\phi_m(x_0)&=g(x_0)\\ \cdots\cdots\cdots\cdots\cdots\cdots\cdots\cdots\cdots\\ c_0\phi_0(x_i)+c_1\phi_1(x_i)+\cdots+c_m\phi_m(x_i)&=g(x_i)\\ \cdots\cdots\cdots\cdots\cdots\cdots\cdots\cdots\cdots\\ c_0\phi_0(x_m)+c_1\phi_1(x_m)+\cdots+c_m\phi_m(x_m)&=g(x_m).\end{aligned} \tag{4.3.117}$$

Applying, for instance, Cramer's rule, we find

$$c_0=\frac{w_g(x_0,x_1,\ldots,x_m)}{w(x_0,x_1,\ldots,x_m)}, \tag{4.3.118}$$

where

$$w(x_0,\ldots,x_m)=\begin{vmatrix}\phi_0(x_0),\phi_1(x_0),\ldots,\phi_m(x_0)\\ \cdots\cdots\cdots\cdots\cdots\cdots\\ \phi_0(x_i),\phi_1(x_i),\ldots,\phi_m(x_i)\\ \cdots\cdots\cdots\cdots\cdots\cdots\\ \phi_0(x_m),\phi_1(x_m),\ldots,\phi_m(x_m)\end{vmatrix} \tag{4.3.119}$$

is the $(n+1)(n+1)$ determinant and $w_g(x_0, x_1, \ldots, x_m)$ is the corresponding determinant in which the first column vector $\phi_0(x) = \{\phi_0(x_0), \ldots, \phi_0(x_m)\}$ is replaced by the right-hand side vector $g(x) = \{g(x_0), g(x_1), \ldots, g(x_m)\}$. With these results at hand let us consider the problem of calculating the integral

$$I_0 = \int \phi_0(x) g(x)\, dx. \tag{4.3.120}$$

Substituting (4.3.115) in the last formula, we have

$$\hat{I}_0 = \int \phi_0(x) g_m(x)\, dx = \int \phi_0(x) \left(\sum_{i=0}^{m} c_i \phi_i(x) \right) dx \approx I_0, \tag{4.3.121}$$

which is an approximation of I_0 and is called an *interpolation quadrature formula* [4] for I_0. Taking into consideration the orthonormality condition (4.3.114), we immediately obtain

$$\hat{I}_0 = c_0. \tag{4.3.122}$$

Therefore the value of integral I_0 is approximately equal to the coefficient c_0 in the interpolation formula (4.3.115) and can be calculated by Cramer's rule (4.3.118).

Ermakov [4] suggested choosing the points $x_i \in D$ in the interpolation formula (4.3.115) according to some probabilistic law rather than determining them in advance.

Assuming that x_i, c_i, or both of them are random variables, they called (4.3.115) a *random quadrature formula*, which is a natural generalization of the same formula (4.3.115) with deterministic x_i and c_i. They proved the following theorem.

Theorem 4.3.3 Let

$$\theta_{11} = \begin{cases} \dfrac{w_g(X)}{w(X)}, & \text{if } X \in B_+ \subset R^{m+1} \\ 0, & \text{if } X \in B_0 \subset R^{m+1} \end{cases} \tag{4.3.123}$$

be a random variable distributed with

$$f_X(x) = \frac{1}{(m+1)!} w^2(x), \tag{4.3.124}$$

where

$$B_0 = \{x : w(x) = 0\}$$

and

$$B_+ = \{x : w(x) \neq 0\}.$$

Then θ_{11} is an unbiased estimator of I_0, that is,

$$E(\theta_{11}) = I_0 \tag{4.3.125}$$

with variance

$$\operatorname{var}\theta_{11} \le \int g^2(x)\,dx - \sum_{i=1}^{m}\left[\left(\int g(x)\phi_i(x)\right)dx\right]^2. \tag{4.3.126}$$

The proof of the theorem as well, as some generalizations and applications can be found in Ermakov's monograph [4].

This method offers great possibilities because of its general character. But it also has some weak points: first, we must define a set of orthonormal functions over the region D; second, we must find an efficient way of sampling $X_0, X_1, \ldots, X_m$ with joint p.d.f. $\dfrac{1}{(m+1)!}[w(x_0, x_1, \ldots, x_m)]^2$. Even then computation of θ_{11} is generally no small matter, and therefore the random quadrature method seems to be of rather limited practicality.

4.3.10 Biased Estimators

Until now we have considered unbiased estimators for computing integrals. Using biased estimators, we can sometimes achieve useful results. Let us estimate the integral

$$I = \int_D g(x)\,dx \tag{4.3.127}$$

by

$$\theta_{12} = \frac{\sum_{i=1}^{N} g(U_i)}{\sum_{i=1}^{N} f(U_i)} \tag{4.3.128}$$

instead of using the usual sample-mean estimator

$$\theta_3 = \frac{1}{N}\sum_{i=1}^{N}\frac{g(X_i)}{f_X(X_i)}.$$

Here U is distributed uniformly in D, that is,

$$h(u) = \begin{cases} \dfrac{1}{V}, & \text{if } u \in D \\ 0, & \text{otherwise} \end{cases}, \quad V = \int_D dx, \tag{4.3.129}$$

and X is distributed according to $f_X(x)$.

It is clear that $E(\theta_{12}) \neq I$, that is, θ_{12} is a biased estimator of I. Let us show that θ_{12} is consistent. To prove consistency let us represent θ_{12} as a ratio of two random variables θ'_{12} and θ''_{12}, that is,

$$\theta_{12} = \frac{\theta'_{12}}{\theta''_{12}} = \frac{V\dfrac{1}{N}\sum_{i=1}^{N} g(U_i)}{V\dfrac{1}{N}\sum_{i=1}^{N} f(U_i)}, \tag{4.3.130}$$

where

$$\theta'_{12} = \frac{V}{N}\sum_{i=1}^{N} g(U_i) \tag{4.3.131}$$

and

$$\theta''_{12} = \frac{V}{N}\sum_{i=1}^{N} f(U_i). \tag{4.3.132}$$

Further,

$$E(\theta'_{12}) = V\int g(u)h(u)\,du = \int g(u)\,du = I \tag{4.3.133}$$

and

$$E(\theta''_{12}) = V\int f(u)h(u)\,du = \int f(u)\,du = 1. \tag{4.3.134}$$

With these results in hand we conclude that θ'_{12} and θ''_{12} converge a.s. to I and 1, respectively, when $N \to \infty$, which also means that

$$\lim_{N\to\infty} \left(\frac{\sum_{i=1}^{N} g(U_i)}{\sum_{i=1}^{N} f(U_i)} \right) \xrightarrow{\text{a.s.}} I, \qquad \text{if } \int |g(x)|\,dx < \infty \tag{4.3.135}$$

and this shows that θ_{12} is a consistent estimator of I. The bias of θ_{12} follows from

$$E(\theta_{12}) = E\left[\frac{\sum_{i=1}^{N} g(U_i)}{\sum_{i=1}^{N} f(U_i)} \right] \neq \frac{E\left[\sum_{i=1}^{N} g(U_i)\right]}{E\left[\sum_{i=1}^{N} f(U_i)\right]} = I. \tag{4.3.136}$$

One major advantage of this method is that the sample is taken from a uniform distribution rather than from a general $f_X(x)$ from which the generation of r.v.'s can be difficult (recall for instance that in importance sampling $f_X(x)$ has to be proportional to $|g(x)|$, and if $g(x)$ is a complicated function, it is difficult to generate from $f_X(x)$).

Powell and Swann [20] called this method *weighted uniform sampling*. They showed that for sufficiently large N this method is $N^{1/2}$ times more efficient than the sample-mean method.

4.3.11 Weighted Monte Carlo Integration

Yakowitz et al. [27] suggested estimating the integral

$$I = \int_0^1 g(x)\,dx$$

using the following Monte Carlo procedure:

1. Generate $U_1, \ldots, U_N$ from $\mathcal{U}(0, 1)$.
2. Arrange $U_1, \ldots, U_N$ in the increasing order $U_{(1)}, \ldots, U_{(N)}$.
3. Estimate the integral by

$$\theta_{13} = \frac{1}{2}\left[\sum_{i=0}^{N} \left(g(U_{(i)}) + g(U_{(i+1)})\right)\left(U_{(i+1)} - U_{(i)}\right)\right], \tag{4.3.137}$$

where $U_{(0)} \equiv 0$ and $U_{(N+1)} \equiv 1$. They proved the following

Proposition 4.3.5 Assume $g(x)$ is a function with a continuous second derivative on [0, 1]. If $\{U_{(i)}\}_{i=1}^{N}$ is the ordered sample associated with N independent uniform observations, then

$$\operatorname{var} \theta_{13} = E(\theta_{13} - I)^2 \leq \frac{k}{N^4}, \tag{4.3.138}$$

where k is some positive constant.

It is also shown in [27] that in the one-dimensional case $\operatorname{var}\theta_{13} = 0(1/N^4)$, which is much less than $\operatorname{var}\theta_3 = 0(1/N)$ in the sample-mean Monte Carlo method and in the two-dimensional case $\operatorname{var}\theta_{13} = 0(1/N^2)$, which is bigger than $\operatorname{var}\theta_{13}$ in the one-dimensional case but less then $\operatorname{var}\theta_3 = 0(1/N)$ for the sample-mean Monte Carlo method. Unfortunately, Yakowitz et al.'s method becomes inefficient as the dimensionality of x increases.

4.3.12 More about Variance Reduction (Queueing Systems and Networks)

In this section we consider two more examples of application of variance reduction techniques, which are taken from Refs. 29, 32, and 33. The first example is a single server queue $GI/G/1$, the second, a network. Some other examples of variance reduction with application to different problems can be found in Refs. 28 through 46.

(a) Single Server Queue $GI/G/1$ [33] Consider a single server queueing system $GI/G/1$, with a general distribution of service and interarrival time. We assume that, if an arriving customer finds the server free, his service commences immediately, and he departs from the system after completion of his service. If the arriving customer finds the server busy, he enters the waiting room and waits for his turn to be served. Customers are served on a first-in–first-out (FIFO) basis.

Let S_i denote the service time of ith customer who arrives at time t_i and let $A_i = t_i - t_{i-1}$, $i \geq 1$, denote the interarrival time (the time between the arrivals of the $(i-1)$th and ith customers).

Assume that the sequences $\{S_i, i \geq 0\}$ and $\{A_i, i \geq 1\}$ each consist of i.i.d. r.v.'s and are themselves independent. Let μ be the mean service rate, and let λ be the mean arrival rate, that is,

$$E(S_i) = \mu^{-1} \qquad \text{and} \qquad E(A_i) = \lambda^{-1}.$$

The parameter $\rho = \lambda/\mu$ is called the *traffic intensity* and measures the congestion of the queueing system. The necessary and sufficient conditions for the system to reach steady-state position (to become stable) is $\rho < 1$.

To measure the performance of the system we can use the mean waiting time of the ith customer (time for arrival to commencement of service); the number of customers in the system at time t; the amount of time in the interval $[0, t]$ that the server is busy; or the total number of customers who have been served in the interval $[0, t]$. As our measure of performance we take the mean waiting time of the ith customer and denote it by $E(W_i)$.

We assume that customer 0 arrives at time $t_0 = 0$ and finds an empty system. The following recursive formula is well known [33]:

$$W_0 = 0$$

$$W_i = \max(W_{i-1} - A_i + S_{i-1}, 0) = (W_{i-1} - A_i + S_{i-1})^+, \qquad i = 1, 2, \ldots . \tag{4.3.139}$$

Usually, for the $GI/G/1$ queueing system it is difficult to find $E(W_i)$ analytically and simulation may be used. In order to estimate $E(W_i)$ we run the queueing system N times, each time starting from $t_0 = 0$, obtain a sequence of service times $\{S_{ik}, i \geq 0, k = 1, \ldots, N\}$ and a sequence of interarrival times $\{A_{ik}, i \geq 1, k = 1, \ldots, N\}$, and estimate $E(W_i)$ by the sample-mean formula

$$\overline{W}_i = \frac{1}{N} \sum_{k=1}^{N} W_{ik}, \tag{4.3.140}$$

where $W_{ik} = (W_{(i-1)k} - A_{ik} + S_{(i-1)k})^+$, $W_{0k} = 0$.

We now explain how the antithetic and control variates methods can be applied for variance reduction, thereby improving the efficiency of the simulation. Both methods are based on reuse of the same random numbers.

Antithetic variates. Let $F_1(x)$ be the c.d.f. of the interarrival time A_i and let $F_2(x)$ be the c.d.f. of the service time S_i. Let us generate two sequences of random numbers $\{U_{ik}^{(1)}, i \geq 0, k = 1, \ldots, N\}$ and $\{U_{ik}^{(2)}, i \geq 0, k = 1, \ldots, N\}$, and obtain two corresponding sequences $A_{ik} = F_1^{-1}(U_{ik}^{(1)})$ and $S_{ik} = F_2^{-1}(U_{ik}^{(2)})$ of interarrival and service times. Introducing the antithetic sequences $\{1 - U_{ik}^{(1)}, i \geq 0, k = 1, \ldots, N\}$ and $\{1 - U_{ik}^{(2)}, i \geq 0, k = 1, \ldots, N\}$, we can define another two sequences $A_{ik}^1 = F_i^{-1}(1 - U_{ik}^{(1)})$ and $S_{ik}^1 = F_1^{-1}(1 - U_{ik}^{(2)})$ of interarrival and service times and estimate the mean waiting time $E(W_i)$ by

$$\overline{W}_i^{(A)} = \frac{1}{N} \sum_{k=1}^{N} \frac{W_{ik} + W'_{ik}}{2}, \tag{4.3.141}$$

where

$$W_{ik} = (W_{(i-1)k} - A_{ik} + S_{(i-1)k})^+ = \left[W_{(i-1)} - F_1^{-1}(U_{ik}^{(1)}) + F_2^{-1}(U_{(i-1)k}^{(2)}) \right]^+$$

$$W'_{ik} = \left[W'_{(i-1)k} - F_1^{-1}(1 - U_{ik}^{(1)}) + F_2^{-1}(1 - U_{(i-1)k}^{(2)}) \right]^+.$$

Now

$$\operatorname{var} \overline{W}_i^{(A)} = \frac{1}{4N} \left[\operatorname{var} W_i + \operatorname{var} W'_i + 2 \operatorname{cov}(W_i, W'_i) \right]$$

$$= \frac{1}{2N} \left[\operatorname{var} W_i + \operatorname{cov}(W_i, W'_i) \right]. \tag{4.3.142}$$

By analogy with (4.3.65) we can conclude that the method of antithetic variates will be more efficient than the sample-mean method if

$$\operatorname{var} \overline{W}_i^{(A)} \leq \tfrac{1}{2} \operatorname{var} \overline{W}_i, \tag{4.3.143}$$

which means that $\operatorname{cov}(W_i, W'_i)$ is negative and $|\operatorname{cov}(W_i, W'_i)| > \frac{1}{2} \operatorname{var} W_i$.

Page [46] suggested estimating $E(W)$ by

$$\tilde{W}_i^{(A)} = \frac{1}{N} \sum_{k=1}^{N} \frac{W_{ik} + W''_{ik}}{2}, \tag{4.3.144}$$

where $W''_{ik} = [W''_{(i-1)k} - F_1^{-1}(U_{ik}^{(2)}) + F_2^{-1}(U_{(i-1)k}^{(1)})]^+$.

Comparing the estimates $\overline{W}_i^{(A)}$ and $\tilde{W}_i^{(A)}$, we can see that antithetic pairs $1 - U_{ik}^{(1)}$ and $1 - U_{ik}^{(2)}$ in $\overline{W}_i^{(A)}$ were replaced, correspondingly, by $U_{ik}^{(2)}$ and $U_{ik}^{(1)}$ in $\tilde{W}_i^{(A)}$.

Mitchel [45] proved that, for any $i > 0$, both estimators $\overline{W}_i^{(A)}$ and $\tilde{W}_i^{(A)}$ are more efficient than the sample-mean estimator.

Control variates. It is suggested in Ref. 33 that

$$C_i = C_{i-1} - A_i + S_{i-1}, \qquad C_0 = 0 \tag{4.3.145}$$

be chosen as a control variate for $W_i = \max(W_{i-1} - A_i + S_{i-1}, 0)$, $W_0 = 0$.

Table 4.3.1 presents $\text{var}(W_i)$ for different methods and for the 200th customer, based on 25 runs.

The service time has an exponential distribution with mean $\mu^{-1} = 1.111$; the interarrival time is assumed to be constant and equal to unity, and at time $t_0 = 0$ there are no customers in the system. We can see that the effect of variance reduction by the antithetic and control variates is substantial.

(b) Networks

i) Antithetic variates

To illustrate the use of antithetic variates for networks, consider the network shown in Fig. 4.3.1.

Suppose we wish to estimate the expected completion time of $T = T_1 + T_2$ by simulation, assuming that T_1 and T_2 are independent.

The procedure of using antithetic variates for estimating $E(T)$ is straightforward and can be written as:

1 Generate two sequences of random numbers $\{U_i^{(1)}, i = 1, \ldots, N\}$ and $\{U_i^{(2)}, i = 1, \ldots, N\}$.

Table 4.3.1 var (W_i) for Different Methods

Method	Sample-Mean	Antithetic Variates	Control Variates ($\beta = 1$)
var (W_i)	10.678	1.770	1.427

Source: Data from Ref. 33.

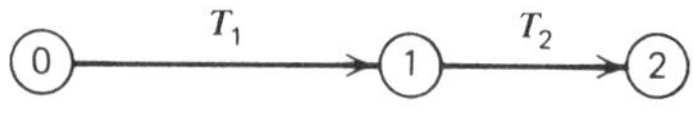

Fig. 4.3.1 Network (from Ref. 29).

2 Compute $T_{1i} = F_1^{-1}(U_i^{(1)})$, $T_{2i} = F_2^{-1}(U_i^{(2)})$, $T'_{1i} = F_1^{-1}(1 - U_i^{(1)})$, and $T'_{2i} = F_2^{-1}(1 - U_i^{(2)})$.

3 Estimate $E(T)$ by

$$\bar{T}_A = \frac{1}{N} \sum_{i=1}^{N} \frac{(T_{1i} + T_{21}) + (T'_{1i} + T'_{2i})}{2}.$$

Let us assume that both T_1 and T_2 are distributed exp(1). Then denoting $T_i = T_{1i} + T_{21}$ and $T'_i = T'_{1i} + T'_{2i}$, we obtain

$$\begin{aligned} \operatorname{var}(\bar{T}_A) &= \frac{1}{4N} \{ \operatorname{var} [(T_{1i} + T'_{1i}) + (T_{2i} + T'_{2i})] \} \\ &= \frac{1}{4N} [\operatorname{var} T_{1i} + \operatorname{var} T'_{1i} + \operatorname{var} T_{2i} + \operatorname{var} T'_{2i} \\ &\quad + 2 \operatorname{cov}(T_{1i}, T'_{1i}) + 2 \operatorname{cov}(T_{2i}, T'_{2i})] \\ &= \frac{1}{4N} [4 + 2E[(T_{1i} - 1)(T'_{1i} - 1)] + 2E[(T_{2i} - 1)(T'_{2i} - 1)] \\ &= \frac{1}{4N} [4 + 2E\{ \ln(U_i^{(1)} + 1)[\ln(1 - U_i^{(1)}) + 1] \} \\ &\quad + 2E\{ (\ln U_i^{(2)} + 1)[\ln(1 - U_i^{(2)}) + 1] \}] \\ &= \frac{1}{N} \left[\int_0^1 \ln u \ln(1 - u)\, du \right] = \frac{1}{N}\left(2 - \frac{\pi^2}{6}\right) = \frac{0.36}{N}. \end{aligned} \tag{4.3.146}$$

On the other hand, in the sample-mean method with $2N$ runs we have

$$\operatorname{var}(\bar{T}) = \frac{1}{2N}. \tag{4.3.147}$$

Thus the variance has been reduced by about one third.

It can be proven by analogy with Proposition 4.3.3 that for any continuous r.v. T_1 and T_2 the method of antithetic variates is more efficient than the sample-mean method.

This simple example has been chosen solely to simplify the presentation. The method of antithetic variates can be successfully employed for any more composed network.

Control variates. Consider the network shown in Fig. 4.3.2. We are interested in finding $E(T_{AB})$, the mean completion time of the network. We

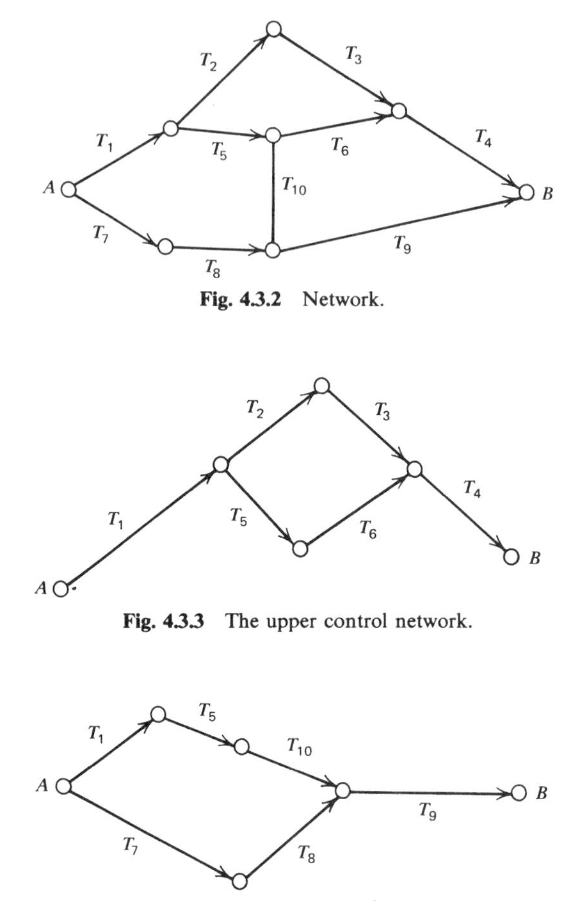

Fig. 4.3.2 Network.

Fig. 4.3.3 The upper control network.

Fig. 4.3.4 The lower control network.

assume that all $T_i, i = 1, \ldots, 10$, are independent exponentially distributed r.v.'s with the same mean $\mu^{-1} = 10$. Even in this case it is difficult to calculate $E(T_{AB})$ because of the "crossing" link of duration T_{10}. It is suggested in Ref. 29 that the control networks be chosen as a subnetwork of the original complex network, formed by deleting links with low probabilities of falling within the critical part. Two such control networks are shown in Fig. 4.3.3 and 4.3.4: the upper and lower control networks, respectively.

For these two control networks the mean completion times are available analytically. Table 4.3.2 presents simulation results for the expected value and the variance of the completion time for the network in Fig. 4.3.2. The

Table 4.3.2 Simulation results for the Network in Fig. 4.3.2

Method	Sample Mean	Antithetic Variates	Control Variates	
			Upper Network	Lower Network
Expected Value $E(T_{AB})$	55.1	54.1	54.3	53.8
Variance var (T_{AB})	6.2	4.0	3.8	3.1

Source: Data from Ref. 29.

following methods are considered: sample-mean, antithetic variates, and control variates, using both the upper and the lower control networks. The simulation results are based on 50 runs. It is clear that the degree of variance reduction depends on our skill in selecting the control networks, which is not an easy problem.

EXERCISES

1 Apply Chebyshev's rule to find the minimal sample size N for which the following formula will hold:

$$P(|\theta_2 - I| \leq \epsilon) = \alpha,$$

where

$$\theta_2 = (b-a)\frac{1}{N}\sum_{i=1}^{N} g(X_i) \qquad \text{and} \qquad X \sim \mathcal{U}(a,b).$$

2 Assuming that for sufficiently large N

$$\hat{\theta}_2 = \frac{\theta_2 - I}{\sqrt{\operatorname{var}\theta_2}} \sim N(0,1),$$

find the confidence interval for I with the level of significance α.

3 Prove Theorem 4.3.2. *Hint:* apply Bellman's dynamic programming recursive equation, or Lagrangian multipliers.

4 Let $I = \sum_{i=1}^{n} a_i I_i$, where $I_i = \int g_i(x)\,dx$ and a_i are known coefficients. An unbiased estimator of I is

$$\eta = \sum_{i=1}^{n} a_i \frac{g_i(X)}{f_X(X)},$$

where $f_X(x)$ is a multidimensional distribution.

(a) Prove that $\min_{f_X(x)} \operatorname{var}(\eta)$ is achieved when $f_X(x) = |Q(x)| / \int |Q(x)|\,dx$, where $Q(x) = \sum_{i=1}^{n} a_i g_i(x)$ and is equal to

$$\operatorname{var}\eta = \left(\int |Q(x)|\,dx\right)^2 - I^2 = \left(\int \Big|\sum_{i=1}^{n} a_i g_i(x)\Big|\,dx\right)^2 - \left(\int a_i g_i(x)\,dx\right)^2.$$

(b) Prove that, if $X_1, \ldots, X_n$ are independent, then $\min_{f_X(x)} \operatorname{var}(\eta)$ is achieved when

$$f_X(x) = \frac{1}{\bar{I}}\left(\sum_{i=1}^{n} a_i^2 g_i^2(x)\right)^{1/2}$$

and is equal to

$$\bar{I}^2 - \sum_{i=1}^{n} a_i^2 I_i^2,$$

where

$$\bar{I} = \int\left(\sum_{i=1}^{n} a_i^2 g_i^2(x)\right)^{1/2} dx.$$

From Evans [5].

5 Consider the integral

$$I = \int_a^b g(x) f_X(x)\,dx,$$

which can be estimated by both the sample-mean Monte Carlo method

$$\theta_4 = \frac{1}{N}\sum_{i=1}^{N} g(X_i)$$

and by the antithetic variates method

$$\theta_7 = \frac{1}{2N}\sum_{i=1}^{N}\left[g(X_i) + g(b + a - X_i)\right],$$

where the sample $X_i, i = 1, \ldots, N$, is taken from $\mathcal{U}(a,b)$. By the assumptions of Proposition 4.3.2 prove

$$\operatorname{var}\theta_7 \le \tfrac{1}{2}\operatorname{var}\theta_4.$$

6 Let $m = 2$, $N_1 = N_2 = N/2$, in the stratified sampling method. According to Proposition 4.3.1, $P_1 = \frac{1}{2}$ and $P_2 = \frac{1}{2}$. Prove that if we choose $P_1 = \frac{3}{4}$ and $P_2 = \frac{1}{4}$, then for any $g(x) \in L_2(X, f)$, $\operatorname{var}\theta_6 > \operatorname{var}\theta_4$, that is, the stratified sampling method is worse than the sample mean method. From Ermakov [4].

7 Prove by induction on m that

$$\underbrace{\int \cdots \int}_{m+1} w^2(x_0, x_2, \ldots, x_m)\,dx_0, dx_1, \ldots, dx_m = (m+1)!,$$

where $w(x_0, x_1, \ldots, x_m)$ is defined in (4.3.119). From Sobol [22].

8 Find an estimator for

$$I = \int_0^\infty g(x) c^{-kx}\, dx, \qquad k > 0,$$

assuming that the sample is taken from the exponential distribution $f_X(x) = \lambda e^{-\lambda x}$, $\lambda \geq 0$. Prove that, for $g(x) = cx_i^n$ $c > 1$. The minimum variance of the estimator will be achieved when $\lambda = k/(n + 1)$. From Sobol [22].

9 Let U be a random number and let $X = aU + b$ and $X' = a(1 - U) + b$. Show that the correlation coefficient between X and X' is equal to -1.

10 Consider the following network:
Assume that T_i, $i = 1, 2, 3$, are i.i.d. r.v.'s distributed $F_T(t)$. Write two formulas for estimating the expected completion time $E(T_{AB})$, using the following methods:

(a) Sample-mean Monte Carlo method.
(b) Antithetic variates.

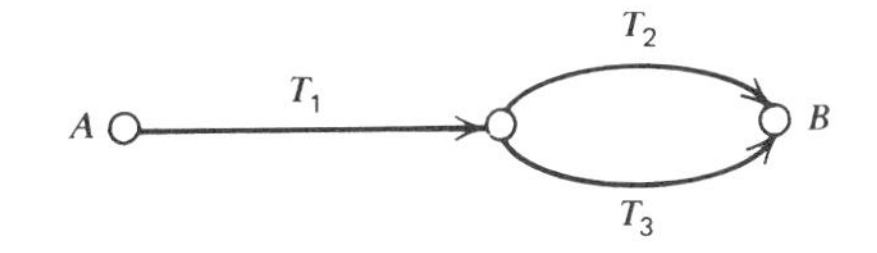

11 Prove that while integrating in situation of noise (see Section 4.2.4) both $\tilde{\theta}_1$ and $\tilde{\theta}_2$ converge a.s. and in mean square to $I = \int g(x)\, dx$ and that $\operatorname{var} \tilde{\theta}_2 < \operatorname{var} \tilde{\theta}_1$.

12 Let $I = \int g(x) h(x)\, dx = E(g(X))$, where $h(x)$ is a p.d.f. Let $f_X(x)$ be another p.d.f. An unbiased estimator of I is

$$\eta = \frac{1}{N} \sum_{i=1}^{n} g(X_i) \frac{h(X_i)}{f_X(X_i)}.$$

Prove that $\min_{f_X(x)} \operatorname{var}(\eta)$ is achieved when

$$f_X(x) = \frac{|g(x)| h(x)}{\int |g(x)| h(x)\, dx}$$

and is equal to

$$\operatorname{var} \eta = \frac{1}{N} \left\{ \left[\int |g(x)| h(x)\, dx \right]^2 - I^2 \right\}.$$

13 Show that the method of antithetic variates is a particular case of the method of control variates.

REFERENCES

1. Burt, I. M. and M. B. Garman, Conditional Monte Carlo: A simulation technique for stochastic network analysis, *Manage. Sci.*, **18**, 1971, 207–217.
2. Clark, C. E., Importance sampling in Monte Carlo Analyses, *Oper. Res.*, **9**, 1961, 603–620.
3. Cochran, W. S., *Sampling Techniques*, 2nd ed., Wiley, New York, 1966.

4 Ermakov, J. M., *Monte Carlo Method and Related Questions*, Nauka, Moscow, 1976 (in Russian).

5 Evans, O. H., Applied multiplex sampling, *Technometrics*, **5**, No. 3, 1963, 341–359.

6 Garman, M. B., More on conditional sampling in the simulation of stochastic networks, *Manage. Sci.*, **17**, 1972, 90–95.

7 Gray, K. G., and K. I. Travers, *The Monte Carlo Method*, Stipes, Champaign, Illinois, 1978.

8 McGrath, E. I., *Fundamentals of Operations Research*, West Coast University, 1970.

9 Halton, I. H., A retrospective and prospective survey of Monte Carlo method, *Soc. Indust. Appl. Math. Rev.*, **12**, 1970, 1–63.

10 Hammersley, J. M. and D. C. Handscomb, *Monte Carlo Methods*, Wiley, New York, 1964.

11 Hammersley, J. M. and K. W. Morton, A new Monte Carlo technique antithetic variates, *Proc. Cambridge Phil. Soc.*, **52**, 1956, 449–474.

12 Kahn, M. and A. W. Marshall, Methods of reducing sample size in Monte Carlo computations, *Oper. Res.*, **1**, 1953, 263–278.

13 Lavenberg, S. S. and Welch P. D., A perspective on the use of control variables to increase the efficiency of Monte Carlo simulations. Research Report RC8161, IBM Corporation, Yorktown Heights, New York, 1980.

14 Marshall, A. W., The use of multi-stage sampling schemes in Monte Carlo computations. in *Symposium on Monte Carlo Methods*, edited by M. A. Meyer, Wiley, New York, 1956, pp. 123–140.

15 Michailov, G. A., *Some Problems in the Theory of the Monte Carlo Method*, Nauka, Novosibirsk, U.S.S.R., 1974 (in Russian).

16 Mitchell, B., Various Reduction by Antithetic Variates in $G1/G/1$ Queueing Simulation, *Oper. Res.*, **21**, 1973, 988–997.

17 Mood, A. M., F. A. Graybill, and D. C. Boes, *Introduction to the Theory of Statistics*, 3rd ed., McGraw-Hill, New York, 1974.

18 Moshman, I., The application of sequential estimation to computer simulation and Monte Carlo procedures, *J. Assoc. Comp. Mach.*, **5**, 1968, 343–352.

19 Neuts, M., *Probability*, Allyn and Bacon, 1972.

20 Powell, M. I. D. and I. Swann, Weighted uniform sampling—A Monte Carlo technique for reducing variance, *J. Inst. Math. App.*, **2**, 1966, 228–238.

21 Shreider, Yu. A. (Ed.), *The Monte Carlo Method* (*the Method of Statistical Trials*), Pergamon, Elmsford, New York, 1966.

22 Sobol, J. M., *Computational Methods of Monte Carlo*, Nauka, Moscow, 1973 (in Russian).

23 Spanier, J., An analytic approach to variance reduction, *Soc. Indust. Appl. Math. J. Appl. Math.*, **18**, 1972, 172–192.

24 Trotter, M. F. and Tukey, J. W., Conditional Monte Carlo for normal samples, in *Symposium on Monte Carlo Methods*, edited by M. A. Meyer, Wiley, New York, 1956, pp. 64–79.

25 Wendel, J. G., Groups and conditional Monte Carlo, *Ann. Math. Stat.*, **28**, 1957, 1048–1052.

26 Yakowitz, S. J., *Computational Probability and Simulation*, Addison-Wesley, Reading, Massachusetts, 1977.

27 Yakowitz, S. et al., Weighted Monte Carlo integration, *Soc. Indust. Appl. Math. J. Numer. Anal.*, **15**, No. 6, 1978, 1289–1300.

ADDITIONAL REFERENCES (SECTION 4.3.12)

28. Burt, J. M., Jr. and M. Garman, Monte Carlo techniques for stochastic network analysis, in *Proceedings of the Fourth Conference on the Applications of Simulation*, December 9–11, pp. 146–153.

29 Burt, J. M., Jr., D. P. Gaver, and M. Perlas, Simple stochastic network: Some problems and procedures, *Naval Res. Logist. Quart.* **17**, 1970, 439–460.

30 Carter, G. and E. Ignall, A simulation model of fire department operations, *Inst. Elec. Electron. Eng. Trans. Syst., Man, Cybern.*, **6**, 1970, 282–292.

31 Carter, G. and E. Ignall, Virtual measures for computer simulation experiments, Report P-4817, The Rand Corporation, Santa Monica, California, April 1972.

32 Gaver, D. P. and G. S. Shedler, Control variable methods in the simulation of a model of a multiprogrammed computer system, *Nav. Res. Logist. Quart.*, **18**, 1971, 435–450.

33 Gaver, D. P. and G. L. Thompson, *Programming and Probability Models in Operations Research*, Brooks/Cole, Monterey, California, 1973.

34 Iglehart, D. L., Functional limit theorems for the queue $GI/G/1$ in light traffic, *Adv. Appl. Prob.*, **3**, 1971, 269–281.

35 Iglehart, D. L. and P. A. W. Lewis, Variance reduction for regenerative simulations, I: Internal control and stratified sampling for queues, Technical Report 86–22, Control Analysis Corporation, Palo Alto, California, 1976.

36 Lavenberg, S. S., Efficient estimation of work rates in closed queueing networks, in *Proceedings in Computational Statistics*, Physica Verlag, Vienna, 1974, pp. 353–362.

37 Lavenberg, S. S., Regenerative simulation of queueing networks, Research Report RC 7087, IBM Corporation, Yorktown Heigths, New York, 1978.

38 Lavenberg, S. S., T. L. Moeller, and C. H. Sauer, Concomitant control variables applied to the regenerative simulation of queueing systems, Research Report RC 6413, IBM Corporation, 1977.

39 Lavenberg, S. S., T. L. Moeller, and P. D. Welch, Control variables applied to the simulation of queueing models of computer systems, in *Computer Performance*, North Holland, Amsterdam, 1977, pp. 459–467.

40 Lavenberg, S. S., T. L. Moeller, and P. D. Welch, Statistical results on multiple control variables with application to variance reduction in queueing network simulation, Research Report, IBM Corporation, Yorktown Heights, New York, 1978.

41 Lavenberg, S. S. and C. H. Sauer, Sequential stopping rules for the regenerative method of simulation, *IBM J. Res. Develop.*, **21**, 1977, 545–558.

42 Lavenberg, S. S. and G. S. Shedler, Derivation of confidence intervals for work rate estimators in closed queueing network, *Soc. Indust. App. Math. J. Comp.*, **4**, 1975, 108–124.

43 Lavenberg, S. S. and D. R. Slutz, Introduction to regenerative simulation, *IBM J. Res. Develop.*, **19**, 1975, 458–462.

44 Lavenberg, S. S. and D. R. Slutz, Regenerative simulation of an automated tape library, *IBM J. Res. Develop.*, **19**, 1975, 463–475.

45 Mitchell, B., Variance reduction by antithetic variates in $G1/G/1$ queueing simulations, *Oper. Res.*, **21**, 1971, 988–997.

46 Page, E. S., On Monte Carlo methods in congestion problems, *Oper. Res.*, **13**, 1965, 300–305.

CHAPTER 5

Linear Equations and Markov Chains

In this chapter we show how Monte Carlo methods can be used to solve linear algebraic, integral, and differential equations. As a rule Monte Carlo methods are not competitive with classical numerical methods for solving systems of linear equations (some special cases where Monte Carlo methods can be used are considered at the end of Section 5.1.3). We discuss the Monte Carlo methods, however, because they serve to introduce analogous Monte Carlo methods for solving integral equations. These methods are widely used, since numerical methods are not efficient in this latter case.

This chapter is constructed as follows: In Section 5.1 we solve a system of linear equations and find the elements of the inverse matrix in the system by simulating discrete-time Markov chains. The problem of finding a solution of integral equations by simulating continuous-time Markov chains is the subject of Section 5.2. Finally, in Section 5.3 we construct a Markov chain for solving the Dirichlet problem.

5.1 SIMULTANEOUS LINEAR EQUATIONS AND ERGODIC MARKOV CHAINS

A Monte Carlo solution to a system of linear equations is based on one proposed by von Neumann and Ulam and extended by Forsythe and Leibler [4].

Let us consider a system of simultaneous linear equations written in vector form

$$Bx = f, \tag{5.1.1}$$

where the vector $x^t = (x_1, \dots, x_n)$ is to be found and the matrix $B = \|b_{ij}\|_1^n$ and the vector $f^t = (f_1, \dots, f_n)$ are given; t denotes the transpose operation.

Introducing $I - A = B$, where I is an identity matrix, system (5.1.1) can be rewritten as

$$x = Ax + f. \tag{5.1.2}$$

Suppose

$$\max_i \sum_{j=1}^{n} |a_{ij}| < 1. \tag{5.1.3}$$

Under this assumption we can solve (5.1.2) by applying the following recursive equation:

$$x^{(k+1)} = Ax^{(k)} + f. \tag{5.1.4}$$

Assuming $x^0 \equiv 0$ and $A^0 \equiv \mathrm{I}$, we have

$$\begin{aligned} x^{(k+1)} &= (I + A + \cdots + A^{k-1} + A^k)f \\ &= \sum_{m=0}^{k} A^m f. \end{aligned} \tag{5.1.5}$$

Taking the limit, for B nonsingular,

$$\lim_{k\to\infty} x^{(k)} = \lim_{k\to\infty} \sum_{m=0}^{k} A^m f = (I - A)^{-1} f = B^{-1} f = x, \tag{5.1.6}$$

we obtain the exact solution of x. The jth coordinate of the vector x^{k+1} is equal to

$$\begin{aligned} x_j^{(k+1)} &= f_j + \sum_{i_1} a_{ji_1} f_{i_1} + \sum_{i_1 i_2} a_{ji_1} a_{i_1 i_2} f_{i_2} \\ &\quad + \cdots + \sum_{i_1, i_2, \dots, i_k} a_{ji_1} a_{i_1 i_2} \cdots a_{i_{k-1} i_k} f_{i_k}. \end{aligned} \tag{5.1.7}$$

We also consider the problem of finding the inner product

$$\langle h, x \rangle = h_1 x_1 + \cdots + h_n x_n, \tag{5.1.8}$$

where h is a given vector and x is a solution of (5.1.2).

It is readily seen that by setting

$$h^t = (\underbrace{0, \dots, 0, 1}_{j}, 0, \dots, 0) \tag{5.1.9}$$

we obtain x_j.

In order to solve (5.1.2) let us introduce an arbitrary ergodic Markov chain (M.C.)

$$P = \| P_{ij} \|_1^n \tag{5.1.10}$$

$$\sum_{i=1}^{n} p_i = 1, \sum_{j=1}^{n} P_{ij} = 1, \qquad p_i \geq 0, P_{ij} \geq 0, i,j = 1, \ldots, n,$$

such that*

$$\begin{aligned} &\mathbf{1} \quad p_i > 0, \qquad \text{if } h_i \neq 0 \\ &\mathbf{2} \quad P_{ij} > 0 \qquad \text{if } a_{ij} \neq 0, i,j = 1, \ldots, n, \end{aligned} \tag{5.1.11}$$

where p_i and P_{ij} are, respectively, the initial distribution and the transition probabilities of the Markov chain.

We first consider the problem of estimation $\langle h, x^{(k+1)} \rangle$, which approximates $\langle h, x \rangle$. Let k be a given integer and let us simulate the Markov chain (5.1.10), (5.1.11) k units of time. We associate with the Markov chain a particle that passes through the sequence of states $i_0, i_1, \ldots, i_k$.

Define

$$W_m = \frac{a_{i_0 i_1} a_{i_1 i_2} \cdots a_{i_{m-1} i_m}}{P_{i_0 i_1} P_{i_1 i_2} \cdots P_{i_{m-1} i_m}}, \tag{5.1.12}$$

which can be written recursively

$$W_m = W_{m-1} \frac{a_{i_{m-1} i_m}}{P_{i_{m-1} i_m}}, \qquad W_0 \equiv 1. \tag{5.1.13}$$

We also define the random variable (r.v.)

$$\eta_k(h) = \frac{h_{i_0}}{p_{i_0}} \sum_{m=0}^{k} W_m f_{i_m} \tag{5.1.14}$$

associated with the sample path $i_0 \to i_1 \to \cdots \to i_k$, which has probability $p_{i_0} P_{i_0 i_1} P_{i_1 i_2} \cdots P_{i_{k-1} i_k}$. Now we are able to prove the following

Proposition 5.1.1

$$E[\eta_k(h)] = \left\langle h, \sum_{m=1}^{k} A^m f \right\rangle = \langle h, x^{(k+1)} \rangle, \tag{5.1.15}$$

that is, $\eta_k(h)$ is an unbiased estimator of the inner product $\langle h, x^{(k+1)} \rangle$.

*The Markov chain need not be homogeneous; we are considering the homogeneous case for simplicity only.

Proof Each path $i_0 \to i_1 \to \cdots \to i_k$ will be realized with probability

$$P(i_0, i_1, \ldots, i_k) = p_{i_0} P_{i_0 i_1} P_{i_1 i_2} \cdots P_{i_{k-1} i_k}. \tag{5.1.16}$$

While simulating the M.C. (5.1.10)–(5.1.11), since the r.v. $\eta_k(h)$ is defined along the path $i_0 \to i_1 \to \cdots \to i_k$, we have

$$E[\eta_k(h)] = \sum_{i_0=1}^{n} \cdots \sum_{i_k=1}^{n} \eta_k(h) p_{i_0} P_{i_0 i_1} \cdots P_{i_{k-1} i_k}, \tag{5.1.17}$$

which, together with (5.1.12) through (5.1.14), gives

$$\begin{aligned} E[\eta_k(h)] &= E\left(\frac{h_{i_0}}{p_{i_0}} \sum_{m=0}^{k} W_m f_{i_m} \right) \\ &= \sum_{i_0=1}^{n} \cdots \sum_{i_k=1}^{n} h_{i_0} \sum_{m=0}^{k} a_{i_0 i_1} a_{i_1 i_2} \cdots a_{i_{m-1} i_m} f_{i_m} P_{i_m i_{m+1}} \cdots P_{i_{k-1} i_k}. \end{aligned} \tag{5.1.18}$$

Using the property $\sum_{j=1}^{n} P_{ij} = 1$, the last formula can be written as

$$E[\eta_k(h)] = \sum_{m=0}^{k} \sum_{i_0=1}^{n} \cdots \sum_{i_m=1}^{n} h_{i_0} a_{i_0 i_1} a_{i_1 i_2}, \ldots, a_{i_{m-1} i_m} f_{i_m}. \tag{5.1.19}$$

Taking into account that

$$A^2 = \| a_{i_0 i_1} \|_1^n \times \| a_{i_1 i_2} \|_1^n = \left\| \sum_{i_1=1}^{n} a_{i_0 i_1} a_{i_1 i_2} \right\|_{i_0, i_2 = 1}^{n},$$

and

$$A^m = \left\| \sum_{i_1=1}^{n} \cdots \sum_{i_{m-1}=1}^{n} a_{i_0 i_1} \cdots a_{i_{m-1} i_m} \right\|_{i_0, i_m = 1}^{n},$$

we immediately obtain

$$E[\eta_k(h)] = \left\langle h, \sum_{m=0}^{k} A^m f \right\rangle = \langle h, x^{(k+1)} \rangle.$$

Q.E.D.

To estimate $\langle h, x^{(k+1)} \rangle$ we simulate N random paths $i_0^{(s)} \to i_1^{(s)} \to \cdots \to i_k^{(s)}$, $s = 1, 2, \ldots, N$, of length k each and then find the sample mean

$$\theta_k = \frac{1}{N} \sum_{s=1}^{N} \eta_k^{(s)}(h) \approx \langle h, x^{(k+1)} \rangle. \tag{5.1.20}$$

The Procedure for Estimating $\langle h, x^{(k+1)}\rangle$

1 Choose any integer $k > 0$.

2 Simulate N independent random paths $i_0^{(s)} \to i_1^{(s)} \to \cdots \to i_k^{(s)}$, $s = 1, \ldots, N$, of the Markov chain (5.1.10)–(5.1.11).

3 Find

$$\eta_k^{(s)}(h) = \frac{h_{i_0}}{p_{i_0}^{(s)}} \sum_{m=0}^{k} W_m^{(s)} f_{i_m}, \qquad s = 1, \ldots, N, \tag{5.1.21}$$

where

$$W_m^{(s)} = \frac{a_{i_0^{(s)} i_1^{(s)}} a_{i_1^{(s)} i_2^{(s)}} \cdots a_{i_{m-1}^{(s)} i_m^{(s)}}}{P_{i_0^{(s)} i_1^{(s)}} P_{i_1^{(s)} i_2^{(s)}} \cdots P_{i_{m-1}^{(s)} i_m^{(s)}}}, \qquad W_0^{(s)} \equiv 1. \tag{5.1.22}$$

4 Calculate

$$\theta_k = \frac{1}{N} \sum_{s=1}^{N} \eta_k^{(s)}(h), \tag{5.1.23}$$

which is an unbiased estimator of the inner product $\langle h, x^{(k+1)}\rangle$.

Taking the limit of (5.1.15), we obtain

$$\lim_{k \to \infty} E[\eta_k(h)] = E[\eta_\infty(h)] = \langle h, x\rangle. \tag{5.1.24}$$

Thus provided that the von Neumann series $A + A^2 + \cdots$ converges and the path $i_0 \to i_1 \to \cdots \to i_k \cdots$ is infinitely long, we obtain an unbiased estimator of $\langle h, x\rangle$.

The sample-mean is then of the form

$$\theta_\infty = \frac{1}{N} \sum_{s=1}^{N} \eta_\infty^{(s)}(h) \tag{5.1.25}$$

where

$$\eta_\infty^{(s)}(h) = \frac{h_{i_0}}{p_{i_0}^{(s)}} \sum_{m=0}^{\infty} W_m^{(s)} f_{i_m}, \qquad s = 1, \ldots, N \tag{5.1.26}$$

and

$$W_m^{(s)} = \frac{a_{i_0^{(s)} i_1^{(s)}} a_{i_1^{(s)} i_2^{(s)}} \cdots a_{i_{m-1}^{(s)} i_m^{(s)}}}{P_{i_0^{(s)} i_1^{(s)}} P_{i_1^{(s)} i_2^{(s)}} \cdots P_{i_{m-1}^{(s)} i_m^{(s)}}}. \tag{5.1.27}$$

We note that the inner products $\langle h, \sum_{m=0}^{k} A^m f\rangle$ for different h and f can be found from (5.1.23) by using the same random paths $i_0^{(s)} \to i_1^{(s)} \to \cdots i_k^{(s)}$, $s = 1, \ldots, N$ of the M.C. (5.1.10)–(5.1.11).

Remark In the particular case where $A = \alpha P$, $0 < \alpha < 1$, we have $W_m = \alpha^m$ and

$$\eta_k(h) = \frac{h_{i_0}}{P_{i_0}} \sum_{m=0}^{K} \alpha^m f_{i_m}.$$

5.1.1 Adjoint System of Linear Equations

Let us define for the system of linear equations (5.1.2) an associated system of linear equations

$$x^* = A^t x^* + h \tag{5.1.28}$$

where $A^t = \| a_{ij}^t \|_1^n$ is the transpose of A. It is readily seen that

$$\langle h, x \rangle = \langle x^*, f \rangle. \tag{5.1.29}$$

Indeed, we have from (5.1.2) and (5.1.28) $\langle x^*, x \rangle = \langle x^*, Ax \rangle + \langle x^*, f \rangle$ and $\langle x, x^* \rangle = \langle A^t x^*, x \rangle + \langle h, x \rangle$, respectively. Now (5.1.29) follows because $\langle A^t x^*, x \rangle = \langle x^*, Ax \rangle$. We call the pair (5.1.2) and (5.1.28) adjoint systems.

A direct consequence of (5.1.29) is that there exists another unbiased estimator of $\langle h, x \rangle$, which can be written as

$$\eta_k^*(h) = \frac{f_{i_0}}{p_{i_0}^*} \sum_{m=0}^{k} W_m^* h_{i_m}, \tag{5.1.30}$$

where

$$W_0^* = 1, \qquad \text{and} \qquad W_m^* = W_{m-1}^* \frac{a_{i_{m-1} i_m}^t}{P_{i_{m-1} i_m}^*}$$

are defined on the sample path $i_0 \to i_1 \to \cdots \to i_k$, which is obtained from the Markov chain defined by the following:

$$P^* = \| P_{ij}^* \|_1^n$$

$$\sum_{i=1}^{n} p_i^* = 1, \qquad \sum_{j=1}^{n} P_{ij}^* = 1, \qquad p_i^* \geq 0, \qquad P_{ij}^* \geq 0, \qquad i,j = 1, \ldots, n,$$

such that

1. $p_i^* > 0, \quad$ if $f_i \neq 0$
2. $P_{ij}^* > 0, \quad$ if $a_{ij}^t \neq 0$, $i,j = 1, \ldots, n$.

In the particular case for which P in (5.1.10)–(5.1.11) is a doubly stochastic matrix, that is,

$$\sum_{j=1}^{n} P_{ij}=1 \qquad \text{and} \qquad \sum_{i=1}^{n} P_{ij}=1, \tag{5.1.31}$$

P^* can be chosen equal to P'. Assuming also $A'=A$, then together with (5.1.31) we obtain $P^*=P$, and (5.1.30) becomes

$$\eta_k^*(h)=\frac{f_{i_0}}{p_{i_0}}\sum_{m=0}^{k} W_m h_{i_m}. \tag{5.1.32}$$

Comparing (5.1.14) with (5.1.32), we can see even in this case, that is, when $A'=A$ and $P^*=P$, $\eta_k^*(h)\neq\eta_k(h)$. The difference between $\eta_k^*(h)$ and $\eta_k(h)$ is in terms of f_{i_0} and h_{i_m}, which are interchanged.

We return now to the original problem (5.1.2) of estimating all coordinates x_j of the vector x. In order to estimate the jth coordinate x_j of x we assume

$$h'=e_j=(\underbrace{0,\ldots,0,1}_{j},0,\ldots,0)$$

and start simulating the M.C. from the state j, that is, $p_{i_0}=p_j=1$. The corresponding path is then $j\to i_1\to i_2\to\cdots\to i_k$.

Denoting

$$\eta_k(e_j)=\sum_{m=0}^{k} W_m f_{i_m}, \tag{5.1.33}$$

where

$$W_m=\frac{a_{ji_1}a_{i_1i_2}\cdots a_{i_{m-1}i_m}}{P_{ji_1}P_{i_1i_2}\cdots P_{i_{m-1}i_m}}, \qquad W_0\equiv 1,$$

we immediately obtain the *corollary*

$$E\left[\eta_k(e_j)\right]=x_j^{(k+1)}, \tag{5.1.34}$$

and also

$$\theta_k(e_j)=\frac{1}{N}\sum_{s=1}^{N}\eta_k^{(s)}(e_j)\approx x_j^{(k+1)}. \tag{5.1.35}$$

It follows from (5.1.33) that, in order to estimate all the components x_j, $j=1,2,\ldots,n$, of the vector x, we have to simulate n random paths $j\to i_1\to i_2\to\cdots\to i_k$, $j=1,2,\ldots,n$, of the Markov chain (5.1.10)–(5.1.11), each time starting from a new state $i_0=j$.

Looking carefully at (5.1.33), we find that all $\eta_k(e_j)$, $j=1,2,\ldots,n$, are similar. They differ only in the initial terms $a_{i_0 i_1}/P_{i_0 i_1}$ and f_{i_0}, which are associated with the choice of the initial state i_0. Thus for $\eta_k(e_j)$ and $\eta_k(e_r)$ we have a_{ji_1}/P_{ji_1}, f_j and a_{ri_1}/P_{ri_1}, f_r, respectively.

We now turn to the question of whether or not all the components x_j of x can be estimated simultaneously by simulating one path. The answer is affirmative. We start this topic with the following

Definition The path $i_0 \to i_1 \to \cdots \to i_T$ will be called *covering* if it has visited each state $j=1,\ldots,n$ at least once.

Let $i_0 \to i_1 \to \cdots \to i_T \to \cdots$ be an infinite realization from the Markov chain (5.1.10)–(5.1.11). Because our Markov chain is ergodic, each state will be visited infinitely many times and the first hitting time to the state j, $T_j = \min\{t : i_t = j\}$ is finite almost surely (a.s.). With this result in hand the procedure for finding all the estimates $\eta_k(e_j)$, $j=1,2,\ldots,n$, from one realization can be written in the following way:

1 Simulate a covering path

$$i_0 \to i_1 \to \cdots \to i_{T_j} \to \cdots \to i_T \to \cdots \to i_{T+k}, \tag{5.1.36}$$

where $T = \max_j\{T_j\} = \max\min_j\{t : i_t = j\}$, $j=1,\ldots,n$, and k is some fixed number.

2 Find the first hitting time $T_j = \min\{t : i_t = j\}$ for each state $j = 1,\ldots,n$, separately.

3 Take the subpath $i_{T_j} \to i_{T_{i+l}} \to \cdots \to i_{T_{j+k}}$ (which is the part of the generated path) for each state $j=1,\ldots,n$, separately.

4 Calculate all

$$\tilde{\eta}_k(e_j) = \sum_{m=T_j}^{T_j+k} W_m f_{i_m}, \qquad j=1,\ldots,n \tag{5.1.37}$$

where

$$W_m = \frac{a_{i_{T_j} i_{T_j+1}} a_{i_{T_j+1} i_{T_j+2}} \cdots a_{i_{m-1} i_m}}{P_{i_{T_j} i_{T_j+1}} P_{i_{T_j+1} i_{T_j+2}} \cdots P_{i_{m-1} i_m}} \quad \text{and} \quad W_{T_j} = 1 \tag{5.1.38}$$

are defined on the same path (5.1.36) starting at different points T_j associated with the first hitting time. Each subpath $i_{T_j} \to i_{T_{j+l}} \to \cdots \to \cdots \to i_{T_{j+k}}$ is of the same length k. Thus $i_0 \to i_1 \to \cdots \to i_T$ will be a covering path of minimal length (in a given realization).

5 Simulate N such independent random paths

$$i_0^{(s)} \to i_1^{(s)} \to \cdots \to i_{T_j^{(s)}}^{(s)} \to \cdots \to i_{T^{(s)}}^{(s)} \to \cdots \to i_{T^{(s)}+k}^{(s)}$$

and find

$$\tilde{\theta}_k(e_j) = \frac{1}{N} \sum_{S=1}^{N} \tilde{\eta}_k^{(s)}(e_j), \qquad j = 1, \ldots, n, \tag{5.1.39}$$

which estimates x_j.

Therefore all r.v.'s $\eta_k(e_j)$, $j = 1, \ldots, n$, are defined on the same path and calculated according to the same formula (5.1.37). The only difference between them is the starting point, which is determined by the first hitting time T_j and is a random variable.

Proposition 5.1.2

$$E\left[\tilde{\eta}_k(e_j)\right] = x_j^{(k+1)}. \tag{5.1.40}$$

Proof The proof of this proposition is based on the strong Markov property, which is given in Ref. 2, Proposition 1.22, p. 117, which states that for any homogeneous Markov chain and any bounded function g defined on the state space, we have

$$E\left[g(i_t, i_{t+1}, \ldots,) | i_t = j\right] = E\left[g(i_0, i_1, \ldots) | i_0 = j\right].$$

In our notations

$$E\left[\tilde{\eta}_k(e_j) | T_j = t\right] = E\left[g(i_t, i_{t+1}, \ldots, i_{t+k}) | i_t = j\right] = E\left[g(i_0, \ldots, i_k) | i_0 = j\right].$$

By Proposition 5.1.1 $E[g(i_0, \ldots, i_k) | i_0 = j] = x_j^{(k+1)}$. Since $E[\eta_k(e_j) | T_j = t]$ does not depend on t, we have $E[\eta_k(e_j) | T_j = t] = E[\tilde{\eta}_k(e_j)] = x_j^{(k+1)}$.

Q.E.D.

Corollary $\lim_{k \to \infty} E[\tilde{\eta}_k(e_j)] = x_j$.

Proposition 5.1.3

$$\operatorname{var}\left[\theta_k(e_j)\right] = \operatorname{var}\left[\tilde{\theta}_k(e_j)\right]. \tag{5.1.41}$$

Proof

$$\operatorname{var}\left[\theta_k(e_j)\right] = \operatorname{var}\left[\frac{1}{N} \sum_{s=1}^{N} \eta_k^{(s)}(e_j)\right]$$

$$= \frac{1}{N}\left\{E\left[\eta_k(e_j)\right]^2 - \left[x_j^{(k+1)}\right]^2\right\} = \frac{1}{N} \operatorname{var} \eta_k(e_j).$$

Similarly,

$$\operatorname{var}\left[\tilde{\theta}_k(e_j)\right] = \operatorname{var}\left[\frac{1}{N}\sum_{s=1}^{N}\tilde{\eta}_k^{(s)}(e_j)\right]$$

$$= \frac{1}{N}\left\{E\left[\tilde{\eta}_k(e_j)\right]^2 - \left[x_j^{(k+1)}\right]^2\right\} = \frac{1}{N}\operatorname{var}\tilde{\eta}_k(e_j).$$

Now again using Proposition 1.22 of Ref. 2 (p. 117), we have

$$E\left\{\left[\tilde{\eta}_k(e_j)\right]^2 | T_j = t\right\} = E\left[\hat{g}(i_t,\ldots,i_{t+k})|i_t = j\right]$$

$$= E\left[\hat{g}(i_0,\ldots,i_k)|i_0 = j\right] = E\left\{\left[\eta_k(e_j)\right]^2\right\}.$$

Therefore $\operatorname{var}[\theta_k(e_j)] = \operatorname{var}[\tilde{\theta}_k(e_j)]$. Q.E.D.

To compare the efficiencies of the two methods we use (4.2.28), which can be written

$$\varepsilon = \frac{t \operatorname{var}\theta_k(e_j)}{\tilde{t}\operatorname{var}\tilde{\theta}_k(e_j)}, \tag{5.1.42}$$

and assume without loss of generality $N = 1$. Since $\operatorname{var}\theta_k(e_j) = \operatorname{var}\tilde{\theta}_k(e_j)$, we have $\varepsilon = t/\tilde{t}$. In the first case we have n trajectories each of length k, so the total length of these trajectories is nk. In the second case we have one trajectory of length $\max_{j=1,\ldots,n}\{T_j\} + k$, with mean $E(\max_{j=1,\ldots,n}\{T_j\}) + k$.

It is obvious that the second algorithm is on the average more efficient when $n > 1$ and $k \gg (n-1)^{-1}E[T = \max_{j=1,\ldots,n}\{T_j\}]$. Because the first hitting time $T_j, j = 1,\ldots,n$, to each state is finite a.s., it can be proven that

$$\varepsilon = \frac{t}{\tilde{t}} \to \frac{1}{n} \text{ a.s.} \qquad \text{as } k \to \infty, \tag{5.1.43}$$

that is asymptotically the method of covering path is n times more efficient than the standard Monte Carlo method.

The efficiency of the second method can be improved if we can find $i_0 = l$ such that

$$E\left[\max_{j=1,\ldots,n}\{T_j\}|i_0 = l\right] = \min_{i_0=1,\ldots,n} E\left[\max\{T_j\}|i_0 = l\right]$$

and then take this $i_0 = l$ as a starting point of the path or, equivalently, choose the initial distribution as

$$p_{i_0} = \begin{cases} 1, & \text{if } i_0 = l \\ 0, & \text{if } i_0 \neq l. \end{cases}$$

5.1.2 Computing the Inverse Matrix

It follows from (5.1.6) that

$$x = \sum_{m=0}^{\infty} A^m f = B^{-1} f$$

where $B^{-1} = \| b_{jr}^{-1} \|_{j=1}^{n} = 1 + A + A^2 + \cdots$. The jth coordinate of x is

$$x_j = \sum_{r=1}^{n} b_{jr}^{-1} f_r.$$

Setting

$$f_r' = e_r = (\underbrace{0, \ldots, 0, 1}_{r}, 0, \ldots, 0), \tag{5.1.44}$$

we obtain

$$x_j = b_{jr}^{-1}, \tag{5.1.45}$$

and the estimator $\eta_k(x_j)$ in (5.1.33) becomes

$$\eta_k\left(b_{jr}^{-1}\right) = \sum_{m/i_m = r} W_m. \tag{5.1.46}$$

Here the summation with respect to W_m is taken over the indices $i_m = r$, that is, when the particle visits the state r.

The sample mean is then

$$\theta_k\left(b_{jr}^{-1}\right) = \frac{1}{N} \sum_{s=1}^{N} \eta_k^{(s)}\left(b_{jr}^{-1}\right) \approx b_{jr}^{-1}, \tag{5.1.47}$$

where $s = 1, 2, \ldots, N$ is the path number.

Thus setting

$$h' = h_j = e_j = (\underbrace{0, \ldots, 0, 1}_{j}, 0, \ldots, 0) \tag{5.1.48}$$

and

$$f = f_r = e_r = \underbrace{0, \ldots, 0, 1}_{r}, 0, \ldots, 0,$$

we can estimate all the elements b_{jr}^{-1} of the inverse matrix B^{-1} by (5.1.47).

Inasmuch as the problem of determining b_{jr}^{-1} is a particular case of the problem of finding x_j, we can estimate all the elements b_{jr}^{-1} of the jth row of the inverse matrix B^{-1} simultaneously with x_j. Thus the Monte Carlo method provides a way of estimating a single element or any collection of the elements of B^{-1}. This desirable feature differentiates the Monte Carlo method from other numerical methods in which, as a rule, all the elements of B^{-1} are computed simultaneously.

By solving the adjoint system we can estimate simultaneously all the elements b_{jr}^{-1} of the rth column of the inverse matrix B^{-1}. It follows also from (5.1.36) through (5.1.39) that all the elements b_{jr}^{-1} can be estimated simultaneously with the x_j's from the covering path.

Before leaving this section we want to turn the readers' attention to the analogy that exists in calculating integrals and solving systems of linear equations by Monte Carlo methods.

Calculating the integral

$$I=\int g(x)\,dx,$$

we introduce any p.d.f. $f_X(x)$ such that

$$I=\int \frac{g(x)}{f_X(x)} f_X(x)\,dx = E\left[\frac{g(X)}{f_X(X)}\right]$$

where X is distributed with p.d.f. $f_X(x)$ and $f_X(x)>0$ when $g(x)\neq 0$. Then taking a sample N from $f_X(x)$, we estimate the integral I by (see (4.3.4))

$$I\approx\theta_3=\frac{1}{N}\sum_{i=1}^{N}\frac{g(X_i)}{f_X(X_i)}.$$

While solving the system of linear equations we introduce any ergodic Markov chain (5.1.10)–(5.1.11). Then simulating our Markov chain, we obtain the path $i_0\to i_1\to\cdots\to i_k$ with probability $P(i_0,i_1,\ldots,i_k)=p_{i_0}P_{i_0i_1},\ldots,P_{i_{k-1}i_k}$.

The element $x_j^{(k+1)}$ of the vector $x^{(k+1)}$ can be written (see (5.1.7)) as

$$\begin{aligned}
x_j^{(k+1)} &= f_j+\sum_{i_1}a_{ji_1}f_{i_1}+\sum_{i_1,i_2}a_{ji_1}a_{i_1i_2}f_{i_2}+\cdots\\
&\quad+\sum_{i_1,i_2,\ldots,i_k}a_{ji_1}a_{i_1i_2}\cdots a_{i_{k-1}i_k}f_{i_k}\\
&= f_i+\sum_{i_1}\frac{a_{ji_1}}{P_{ji_1}}f_{i_1}P_{ji_1}+\sum_{i_1,i_2}\frac{a_{ji_1}a_{i_1i_2}}{P_{ji_1}P_{i_1i_2}}f_{i_2}P_{ji_1}P_{i_1i_2}\\
&\quad+\sum_{i_1,i_2,\ldots,i_k}\frac{a_{ji_1}a_{i_1i_2}\cdots a_{i_{k-1}i_k}}{P_{ji_1}P_{i_1i_2}\cdots P_{i_{k-1}i_k}}f_{ik}P_{ji_1}P_{i_1i_2}\cdots P_{i_{k-1}i_k}\\
&= E\left[\sum_{m=0}^{k}W_m f_{i_m}\right]=E\left[\eta_k(e_j)\right],
\end{aligned}\tag{5.1.49}$$

where η_k is distributed according to $P_{ji_1}P_{i_1i_2}, \ldots, P_{i_{k-1}i_k}$. Here $i_0 = j$ and $p_j = 1$.

Now considering N random paths $j^{(s)} \to i_1^{(s)} \to i_2^{(s)} \to \cdots \to i_k^{(s)}$, $s = 1, \ldots, N$, we can estimate $x_j^{(k)}$ by (5.1.39).

Comparing (4.3.4) and (5.1.35), we realize that both problems of calculating the integral and solving the system of linear equations can be reduced to the problem of estimating the expected value of some random function. In our case the random functions are $g(X)/f_X(X)$ and $\eta_k(e_j)$, respectively.

These results allow us to suggest a *general Monte Carlo procedure* for solving different problems, which can be written as:

1 Find a suitable distribution associated with the problem.
2 Take a sample from this distribution.
3 Substitute the values from the sample in a proper formula, which estimates the solution.

5.1.3. Solving a System of Linear Equations by Simulating a Markov Chain with an Absorbing State

Another possibility of estimating $\langle h, x \rangle$ is by simulating a Markov chain with an absorbing state, as was suggested by Forsythe and Leibler [4].

$$P = \begin{vmatrix} P_{11} & P_{12} & \cdots & P_{1n} & P_{1n+1} \\ P_{21} & P_{22} & \cdots & P_{2n} & P_{2n+1} \\ \vdots & & & & \\ P_{n1} & P_{n2} & \cdots & P_{nn} & P_{nn+1} \\ 0 & 0 & \cdots & 0 & 1 \end{vmatrix} \qquad (5.1.50)$$

with

$$P_{ij} \geq 0, \qquad i,j = 1, \ldots, n, \; P_{i,n+1} = g_i = 1 - \sum_{j=1}^{n} P_{ij} \geq 0$$

$$\sum_{i=1}^{n} p_i = 1, \qquad p_i \geq 0, \; i = 1, 2, \ldots, n,$$

which is essentially an augmented (5.1.10) matrix. Here p_i and P_{ij} are, respectively, the initial and the transition probabilities.

Assume also:

$$\begin{aligned} &\mathbf{1} \quad p_i > 0, && \text{if } h_i \neq 0, \\ &\mathbf{2} \quad P_{ij} > 0, && \text{if } a_{ij} \neq 0, \; i,j = 1, 2, \ldots, n. \end{aligned} \qquad (5.1.51)$$

The state $n+1$ is called an absorbing state of the Markov chain (5.1.50)–(5.1.51). It is well known (Činlar [2]) that, if there exists a state i, $i=1,\ldots,n$, such that $P_{i,n+1}>0$, then all the random paths $i_0 \to i_1 \to \cdots \to i_{(\nu)}$ terminate in state $n+1$ a.s. and the expected time of termination of each random path is finite, that is, $E(\nu)<\infty$.

We start to simulate our Markov chain (5.1.50)–(5.1.51) by choosing the initial state i_0 according to the probability p_{i_0}, $i_0=1,2,\ldots,n$, where $\Sigma_{i_0} p_{i_0}=1$. Consider now a particle that is in state i_0. The particle either will be absorbed with probability g_{i_0} in state i_0 or will pass to another state i_1 with probability $P_{i_0 i_1}$. Generally, if at time $m-1$ the particle arrived at the state i_{m-1}, then it will either be absorbed from there with probability $g_{i_{m-1}}$ or will continue along the random path to the next state i_m with probability P_{i_{m-1}, i_m}. The random path $i_0 \to i_1 \to \cdots \to i_{(\nu)}$ has probability

$$p_{i_0} P_{i_0 i_1} P_{i_1 i_2} \cdots P_{i_{\nu-1} i_\nu} g_{i_\nu}, \qquad \text{where } g_{i_\nu} = P_{i_\nu n+1} = 1 - \sum_{j=1}^{n} P_{i_\nu j}$$

is the probability of absorption from state i_ν.

Consider any r.v. η, which is defined on the parth $i_0 \to i_1 \to \cdots \to i_{(\nu)}$. The expectation of η is

$$E(\eta) = \sum_{k=0}^{\infty} \sum_{i_0=1}^{n} \cdots \sum_{i_k=1}^{n} \eta_k p_{i_0} P_{i_0 i_1} \cdots P_{i_{k-1} i_k} g_{i_k},$$

where η_k is defined on the path that terminates exactly after k units of time.

Let

$$\eta_{(k)}(h) = \frac{h_{i_0}}{p_{i_0}} W_k \frac{f_{i_k}}{g_{i_k}}, \tag{5.1.52}$$

where W_k is the same as in (5.1.12).

Proposition 5.1.4

$$E\left[\eta_{(k)}(h)\right] = \langle h, x \rangle, \tag{5.1.53}$$

that is, $\eta_{(k)}(h)$ is an unbiased estimator of the inner product $\langle h, x \rangle$, provided $E(k)<\infty$.

Proof The r.v. $\eta_{(k)}(h)$ has the same probability as η_k, that is, $p_{i_0} P_{i_0 i_1} P_{i_1 i_2} \cdots P_{i_{k-1} i_k} g_{i_k}$. Therefore

$$E\left[\eta_{(k)}(h)\right] = \sum_{k=0}^{\infty} \sum_{i_0=1}^{n} \cdots \sum_{i_k=1}^{n} \eta_{(k)}(h) p_{i_0} P_{i_0 i_1} P_{i_1 i_2} \cdots P_{i_{k-1} i_k} g_{i_k}. \tag{5.1.54}$$

Substituting (5.1.22) in (5.1.54) and taking (5.1.12) into account, we obtain

$$E\left[\eta_{(k)}(h)\right]=\sum_{k=0}^{\infty}\sum_{i_0=1}^{n}\sum_{i_k=1}^{n} h_{i_0}a_{i_0i_1}a_{i_1i_2}\cdots a_{i_{k-1}i_k}f_{i_k}. \tag{5.1.55}$$

Now comparing (5.1.55) with (5.1.19), we immediately obtain (5.1.53). Q.E.D.

The procedure for estimating $\langle h,x\rangle$ is:

1 Simulate N independent random paths $i_0^{(s)}\to i_1^{(s)}\to\cdots\to i_{(k)}^{(s)}$, $s=1,\ldots,N$, from the Markov chain (5.1.50)–(5.1.51).

2 Determine

$$\eta_{(k)}^{(s)}(h)=\frac{h_{i_0}}{p_{i_0}^{(s)}}W_k^{(s)}\frac{f_{i_k}}{g_{i_k}},\qquad s=1,\ldots,N,$$

where $W_k^{(s)}$ is the same as in (5.1.22).

3 Estimate $\langle h,x\rangle$ by

$$\theta_{(k)}=\frac{1}{N}\sum_{s=1}^{N}\eta_{(k)}^{(s)}(h)\approx\langle h,x\rangle.$$

In the particular case where $a_{ij}\geq 0$ and $\sum_{i=1}^{n}a_{ij}<1$, the matrix P in (5.1.50) can be chosen as

$$P=\begin{vmatrix} a_{11} & \cdots & a_{1n} & p_{1n+1}\\ \vdots & & & \\ a_{n1} & \cdots & a_{nn} & p_{nn+1}\\ 0 & \cdots & 0 & 1 \end{vmatrix}$$

that is, $p_{ij}=a_{ij}$, $i,j=1,\ldots,n$. In this case $W_k=1$ and

$$\eta_k(h)=\frac{h_{i_0}}{p_{i_0}}\frac{f_{i_k}}{g_{i_k}}.$$

There are, however, few applications of these techniques. The reason is that the Monte Carlo method is not competitive with classical numerical analysis in solving systems of linear equations. Still, there are some situations where the Monte Carlo method can be successfully used:

1 The size of the matrix $A=\|a_{ij}\|_1^n$ is sufficiently large ($n>10^3$), and a very rough approximation is required.

2 It is necessary to find $\langle h,x\rangle$ for different h and f, where $x=Ax+f$. As mentioned above, such problems can be solved (estimated) simultaneously by simulating only one Markov chain.

5.2 INTEGRAL EQUATIONS

One of the most fruitful applications of Monte Carlo methods is in solution of integral equations. The reason is that such equations cannot be solved efficiently by classical numerical analysis.

The idea of solving integral equations by a Monte Carlo method is similar to that of solving simultaneous linear equations. Both methods use Markov chains for simulation.

There exists ample literature on solving integral equations by Monte Carlo methods (see [3, 7–9]). Its history is connected with the problem of neutron transport, which is described in Spanier and Gelbard's monograph [9]. One of the earliest methods for solving integral equations by a Monte Carlo method was proposed by Albert [1] and was later developed in Refs. 3, 7, and 8.

Before proceeding with this topic we need some background on integral transforms.

5.2.1 Integral Transforms

Throughout this section we follow Sobol [8]. Let K be an integral operator such that

$$K\psi(x) = \int K(x, x_1)\psi(x_1)\,dx_1, \qquad x_1 \in D, \tag{5.2.1}$$

which maps the function $\psi(t)$ into $K\psi(x)$. $K\psi(x)$ is usually called the first iteration of ψ with respect to the kernel K.

The second iteration is

$$K[K\psi](x) = K^2\psi(x) = \int\int K(x, x_1)K(x_1, x_2)\psi(x_2)\,dx_1\,dx_2. \tag{5.2.2}$$

Proceeding recursively we obtain

$$K[K^{k-1}\psi](x) = K^k\psi(x) = \int K(x, x_k)K^{k-1}\psi(x_k)\,dx_k, \tag{5.2.3}$$

the kth iteration of ψ with respect to the kernel K.

We can estimate such integrals by quadrature methods or by Monte Carlo methods, as described in Chapter 4. However, there exists another Monte Carlo method of estimating such integrals, a method that is similar to the method of solving systems of simultaneous linear equations and that based on simulating a Markov chain.

Before describing the method let us introduce some notations and make some assumptions.

For any two functions $h(x)$ and $\psi(x)$ their inner product is denoted by $\langle h, \psi \rangle$, where

$$\langle h, \psi \rangle = \int h(x)\psi(x)\,dx. \tag{5.2.4}$$

Assume also that

$$\psi(x) \in L_2(D) \tag{5.2.5}$$

$$h(x) \in L_2(D) \tag{5.2.6}$$

and

$$K(x,y) \in L_2(D \times D), \tag{5.2.7}$$

which is the same as

$$\int h^2\,dx < \infty \tag{5.2.8}$$

$$\int \psi^2\,dx < \infty \tag{5.2.9}$$

and

$$\int\int K^2\,dx\,dy < \infty, \tag{5.2.10}$$

respectively.

It is easy to prove, using the Cauchy-Schwarz inequality, that, if conditions (5.2.8) and (5.2.9) are met, then $|\langle h, \psi \rangle| < \infty$. Indeed

$$|\langle h, \psi \rangle| = \left| \int h\psi\,dx \right| \le \left\{ \int h^2\,dx \int \psi^2\,dx \right\}^{1/2} < \infty. \tag{5.2.11}$$

In exercise 2 the reader is asked to prove $K\psi(x) \in L_2(D)$, given (5.2.5) and (5.2.7).

With these results we can return to our problem of evaluating $K^k\psi$. As we mentioned before, the method of evaluating $K^k\psi$ is similar to those for solving the system of linear equations described in Section 5.1.1. From now we consider the problem of finding the inner product $\langle h, K^k\psi \rangle$, which is similar to the problem $\langle h, \Sigma_{m=0}^k A^m f \rangle$. The reader is asked to keep this similarity in mind.

By analogy with (5.1.10) and (5.1.11) let us introduce any continuous Markov chain

$$P = \| P(x,y) \|, \qquad x,y \in D \tag{5.2.12}$$

satisfying $\int P(x,y)\,dy = 1$, $\int p(x)\,dx = 1$, such that

$$\begin{aligned} &1 \quad p(x) > 0, && \text{if } h(x) \neq 0 \\ &2 \quad P(x,y) > 0 && \text{if } K(xy) \neq 0, \end{aligned} \tag{5.2.13}$$

where $p(x)$ and $P(x,y)$ are, respectively, the initial and the transition densities of the Markov chain (5.2.12)–(5.2.13).

By analogy with Proposition 5.1.1 we can readily prove the following

Proposition 5.2.1

$$E[\eta_k(h)] = \langle h, K^k\psi \rangle \tag{5.2.14}$$

where the r.v.

$$\eta_k(h) = \frac{h(x_0)}{p(x_0)} W_k \psi(x_k) \tag{5.2.15}$$

has densities $p(x_0)P(x_0, x_1)P(x_1, x_2) \cdots P(x_{k-1}, x_k)$ and

$$W_k = W_{k-1} \frac{K(x_{k-1}, x_k)}{P(x_{k-1}, x_k)}, \qquad W_0 \equiv 1. \tag{5.2.16}$$

Assuming for some given y that $h(x) = p(x) = \delta(x-y)$, where $\delta(\cdot)$ is Dirac's delta function, we immediately obtain $E[\eta_k(h)] = K^k\psi$.

The procedure for estimating the inner product $\langle h, K^k f \rangle$, where $K^k\psi$ is defined in (5.2.3), can be written by analogy with the procedure for estimating $\langle h, x^{(k+1)} \rangle$ in Section 5.1.1 as follows:

1 Choose any integer $k > 0$.

2 Simulate N independent random paths $x_0^{(s)} \to x_1^{(s)} \to \cdots \to x_k^{(s)}$, $s = 1, 2, \ldots, N$, from Markov chain (5.2.12)–(5.2.13).

3 Find

$$\eta_k^{(s)}(h) = \frac{h(x_0)}{p(x_0)} W_k^{(s)} \psi(x_k), \qquad s = 1, \ldots, N, \tag{5.2.17}$$

where

$$W_k^{(s)} = \frac{K(x_0^{(s)}, x_1^{(s)})K(x_1^{(s)}, x_2^{(s)}) \cdots K(x_{k-1}^{(s)}, x_k^{(s)})}{p(x_0^{(s)}, x_1^{(s)})P(x_1^{(s)}, x_2^{(s)}) \cdots P(x_{k-1}^{(s)}, x_k^{(s)})}, \qquad W_0^{(s)} \equiv 1. \tag{5.2.18}$$

4 Calculate

$$\theta_k = \frac{1}{N} \sum_{s=1}^{N} \eta_k^{(s)}(h) \approx \langle h, K^k\psi \rangle, \tag{5.2.19}$$

which is an unbiased estimator of the inner product $\langle h, K^k\psi \rangle$.

5.2.2 Integral Equations of the Second Kind

Consider the following integral equation of the second kind:

$$z(x) = \int_D K(x, x_1) z(x_1)\, dx_1 + f(x), \tag{5.2.20}$$

which can be written as

$$z = Kz + f. \tag{5.2.21}$$

Let us assume that $f(x) \in L_2(D)$, $K(x, x_1) \in L_2(D \times D)$, and

$$|K| = \sup_D \int_D |K(x,y)|\, dy < 1. \tag{5.2.22}$$

Under these assumptions by analogy with (5.1.4) we can estimate (5.2.20), applying the following recursive equation:

$$z^{(k+1)} = Kz^{(k)} + f. \tag{5.2.23}$$

Setting $Z^0 \equiv 0$ and $K^0 \equiv 0$, we get

$$z^{(k+1)} = f + Kf + \cdots + K^k f = \sum_{m=0}^{k} K^m f. \tag{5.2.24}$$

Taking the limit

$$\lim_{k\to\infty} z^{(k)} = \lim_{k\to\infty} \sum_{m=0}^{k} K^m f = z,$$

we obtain the exact solution of z provided the von Neumann series converges.

One way of estimating $\langle h, z \rangle$ is via simulation of a continuous Markov chain similar to (5.1.10)–(5.1.11).

The following proposition can be readily proved [8] by analogy with Proposition 5.1.1.

Proposition 5.2.2 For any given vector h

$$E[\eta_k(h)] = \left\langle h, \sum_{m=0}^{k} K^m f \right\rangle, \tag{5.2.25}$$

where the r.v. $\eta_k(h)$ is defined on the path $x_0 \to x_1 \to \cdots \to x_k$, such that

$$\eta_k(h) = \frac{h(x_0)}{p(x_0)} \sum_{m=0}^{k} W_m f(x_m) \tag{5.2.26}$$

$$W_m = \frac{K(x_0, x_1) K(x_1, x_2) \cdots K(x_{m-1}, x_m)}{P(x_0, x_1) P(x_1, x_2) \cdots P(x_{m-1}, x_m)}. \tag{5.2.27}$$

The sample mean

$$\theta_k = \frac{1}{N} \sum_{s=1}^{N} \eta_k^{(s)}(h) \approx \left\langle h, \sum_{m=0}^{k} K^m f \right\rangle \tag{5.2.28}$$

estimates the inner product $\langle h, \Sigma_{m=0}^{k} K^m f \rangle$.

Assuming again $h(x) = p(x) = \delta(x - y)$, we obtain (5.2.23). Considering an infinite path $x_0 \to x_1 \to \cdots \to x_k \to \cdots$, we define the random variable

$$\eta_\infty(h) = \frac{h(x_0)}{p(x_0)} \sum_{m=0}^{\infty} W_m f(x_m). \tag{5.2.29}$$

It can be shown that for $\eta_\infty(h)$ to be an unbiased estimator of $\langle h, z \rangle$, that is,

$$\lim_{k \to \infty} E[\eta_k(h)] = E[\eta_\infty(h)] = \langle h, z \rangle, \tag{5.2.30}$$

it is not enough to assume the convergence of the von Neumann series $\Sigma_{m=0}^{\infty} K^m f$, that is,

$$\sum_{m=0}^{\infty} K^m f < \infty. \tag{5.2.31}$$

The reader is asked in exercise 5 to prove (5.2.30), provided

$$\sum_{m=0}^{\infty} |K^m f| < \infty. \tag{5.2.32}$$

It is obvious that, when $K(x,y) \geq 0$ and $f(x) \geq 0$, both (5.2.31) and (5.2.32) coincide.

Another way of estimating $\langle h, z \rangle$ is via simulation of a continuous Markov chain with an absorbing state similar to that of (5.1.50)–(5.1.51). Consider the random path $x_0 \to x_1 \to \cdots \to x_{(k)}$ with the absorption time k, which is a random variable such that $E(k) < \infty$. Define on this path the r.v. (compare with (5.1.52))

$$\eta_{(k)}(h) = \frac{h(x_0)}{p(x_0)} W_k \frac{f(x_k)}{g(x_k)}, \tag{5.2.33}$$

where $g(x)$ is the absorption probability, $p(x)$ is the initial distribution, and

$$W_k = \frac{K(x_0, x_1) K(x_1, x_2) \cdots K(x_{k-1}, x_k)}{P(x_0, x_1) P(x_1, x_2) \cdots P(x_{k-1}, x_k)}. \tag{5.2.34}$$

Then by analogy with Proposition 5.1.4 we can readily prove

$$E[\eta_{(k)}(h)] = \langle h, z \rangle, \tag{5.3.35}$$

provided $\Sigma_{m=0}^{\infty} |K^m f| < \infty$.

To estimate $\langle h, z \rangle$ we simulate N random paths $x_0^{(s)} \to x_1^{(s)} \to \cdots \to x_{(k)}^{(s)}$ with absorption state and find

$$\theta_{(k)} = \frac{1}{N} \sum_{s=1}^{N} \eta_{(k)}^{(s)}(h) \approx \langle h, z \rangle. \tag{5.2.36}$$

The problem $x = Ax + f$ can be considered as a particular case of the problem $z = Kz + f$. Indeed, let us partition the region D into n mutually disjoint subregions D_i, $i = 1, 2, \ldots, n$, such that $D = \cup_{i=1}^{n} D_i$, and let us assume that $f(x)$ and $K(x, x_1)$ are constant functions in each subregion D_i, $i = 1, 2, \ldots, n$, that is,

$$\begin{aligned} f(x) &= f_i, & x &\in D_i \\ K(x, x_1) &= a_{ij}, & x &\in D_i,\ x_1 \in D_j. \end{aligned} \tag{5.2.37}$$

Then, for any $x \in D_i$,

$$\begin{aligned} z(x) &= \sum_{j=1}^{n} \int_{D_j} K(x, x')z(x_1)\, dx_1 + f_i \\ &= \sum_{j=1}^{n} a_{ij} \int_{D_j} z(x_1)\, dx_1 + f_i. \end{aligned} \tag{5.2.38}$$

Inasmuch as $z(x)$ does not depend on x, the last formula can be written as

$$z_i = \sum_{j=1}^{n} a_{ij} z_j + f_i. \tag{5.2.39}$$

Thus by partitioning the region D into n disjoint subregions, we can find the solution of the integral equation (5.2.19) by solving the system of linear equation (5.2.39).

5.2.3 Eigenvalue Problem

Consider the following homogeneous integral equation:

$$z(x) = \lambda \int K(x, x_1) z(x_1)\, dx_1, \tag{5.2.40}$$

which can be written as

$$z = \lambda K z. \tag{5.2.41}$$

If $z \neq 0$, then λ is called the eigenvalue and $z(x)$ is called the eigenfunction corresponding to λ.

Let us assume that the smallest eigenvalue λ_1 is positive and that the kernel $K(x,y) = K(y,x)$ is symmetric and positive definite, that is, $\langle K\psi, \psi \rangle > 0$ if $\psi \neq 0$. Under these assumptions for any two positive functions f and

h we have (see Sobol [8])

$$\lim_{m\to\infty} \frac{\langle h, K^m f\rangle}{\langle h, K^{m+1} f\rangle} = \lambda \tag{5.2.42}$$

and

$$\lim_{m\to\infty} K^m f(x) \langle K^m f, K^m f\rangle^{-1/2} = z(x), \tag{5.2.43}$$

where $z(x)$ is the eigenfunction corresponding to λ.

We can estimate $\langle h, K^m f\rangle$ and $K^m f$ simultaneously by a Monte Carlo method as described in Section 5.2.1.

For further discussion of eigenvalue problems we refer the reader to Hammersley and Handscomb [5] and Sobol [8].

Until now we have not made any special assumption about our Markov chains. We have required only that the estimators $\eta_k(h)$ and $\eta_{(k)}(h)$ be unbiased. It is clear that the variance of both $\eta_k(h)$ and $\eta_{(k)}(h)$ depends on the transition probabilities P_{ij}. Since in solving linear and integral equations we have, respectively, sums and integrals to deal with, it should be possible to use some of the variance reduction techniques of Chapter 4 for better efficiency. In this context the reader is referred to Michailov [7] and Ermakov [3].

5.3 THE DIRICHLET PROBLEM

One of the earliest and most popular illustrations of the Monte Carlo method is the solution of Dirichlet's problem [4].

Dirichlet's problem is to find a continuous and differentiable function u over a given domain D with boundary D^0, satisfying

$$\frac{\partial^2 u}{\partial x^2} + \frac{\partial^2 u}{\partial y^2} = F(x,y), \qquad (x,y) \in D \tag{5.3.1}$$

and

$$u(x,y) = g(x,y), \qquad \text{for } (x,y) \in D^0 \tag{5.3.2}$$

where $g = g(x,y)$ is some prescribed function.

Equation (5.3.1) with $F(x,y) \neq 0$ is called the Laplace equation; with $F(x,y) = 0$ it is known as the Poisson equation.

Generally, there is no analytical solution known to this problem and we have to apply a numerical method. We usually start by covering D with a grid and replacing the differential equation by its finite-difference approximation. Let us denote the closure of D by $\bar{D}$, that is, $D \cup D^0 = \bar{D}$, and the coordinates of the grid by $x_\alpha = \alpha h$ and $y_\beta = \beta h$, where h is the step size. Taking the two-dimensional case for convenience, we call the point $(x_\alpha, y_\beta) \in \bar{D}$ an interior point of $\bar{D}$ if four neighbor points of (x_α, y_β), namely, $(x_\alpha - h, y_\beta)$, $(x_\alpha + h, y_\beta)$, $(x_\alpha, y_\beta - h)$, and $(x_\alpha, y_\beta + h)$ also belong to $\bar{D}$.

We call $(x_\alpha, y_\beta) \in D$ a boundary point if there are not four neighbor points that belong to $\bar{D}$.

Taking this definition into account, we have for any interior point

$$\frac{u_{\alpha+1,\beta} - 2u_{\alpha\beta} + u_{\alpha-1,\beta}}{h^2} + \frac{u_{\alpha,\beta+1} - 2u_{\alpha\beta} + u_{\alpha,\beta-1}}{h^2} = F_{\alpha\beta}, \qquad (x_\alpha, y_\beta) \in D, \tag{5.3.3}$$

which is the finite-difference equation of (5.3.1). Here $u_{\alpha\beta} = u(x_\alpha, y_\beta)$, $F_{\alpha\beta} = F(x_\alpha, x_\beta)$, $u_{\alpha \pm 1,\beta} = u(x_\alpha \pm h, y_\beta)$, and $u_{\alpha,\beta\pm 1} = u(x_\alpha, y_{\beta\pm 1})$. The last equation can be rewritten as

$$u_{\alpha\beta} = \tfrac{1}{4}\left(u_{\alpha-1,\beta} + u_{\alpha+1,\beta} + u_{\alpha,\beta-1} u_{\alpha,\beta+1} - h^2 F_{\alpha\beta}\right). \tag{5.3.4}$$

The boundary condition (5.3.2) is then

$$u_{\alpha,\beta} \equiv g_{\alpha,\beta} \qquad \text{and} \qquad (x_\alpha, y_\beta) \in D^0. \tag{5.3.5}$$

It is not difficult to see that by numbering all the points $(x_\alpha, y_\beta) \in \bar{D}$ in any order we can rewrite (5.3.4) and (5.3.5) as

$$u_i = \sum_{j=1}^{n} a_{ij} u_j + f_i, \qquad i,j = 1, 2, \ldots, n. \tag{5.3.6}$$

Here n is the number of mesh points $(x_\alpha, y_\beta) \in \bar{D}$, which is also equal to the order of the matrix $\|a_{ij}\|_1^n$.

The matrix $\|a_{ij}\|$ has a specific structure: all diagonal elements are equal to zero; each row corresponding to an interior point of D has four elements equal to $\frac{1}{4}$, all other elements being zero; each row corresponding to a boundary point of D^0 contains also elements equal to $\frac{1}{4}$ or zero, but the number of $\frac{1}{4}$ elements is as that of neighboring points, which is always less than 4.

Thus the Dirichlet problem is approximated by a system of linear equations (5.3.6), which can be solved by the Monte Carlo methods described in Section 5.1.3.

EXERCISES

1 Describe an algorithm for simulating an ergodic Markov chain.

2 Prove that $K\psi(x) \in L_2(D)$, given (5.2.5) and (5.2.7).

3 Prove Proposition 5.2.1, that is, $E[\eta_k(h)] = \langle h, K^k\psi \rangle$.

4 Let $K^i\psi(x) = \Sigma_{k=1}^{m} C_{ik}\phi_k(x)$, where

$$\langle \phi_k, \phi_j \rangle = \begin{cases} 1, & \text{if } k = j \\ 0, & \text{if } k \neq j. \end{cases}$$

Prove that

$$E[\eta_i(\phi_k)] = K^i\psi, \phi_k = C_{ik}.$$

5 Prove (5.2.30), given $\Sigma_{m=0}^{\infty} |K^m f| < \infty$.

6 Prove (5.2.35), given $\Sigma_{m=0}^{\infty} |K^m f| < \infty$.

7 Consider the recursive formula (5.2.23)

$$z^{k+1} = Kz^k + f.$$

Assume $z^0 = \phi(x)$, where $\phi(x)$ is any function. Then

$$z^{k+1} + \sum_{m=0}^{k} K^m f + K^{k+1}\phi.$$

Define

$$\eta_k(h) = \frac{h(x_0)}{p(x_0)} \sum_{m=0}^{k} W_m f(x_m) + W_{k+1} f(x_{k+1})$$

and prove that

$$E[\eta_k(h)] = \langle h, z^{(k+1)} \rangle.$$

8 Prove (5.1.43), that is, prove that asymptotically the method of covering path is n times more efficient than the standard Monte Carlo method.

9 Consider the systems of linear equations $x = Ax + f$, where

$$A = \begin{vmatrix} a_{11} & a_{12} \\ a_{21} & a_{22} \end{vmatrix} = \begin{vmatrix} 0.5 & 0.2 \\ 0.3 & 0.4 \end{vmatrix} \quad \text{and} \quad f = (f_1, f_2) = (2, 3).$$

The exact solution of this is $x = (x_1, x_2) = (7.5, 8.75)$.

Simulate the following Markov chain with an absorbing state:

$$P = \begin{vmatrix} 0.5 & 0.2 & 0.3 \\ 0.3 & 0.4 & 0.3 \\ 0 & 0 & 1 \end{vmatrix}$$

and estimate the exact solution $x = (x_1, x_2) = (7.5, 8.75)$ by making a run of the 1000th replication of the Markov chain.

REFERENCES

1 Albert, G. E., A general theory of stochastic estimates of the Neumann series for solution of certain Fredholm integral equations and related series, in *Symposium of Monte Carlo Methods*, edited by M. A. Meyer, Wiley, New York 1956, pp. 37–46.

2 Činclar, E., *Introduction to Stochastic Processes*, Prentice-hall, Englewood Cliffs, New Jersey, 1975.

3 Ermakov, J. M., *Monte Carlo Method and Related Questions*, Nauka, Moscow, 1976 (in Russian).

4 Forsythe, S. E. and R. A. Leibler, Matrix inversion by a Monte Carlo method, *Math. Tables Other Aids Comput.*, **4**, 1950, 127–129.

5 Hammersley, I. M. and D. C. Handscomb, *Monte Carlo Methods*, Methuen, London, 1964.

6 Halton, I. H., A retrospective and prospective survey of Monte Carlo method, *Soc. Indust. Appl. Math. Rev.*, **12**, 1970, 1–63.

7 Michailov, G. A., *Some Problems in the Theory of the Monte Carlo Method*, Nauka, Novosibirsk, U.S.S.R., 1974 (in Russian).

8 Sobol, J. M., *Computational Methods of Monte Carlo*, Nauka, Moscow, 1973 (in Russian).

9 Spanier, J. and E. M. Gelbard, *Monte Carlo Principles and Neutron Transportation Problems*, Addison-Wesley, Reading, Massachusetts, 1969.

CHAPTER 6

Regenerative Method for Simulation Analysis

6.1 INTRODUCTION

It has already been mentioned in Chapter 1 that many real-world problems are too complex to be solved by analytical methods and that the most practical approach to their study is through simulation. In this chapter we consider simulation of a stochastic system, that is, of a system with random elements. Simulation of such systems can be considered as a statistical experiment, in which we seek valid statistical inferences about some unknown parameters associated with the output of the system (or the associated model) being simulated. However, classical methods of statistics are often unsuitable for estimating these parameters. The reason, as we see later, is that the observations made on the simulated system are highly correlated and nonstationary in time; under these circumstances it is difficult (actually impossible) to carry out adequate statistical analyses of the simulated data. To overcome these difficulties a procedure based on regenerative phenomena, called the regenerative method, has recently been developed.

Historically, Cox and Smith [4] were the first to suggest use of regenerative phenomena for simulating a queueing system with Poisson arrivals. This idea was extended by Kabak [39] and Poliak [59]. Quite recently, Crane and Iglehart [6–9] developed a methodology for the regenerative method, based on a unified approach to analyze the output of simulations of those systems that have the property of self-regeneration, that is, of invariably returning (at particular times) to the conditions under which the future of the simulation becomes a probabilistic replica of its past. In other words, if the simulation output is viewed as a stochastic process, the regenerative property means that at those particular times the future behavior of the process is independent of its past behavior, and is governed by the same probability law, that is, at those times the stochastic process "starts afresh probabilistically." Crane and Iglehart showed that a wide variety of problems, such as communication networks, queues, maintenance and inventory control systems, can be cast into a common framework using regenerative phenomena; they then proposed a simple technique for obtaining point estimators and confidence intervals for parameters associated with the simulation output.

The regenerative method also provides answers to the following important problems: how and when to start the simulation, how long to run it, when to begin collecting data, and how to deal with highly correlated data.

The theory and practice of the regenerative method are now in the process of rapid development. The list of references contains about 100 relevant papers known to the author. An excellent introduction to the regenerative method can be found in Crane and Lemoine's book [10]. Iglehart's forthcoming monograph [38] will present a rigorous development of both the theory and practice. Many others recently obtained results, in particular regarding simulation of response time in networks of queues, are to be found in Iglehart and Shedler's monograph [37].

This chapter is organized as follows. The basic ideas of the regenerative method are discussed in Section 6.2. Section 6.3 deals with statistical problems, in particular with the confidence interval for the expected values of some functions defined on the steady-state distribution of the process being simulated. In Section 6.4 the ideas of the regenerative method are illustrated for a single-server queue, a repairman system, and a closed queueing system. Choice of the best among a set of competing systems is the subject of Section 6.5. Section 6.6 deals with a linear programming problem in which the coefficients are unknown and presents the output parameters of regenerative processes. Variance reduction techniques in regenerative simulation are the subject of Section 6.7.

6.2 REGENERATIVE SIMULATION

We start this section with the definition of a regenerative process. Roughly speaking, a stochastic process $\{X(t) : t \geq 0\}$ is called regenerative if there exist certain random times $0 < T_0 < T_1 < T_2 < \cdots$ forming a renewal process* such that at each such time the future of the process becomes a

*A sequence of random variables $\{T_n : n \geq 0\}$ is a renewal process provided that $T_0 = 0$ and $T_n - T_{n-1}$ $(n \geq 1)$ are i.i.d. r.v.'s.

probabilistic replica of the process itself. Informally, this means that at these times the future behavior of the process is independent of its past behavior and is invariably governed by the same law. In other words, the part of the process $\{X(t): T_{i-1} < t \leq T_i\}$ defined between any pair of successive times is a statistically independent probabilistic replica of any other part of the same process defined between any other pair of successive times.

The times $\{T_i : i \geq 0\}$ are called regeneration times and the time between T_{i-1} and T_i is referred to as the length of the ith cycle. Formally [5], a stochastic process $\{X(t): t \geq 0\}$ is regenerative if there exists a sequence $T_0, T_1, \ldots$ of stopping times* such that:

1 $T = \{T_l : l = 0, 1, \ldots\}$ is a renewal process.

2 For any $l, m \in \{0, 1, \ldots\}$, $t_1, \ldots, t_l > 0$, the random vectors $\{X(t_1), \ldots, X(t_l)\}$ and $\{X(T_m + t_1), \ldots, X(T_m + t_l\}$ are identically distributed and the processes $\{X(t): t < T_m\}$ and $\{X(T_m + t): t \geq 0\}$ are independent.

For example, let $\{X_n : n \geq 0\}$ be an irreducible, aperiodic, and positive recurrent Markov chain with a countable state space $I = \{0, 1, \ldots\}$, and let j be a fixed state; then every time at which state j is entered is a time of regeneration.

Let us select a fixed state of the Markov chain (M.C.), say state 0. We then obtain a sequence of stopping times $\{T_i : i \geq 0\}$ such that $0 = T_0 < T_1 < T_2 < \cdots$ and $X_{T_i} = 0$ almost surely (a.s.); that is, once the system enters state 0, the simulation can proceed without any knowledge of its past history.

For another example, let us consider the queue size at time t for a $GI/G/1$ queueing system. Suppose the time origin is taken to be an instant of departure at which time the departing customer leaves behind exactly j customers. Then every time a departure occurs leaving behind j customers, the future of the stochastic process after such a time obeys exactly the same probability law as when the process started at time zero. More examples of regenerative processes are considered in Section 6.4.

It is shown in Ref. 8 that under certain mild regularity conditions the process $\{X(t): t > 0\}$ has a limiting steady-state distribution in the sense that there exists a random vector X such that

$$\lim_{t \to \infty} P\{X(t) \leq x\} = P(X \leq x).$$

*A random variable T taking values in $[0, +, \infty)$ is a stopping time [5] for a stochastic process $\{X(t): t \geq 0\}$, provided that for every finite $t \geq 0$, the occurrence or nonoccurrence of the event $\{T \leq t\}$ can be determined from the history $\{X(s): s \leq t\}$ of the process up to time t.

This type of convergence is known as weak convergence and is denoted $X(t) \Rightarrow X$ as $t \to \infty$. The random vector X is called the steady-state vector.

Let $f: R^k \to R$ be a given real-valued measurable function, and suppose we wish to estimate the value $r = E\{f(X)\}$, where X is the steady-state vector.

For the M.C. $\{X_n : n \geq 0\}$ we have

$$r = E\{f(X)\} = \sum_{i \in I} f(i) P(X = i) = \sum_{i \in I} f(i) \pi_i. \tag{6.2.1}$$

Here, $\pi = \{P(X = i) : i \in I\}$ is the steady-state (stationary) distribution of the regenerative process $\{X_n : n > 0\}$, and $f(i)$ can be interpreted as the penalty (reward) paid in state i. To find r we can solve the following linear system of stationary equations, $\pi = \pi P$, where $P = \{P_{ij} : i, j \in I\}$ is the transition matrix, and then apply (6.2.1).

Let us assume that the values $f(i)$ are known but the transition matrix is unknown. It is clear that the value r cannot be found analytically, since π is determined by P, and simulation must be used. Another case is when P is known but the state space is very large; in this case it may be quite difficult to solve the system $\pi = \pi P$, and we must resort to simulation again.

Possible functions f of interest are the following:

1 If

$$f(i) = \begin{cases} 1, & i = j, \\ 0, & i \neq j, \end{cases} \quad j \in I,$$

then $E\{f(X)\} = \pi_j$.

2 If

$$f(i) = \begin{cases} 1, & i \geq j, \\ 0, & i < j, \end{cases} \quad j \in I,$$

then $E\{f(X)\} = P\{X \geq j\}$.

3 If $f(i) = i^p$, $p > 0$, then $E\{f(X)\} = E\{X^p\}$.

4 If $f(i) = c_i$ = cost of being in state i, then $E\{f(X)\} = \Sigma_{i \in I} c_i P\{X = i\}$ (the stationary expected cost per unit time).

Let τ_i denote the interval between the ith and the $(i+1)$th regeneration times, that is $\tau_i = T_{i+1} - T_i$, $i \geq 0$; τ_i is referred to as the length of the ith cycle. Next, assume $E(\tau_i) < \infty$, and define

$$Y_i = \int_{T_i}^{T_{i+1}} f\{X(t)\}\, dt \tag{6.2.2}$$

or

$$Y_i = \sum_{j=T_i}^{T_{i+1}-1} f(X_j), \tag{6.2.3}$$

depending on whether the process $\{X(t): t \geq 0\}$ is continuous-time or discrete-time. In other words, Y_i is the penalty (reward) during the cycle of length $\tau_i = T_{i+1} - T_i$. Naturally, Y_i is a random variable (r.v.) because so are τ_i and $f(X_i)$.

We now formulate two fundamental propositions that are used extensively in the rest of this chapter.

Proposition 6.2.1. The sequence $\{(Y_i, \tau_i): i \geq 1\}$ consists of independent and identically distributed (i.i.d) random vectors.

Proposition 6.2.2. If τ_1 is aperiodic,* $E(\tau_1) < \infty$, and $E\{|f(X)|\} < \infty$, then

$$r = E\{f(X)\} = \frac{E(Y_1)}{E(\tau_1)}. \tag{6.2.4}$$

There is an analogous ratio formula when τ_1 is periodic. For proof of these propositions the reader is referred to [8].

Proposition 6.2.1 says that the behavior patterns of the system during different cycles are statistically independent and identically distributed. Proposition 6.2.2. enables us to estimate the value $r = E(Y_1)/E(\tau_1)$ (which is the same as $r = E(Y_i)/E(\tau_i)$) by classical statistical methods, and to find point estimators and confidence intervals for r. These two problems are the subject of the next section.

6.3 POINT ESTIMATORS AND CONFIDENCE INTERVALS [22, 28]

In this section we consider several point estimators and confidence intervals for the ratio $E(Y_i)/E(\tau_i)$. The problem we consider is as follows: given the i.i.d. sequence of random vectors $\{(Y_i, \tau_i): i > 1\}$, find point estimators and construct $100(1-\delta)\%$ confidence intervals for the ratio $E(Y_i)/E(\tau_i)$.

*The random variable τ_1 is periodic with period $\lambda > 0$ if, with probability 1, it assumes values in the set $\{0, \lambda, 2\lambda, \ldots\}$ and λ is the largest such number. If there is no such λ, then τ_1 is said to be aperiodic.

Let $Z_i = Y_i - r\tau_i$. It is readily seen that the Z_i's are i.i.d. r.v.'s, since the vectors (Y_i, τ_i) also are. Note also that

$$E(Z_i) = 0 \tag{6.3.1}$$

and

$$\sigma^2 = \operatorname{var}(Z_i) = \operatorname{var}(Y_i) - 2r\operatorname{cov}(Y_i, \tau_i) + r^2 \operatorname{var}(\tau_i). \tag{6.3.2}$$

Denote $\bar{Y} = (1/n)\Sigma_{i=1}^n Y_i$, and $\bar{\tau} = (1/n)\Sigma_{i=1}^n \tau_i$; then by virtue of the central limit theorem, (c.l.t.) we have

$$\frac{n^{1/2}\left[\bar{Y} - r\bar{\tau}\right]}{\sigma} \Rightarrow N(0,1) \qquad \text{as } n \to \infty, \tag{6.3.3}$$

where $\Rightarrow$ denotes weak convergence and it is assumed that $\sigma^2 < \infty$. The last formula can be rewritten as

$$\frac{n^{1/2}(\hat{r} - r)}{\sigma/\bar{\tau}} \Rightarrow N(0,1) \qquad \text{as } n \to \infty, \tag{6.3.4}$$

where $\hat{r} = \bar{Y}/\bar{\tau}$. Inasmuch as σ is unknown, we cannot obtain a confidence interval for r directly from (6.3.4). However, we can estimate σ^2 in (6.3.2) from the sample, that is, by

$$s^2 = s_{11} - 2\hat{r}s_{12} + \hat{r}^2 s_{22}, \tag{6.3.5}$$

where

$$s_{11} = \frac{1}{n-1}\sum_{i=1}^{n}\left(Y_i - \bar{Y}\right)^2, \qquad s_{22} = \frac{1}{n-1}\sum_{i=1}^{n}\left(\tau_i - \bar{\tau}\right)^2$$

and

$$s_{12} = \frac{1}{n-1}\sum_{i=1}^{n}\left(Y_i - \bar{Y}\right)\left(\tau_i - \bar{\tau}\right).$$

It is straightforward to see that $s^2 \to \sigma^2$ a.s. as $n \to \infty$, so (6.3.4) can be rewritten as

$$\frac{n^{1/2}(\hat{r} - r)}{s/\bar{\tau}} \Rightarrow N(0,1), \qquad \text{as } n \to \infty, \tag{6.3.6}$$

and the $100(1-\delta)\%$ confidence interval for $r = E(Y_i)/E(\tau_i)$ is

$$\hat{I} = \left[\hat{r} - \frac{z_\delta s}{\bar{\tau} n^{1/2}}, \hat{r} + \frac{z_\delta s}{\bar{\tau} n^{1/2}}\right], \tag{6.3.7}$$

where $z_\delta = \phi^{-1}(1 - \delta/2)$, ϕ is the standard normal distribution function, and $\hat{r} = \bar{Y}/\bar{\tau}$ is the point estimator of $E(Y_1)/E(\tau_1)$. The procedure for obtaining a $100(1-\delta)\%$ confidence interval for r can be written as follows:

1 Simulate n cycles of the regenerative process.

2 Compute the sequence $\tau_1, \ldots, \tau_n$ and the associated sequence $Y_1, \ldots, Y_n$ (use (6.2.2) and (6.2.3), respectively, for a continuous-time or discrete-time process).

3 Compute $\bar{Y} = (1/n)\Sigma_{n=1}^{n} Y_i$ and $\bar{\tau} = (1/n)\Sigma_{i=1}^{n} \tau_i$ and find the point estimator by

$$\hat{r} = \frac{\bar{Y}}{\bar{\tau}}. \tag{6.3.8}$$

4 Construct the confidence interval by

$$\hat{I} = \left[\hat{r} \pm \frac{z_\delta s}{\bar{\tau} n^{1/2}}\right]$$

where $z_\delta = \phi^{-1}(1-\delta/2)$ and ϕ is the standard normal distribution.

It is readily seen that $\hat{r} = \bar{Y}/\bar{\tau}$, referred to as the *classical estimator* [28], is a biased but consistent estimator of $E(Y_1)/E(\tau_1)$. Iglehart [28] suggested, for the same purpose, the following alternatives:

BEALE ESTIMATOR

$$\hat{r}_b(n) = \frac{\bar{Y}}{\bar{\tau}} \cdot \frac{\left(1 + s_{12}/n\bar{Y}\bar{\tau}\right)}{\left(1 + s_{22}/n\bar{\tau}^2\right)}; \tag{6.3.9}$$

FIELLER ESTIMATOR

$$\hat{r}_f(n) = \frac{\bar{Y}\bar{\tau} - k_\delta s_{12}}{\bar{\tau}^2 - k_\delta s_{22}}, \tag{6.3.10}$$

where

$$k_\delta = \frac{\left[\phi^{-1}(1-\delta/2)\right]^2}{n};$$

JACKKNIFE ESTIMATOR

$$\hat{r}_j(n) = \frac{1}{n} \sum_{n=1}^{n} \theta_i \tag{6.3.11}$$

where

$$\theta_i = n\left(\frac{\bar{Y}}{\bar{\tau}}\right) - (n-1)\left[\frac{\sum_{k \neq i} Y_k}{\sum_{k \neq i} \tau_k}\right];$$

TIN ESTIMATOR

$$\hat{r}_t(n)=\frac{\bar{Y}}{\bar{\tau}}\left[1+\frac{1}{n}\left(\frac{s_{12}}{\bar{Y}\bar{\tau}}-\frac{s_{22}}{\bar{\tau}^2}\right)\right]. \tag{6.3.12}$$

Let us now cite some results from Ref. 28. The four point estimators (6.3.9) through (6.3.12) as well as the classical estimator are biased. Their expected value can be expressed as

$$E[\hat{r}(n)]=r+\frac{c_1}{n}+\frac{c_2}{n^2}+0\left(\frac{1}{n^3}\right). \tag{6.3.13}$$

The point estimators (6.3.9), (6.3.11), and (6.3.12) have been suggested in order to reduce the bias of (6.3.13) up to order $1/n^2$. For the jackknife method $c_1=0$, since

$$\begin{aligned}E[\hat{r}_j(n)]&=n\left[r+\frac{c_1}{n}+\frac{c_2}{n^2}+0\left(\frac{1}{n^3}\right)\right]\\&\quad-(n-1)\left\{r+\frac{c_1}{n-1}+\frac{c_2}{(n-1)^2}+0\left[\frac{1}{(n-1)^3}\right]\right\}\\&=r+0\left(\frac{1}{n^2}\right).\end{aligned}$$

The reader is asked to prove that for both Beale and Tin estimators c_1/n is also equal to zero.

Since both $n^{1/2}(\hat{r}-\hat{r}_b)\to 0$ and $n^{1/2}(\hat{r}-\hat{r}_t)\to 0$ a.s. as $n\to\infty$, we can replace $\hat{r}$ both in (6.3.6) for the c.l.t. and (6.3.7) for the confidence interval without changing the results.

For the jackknife method formulas (6.3.6) and (6.3.7) can be written, respectively, as

$$\frac{\hat{r}_j-r}{n^{1/2}\hat{s}_j}\Rightarrow N(0,1)\qquad \text{as } n\to\infty \tag{6.3.14}$$

$$\hat{I}_j=\left[\hat{r}_j-\frac{z_\delta\hat{s}_j}{n^{1/2}},\hat{r}_j+\frac{z_\delta\hat{s}_j}{n^{1/2}}\right],$$

where

$$\hat{s}_j=\left\{\sum_{i=1}^{n}\frac{(\theta_i-\hat{r}_j)}{(n-1)}\right\}^{1/2}.$$

The Fieller method yields the following $100(1-\delta)\%$ confidence interval:

$$I_f=\left[\hat{r}_f-\frac{D^{1/2}}{(\bar{\tau}^2-k_\delta s_{22})},\hat{r}_f+\frac{D^{1/2}}{(\bar{\tau}^2-k_\delta s_{22})}\right], \tag{6.3.15}$$

where

$$D = \left(\overline{Y\tau} - k_\delta s_{12}\right)^2 - \left(\bar{\tau}^2 - k_\delta s_{22}\right)\cdot\left(\bar{Y}^2 - k_\delta s_{11}\right)$$

and

$$k_\delta = \frac{\left[\phi^{-1}(1-\delta/2)\right]^2}{n}.$$

The performances of these estimators were compared numerically (via simulating several stochastic models), and the following results were obtained [28].

For short runs the jackknife method is recommended both for point estimators and confidence intervals because it produces slightly better statistical results than other methods. Two minor drawbacks of the jackknife method are a large memory requirement and slightly more complex programming. Additional storage addresses of the order of $2n$ are required, where n is the number of cycles observed. Where the storage requirement for the jackknife method is excessive, the Beale or Tin methods are recommended for point estimates and the classical method for the confidence intervals. The Fieller method is recommended for neither point nor confidence intervals. It is found to be heavily biased for short runs and more complicated than the classical method. The above mentioned five point estimators were based on simulating n cycles of regenerative processes. Another possibility is to consider point estimators based on the simulation for a fixed (but large) length of time t. In this case the number of cycles N_t in the interval $[0, t]$ is a random variable given by

$$N_t = \sum_{s=1}^{\infty} 1_{[0,t]}(T_s)$$

where $1_{[0,t]}$ is the indicator function of the interval $[0, t]$. Replacing n by N_t, we can modify all the point estimators (6.3.8) through (6.3.12), preserving their consistency. For example, for the classical estimator we have

$$\hat{r}(N_t) = \frac{\bar{Y}(N_t)}{\bar{\tau}(N_t)}.$$

Thus, asymptotically, there is little difference while considering point estimators based on simulation n regenerative cycles or on simulation for fixed length of time t. The c.l.t. in this case is

$$\frac{t^{1/2}\left[r(N_t) - r\right]}{\sigma\left\{E(\tau_1)\right\}^{1/2}} \Rightarrow N(0,1) \qquad \text{as } t\to\infty \tag{6.3.16}$$

Recently Heidelberger and Meketon [22] considered estimators based on simulations for a relatively short length of time t. They defined estimators

$$\hat{r}(N_t) = \frac{\sum_{i=1}^{N_t} Y_i}{\sum_{i=1}^{N_t} \tau_i} \tag{6.3.17}$$

and

$$\hat{r}(N_t + 1) = \frac{\sum_{i=1}^{N_t+1} Y_i}{\sum_{i=1}^{N_t+1} \tau_i}. \tag{6.3.18}$$

They then showed that

$$E(\hat{r}(N_t)) = r + 0\left(\frac{1}{t}\right) \tag{6.3.19}$$

and

$$E(\hat{r}(N_t + 1)) = r + 0\left(\frac{1}{t^2}\right), \tag{6.3.20}$$

so that a bias reduction is achieved by continuing the simulation until the first regeneration after time t. The bias reduction is comparable to that of the jackknife, Beale, and tin estimators since t is proportional to the number of cycles. Table 6.4.3 lists empirical results from simulations of a closed queueing network model for these estimators.

We turn now to the problem of determining run length. The $100(1-\delta)\%$ confidence interval for a large but fixed number of cycles has a width approximately equal to

$$\frac{2\sigma\phi^{-1}(1-\delta/2)}{E(\tau_1)n^{1/2}}. \tag{6.3.21}$$

In terms of duration time t (6.3.20) can be written as (see [24])

$$\frac{2\sigma\phi^{-1}(1-\delta/2)}{\{E(\tau_1)\}^{1/2}t^{1/2}} \tag{6.3.22}$$

Note that neither σ nor $E(\tau_1)$ are known in advance. Hence it may be worthwhile to take a small sample and obtain rough estimates for σ and $E(\tau_1)$. Such estimates would form a basis for a final decision or run length and level of confidence. We wish to emphasize that all ratio estimators described in this section are designed for simulations with a fixed number of cycles n or a fixed run length t. An alternative possibility would be to consider procedures based on sequential stopping rules.

6.4 EXAMPLES OF REGENERATIVE PROCESSES

In this section we consider three examples of regenerative processes, taken from Refs. 6, 10, and 49: a single server queue, a repair model with spares, and a closed queueing network.

6.4.1 A Single Server Queue *GI/G/1* [6]

This example was described in Section 4.3.12, and will be briefly recapitulated here.

Let W_i and S_i be the waiting time and service time, respectively, of the ith customer in a single server queue. Let A_{i+1} be the time between the arrival of the ith and $(i+1)$th customers. We assume that $\{S_i, i \geq 0\}$ are i.i.d. with $E(S_i) = \mu^{-1}$ and that $\{A_i, i \geq 1\}$ are i.i.d. with $E(A_i) = \lambda^{-1}$. Let the traffic intensity ρ be defined by $\rho = \lambda/\mu$. We assume that customer number 0 arrives at time 0 to an empty system. Let $X_i = S_{i-1} - A_i$ for $i \geq 1$. The waiting time process $\{W_i, i \geq 0\}$ can be defined recursively by

$$W_0 = 0$$

$$W_i = (W_{i-1} + X_i)^+, \qquad i \geq 1.$$

It is known [36] that, if $\rho < 1$, there exists an infinite number of indices i such that $W_i = 0$ and a random variable W such that $W_i \Rightarrow W$, as $i \to \infty$. Thus we choose zero state as our return state and regenerations occur whenever a customer arrives to find an empty queue. We are interested in estimating $E(W)$, which is finite if $E(S_n^2) < \infty$.

Since no analytical results are available for calculating the steady-state waiting time $E(W)$, we estimate it via simulation by making use of the classical estimator (6.3.8). The simulation results are shown in Fig. 6.4.1.

We see that the customers 1, 3, 4, 7, 11, and 16 find the server idle, that is, $W_1 = W_3 = W_4 = W_7 = W_{11} = W_{16} = 0$, while customers 2, 5, 6, 8, 9, 10, 12, 13, 14, and 15 find the server busy and wait in the queue before being served.

It follows from Fig. 6.4.1 that the simulation data contains five complete cycles with the following pairs $\{(Y_i, \tau_i), i = 1, \ldots, 5) : (Y_1, \tau_1) = (10, 2), (Y_2, \tau_2) = (0, 1), (Y_3, \tau_3) = (30, 3), (Y_4, \tau_4) = (50, 4)$, and $(Y_5, \tau_5) = (60, 5)\}$. The sixth cycle will start with the arrival of customer 16. Using the

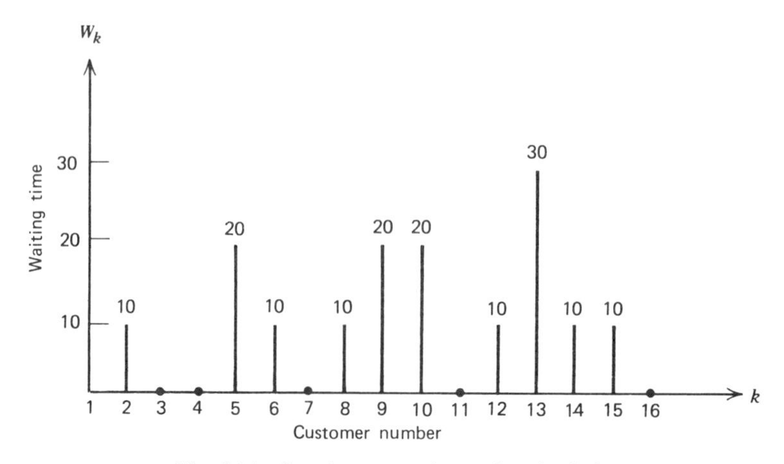

Fig. 6.4.1 Sample output of queueing simulation.

classical estimator $\hat{r} = \sum_{i=1}^{n} Y_i / \sum_{i=1}^{n} \tau_i$, we obtain

$$\hat{r} = \frac{\sum_{i=1}^{5} Y_i}{\sum_{i=1}^{5} \tau_i} = \frac{10+0+30+50+60}{2+1+3+4+5} = \frac{150}{15} = 10.$$

This result can also be obtained by using the sample-mean estimator

$$\bar{r} = \frac{1}{N} \sum_{j=1}^{N} W_j = \frac{1}{15} \sum_{j=1}^{15} W_j = \frac{150}{15} = 10.$$

Here $N = \sum_{i=1}^{5} \tau_i = 15$ is the length of the run and $\sum_{j=1}^{15} W_j = \sum_{i=1}^{5} Y_i$.

A logical question arises. If both points estimators $\hat{r}$ and $\bar{r}$ are equal (we assume that the length of the run N is equal to n complete cycles, $n < N$), why do we need all the ratio estimators (6.3.8) through (6.3.12), (6.3.17) and (6.3.18), based on the regenerative phenomena?

The answer can be found if we consider not only point estimators for $r = E(W)$ but confidence intervals as well. In order to construct confidence intervals in the sense of classical statistics, the simulation data must form a sequence of i.i.d. samples from the same underlying probability distribution. The simulation data from the queueing system is the sequence of waiting times $W_1, \ldots, W_N$. Note, however, that if we start our simulation with an empty queueing system, then the first few waiting times tend to be short, that is, they are correlated, and as a rule, the sample-mean estimator $\bar{r}$ will be a biased estimator of $r = E(W)$.

Table 6.4.1. Simulation Results for the $M/M/1$ Queue

Parameter	Theoretical Value	Point Estimates	Confidence Interval
$r = E(W) = \dfrac{E(Y_1)}{E(\tau_1)}$	0.100	0.110	[0.096, 0.123]
$E(W^2)$	0.040	0.046	[0.035, 0.056]
$E\{\sqrt{(W-0.1)}^{+}\}$	0.120	0.133	[0.116, 0.148]
$E(\tau_1)$	2.000	2.110	[2.012, 2.207]
$\sigma(W)$	0.173	0.182	[0.141, 0.271]

Source: Ref. 6.
Note: Number of cycles $n = 2000$, level of confidence $100(1-\delta) = 90\%$, number of replications $N = 10$, $\lambda = 5$, $\mu = 10$).

To overcome this difficulty we can run the model until it reaches the steady state and then start collecting and updating the simulation data. The problem of determining the steady-state distribution is a difficult one, moreover, requiring considerable computation (CPU) time, but unless we start from it W_i and W_{i+1} will again be correlated (if W_i is short, then W_{i+1} will also tend to be short and vice versa). Since the r.v.'s $W_1, \ldots, W_N$ are correlated, classical statistical methods cannot be applied in constructing confidence intervals for $r = E(W)$. Still, this difficulty can be overcome by using the regenerative property, namely, by grouping the simulation data into independent pairs (blocks) $(Y_1, \tau_1), \ldots, (Y_n, \tau_n)$, which yields different ratio estimators (see (6.3.8) through (6.3.12), (6.3.17) and (6.3.18)) and the associated confidence intervals by means of classical statistics. Table 6.4.1 presents simulation results for the queueing system $M/M/1$ with $\lambda = 5$, $\mu = 10$ based on a run of 2000 cycles. Confidence intervals at the 90% level are given for the parameters $E(W)$, $E(W^2)$, $E\{(\sqrt{W-0.1})^{+}\}$, $E(\tau_1)$, and $\sigma(W)$. The function $E\{\sqrt{(W-0.1)}^{+}\}$ may be interpreted as a penalty for long waiting time.

6.4.2. A Repairman Model with Spares [10]

We now consider a repairman problem with n operating units and m spares (Fig. 6.4.2). Each of the operating units fails with rate λ. A failed unit enters a queue for service from one of s repairmen on a first-in–first-out (FIFO) basis and is replaced by a spare (if available). The distribution of the i.i.d. repair times is exponential with mean μ^{-1} for each repairman. A

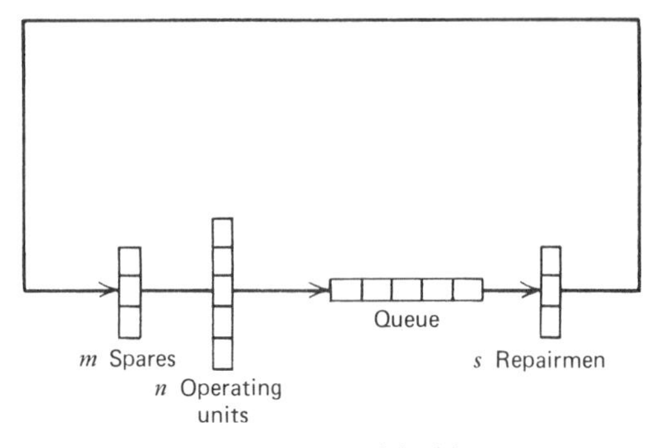

Fig. 6.4.2 Repairman model with spares.

repaired unit enters the pool of spares unless there are fewer than n units in operation, in which case it immediately becomes operational. Denoting by $X(t)$ the number of units in service or waiting in the queue for service, then $\{X(t), t \geq 0\}$ is a birth and death process with state space $I = \{0, 1, \ldots, m + n\}$, and

$$\lambda_i = \begin{cases} n\lambda, & 0 \leq i \leq m \\ (n+m-i)\lambda, & m < i \leq n+m \end{cases}$$

$$\mu_i = \begin{cases} i\mu, & 1 \leq i \leq s \\ s\mu, & s < i \leq m+n \end{cases}.$$

Let us simulate the system for T units of time and have as output the values $X(t)$, $0 \leq t \leq T$, where $X(t)$ is the number of units at the repair facility at time t. The sample mean $(1/T)\int_0^T X(t)\,dt$ is a consistent estimator for $E(X)$ where $E(X)$ is the mean number of units at the repair facility under steady-state conditions. However, unless the value $X(0)$ is obtained by sampling from the steady-state distribution of X, the sample mean will be a biased estimator due to the initial conditions. Moreover, it is seen that, if t_1 is close to t_2, then $X(t_1)$ and $X(t_2)$ will be highly correlated, because the number of units in the repair facility usually does not change quickly.

Due to the initial bias of the estimator and to the correlation of the output data, it is impossible to apply classical statistics in estimating the steady-state value $r = E(X)$. However, by again applying the regenerative approach the difficulty can be overcome. From here on we repeat in essence what was done for the queueing simulation.

The process $\{X(t) : t \geq 0\}$ is a regenerative one in continuous time and

$$P(X(t) = i) \Rightarrow P(X = i) \qquad \text{as } t \to \infty \text{ for all } i \in I$$

Table 6.4.2. Simulation Results for Repairman Model

Parameter	Theoretical Value	Confidence Interval
$E(X)$	5.353	[5.238, 5.432]
$E\{(X-5)^+\}$	1.269	[1.201, 1.325]
$P\{X>5\}$	0.465	[0.444, 0.475]
$P\{X>0\}$	0.988	[0.987, 0.990]
$P\{X=0\}$	0.012	[0.010, 0.013]
$E(\tau_1)$	42.021	[37.459, 47.681]
$E(\hat{\tau}_1)$	73.375	[65.262, 83.342]

Source: Ref. 10.
Note: Run length = 500 cycles; level of confidence = 95%.

Suppose we start the simulation at time $T_1 = 0$ with n operating units and m spares, that is, at $T_1 = 0$ the repair facility is empty; then the sequence is $\{T_i : i \geq 0\}$, where T_i is defined as the regeneration time when the repair facility becomes empty. In other words, the system "starts afresh probabilistically," or regenerates itself, at each time T_i. For any real-valued measurable function f we define

$$Y_i = \int_{T_i}^{T_{i+1}} f(x(t))\, dt;$$

then the pairs $(Y_1, \tau_1), \ldots, (Y_n, \tau_n)$, where $\tau_i = T_{i+1} - T_i$, are i.i.d. Suppose that the simulation time T exactly equals n cycles; then

$$\frac{1}{T}\int_0^T f(X(t))\, dt = \frac{\sum_{i=1}^{n} Y_i}{\sum_{i=1}^{n} \tau_i} = \frac{\frac{1}{n}\sum_{i=1}^{n} Y_i}{\frac{1}{n}\sum_{i=1}^{n} \tau_i} = \hat{r}$$

is a biased but consistent estimator for $r = E(f(X)) = E(Y_1)/E(\tau_1)$.

Table 6.4.2 gives simulation results for some output parameters based on run of 500 cycles. $E(\hat{\tau}_1)$ represents the number of failures over a cycle. It is assumed that $n = 10$, $m = 5$, $s = 4$, and $\mu = 2$. The "lifetime" of an operating unit is exponentially distributed with $\lambda = 5$.

6.4.3 A Closed Queueing Network [49]

Consider a closed queueing system that is a model of the time-sharing computer system in Fig. 6.4.3. The network comprises M service centers with a fixed number N of customers. Service center 1 consists of N

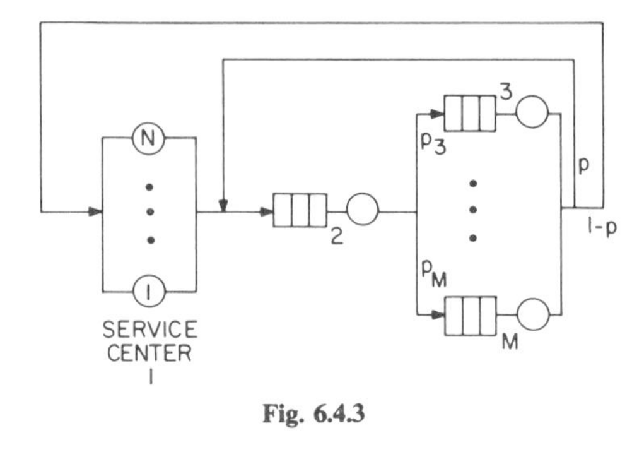

Fig. 6.4.3

terminals (identical parallel servers); hence a customer at this center never has to wait for a server to become free. Service center 2 is a single server processor, that is, all customers receive service immediately, and if there are k customers present each customer is served at $1/k$ of the server's rate. Service centers $3, \ldots, M$ represent peripheral input-output devices (single server queues), each of which is scheduled on a FIFO basis. A customer (device) completing service at service center 1 immediately enters service center 2, and immediately thereafter service center j with probability $p_j > 0$, $j = 3, \ldots, M$, where $\Sigma_{j=3}^{M} p_j = 1$. After completing service at service center j, $j = 3, \ldots, M$, the customer enters service center 1 with probability $1 - p$, or service center 2 with probability p. Service times at service centers $j = 1, 2, \ldots, M$ are i.i.d. and exponentially distributed with mean μ_j^{-1}. It is assumed that routing through the network is Markovian and that all service and routing mechanisms are mutually independent.

Let $Q(t) = (Q_1(t), \ldots, Q_M(t))$, where $Q_j(t)$ is the number of customers at service center j at time t. It can then be shown [49] that $\{Q(t) : t \geq 0\}$ is a continuous-time irreducible Markov chain, and hence a regenerative process. We define a response time as the time interval between a customer's departure from service center 1 and his next return to it, and let W_i be the just completed time of the ith customer arriving there. Then $W = \{W_i, i \geq 0\}$ is regenerative with regeneration occurring whenever a last customer arrives at service center 1 leaving centers $2, \ldots, M$ empty. Again, we are interested in the expected stationary response time $r = E(W)$, which is known to be finite [49]. Let $\hat{p}_i$ be the utilization of service center i, that is, the long run average proportion of time the server there is busy. The particular parameters chosen for this model are listed in Fig. 6.4.3 and yield $\hat{p}_2 = 0.894$, $\hat{p}_i = 0.268$, $r = 8.65$.

Table 6.4.3 presents point estimators and 90% confidence intervals for several ratio estimators discussed in Section 6.4.3.

TABLE 6.4.3. Point Estimates and 90 Confidence Intervals for $E(W) = 8.65$ in Closed Queuing Network

Estimate	$N=5$ $t=220$	$N=10$ $t=440$	$N=30$ $t=1320$	$N=50$ $t=2220$
$\hat{r}(N_t)$	8.28	8.46	8.55	8.59
	± 0.10	± 0.07	± 0.07	± 0.07
$\hat{r}(N_t+1)$	8.64	8.60	8.62	8.63
	± 0.10	± 0.07	± 0.07	± 0.07
Classical $\hat{r}$	8.23	8.50	8.56	8.60
	± 0.17	± 0.09	± 0.07	± 0.08
Jackknife	8.93	8.71	8.61	8.62
	± 0.23	± 0.09	± 0.07	± 0.08

Source: Ref.22.
Note: N = number of cycles simulated; t = number of response times simulated; $R = 200$ replications for $t = 220$, 440; $R = 100$ replications for $t = 1320$; $R = 60$ replications for $t = 2200$.

6.5 SELECTING THE BEST STABLE STOCHASTIC SYSTEM

In this section we consider some techniques for selecting the best system from among m alternative systems according to a certain criteria.

Assume that $N (N \geq 2)$ stochastic systems are being simulated, each giving rise to a regenerative process $\{X^i(t) : t \geq 0\}$, $i = 1, \ldots, N$. For example, N alternative designs are considered for a new system.

Suppose that the measure of performance for the ith system is

$$r_i = E\{f(X^i)\}, \qquad i = 1, \ldots, N \tag{6.5.1}$$

where f is a real-valued bounded measurable function, X^i is the steady-state random variable of the regenerative process $\{X^i(t) : t \geq 0\}$. The problem is to choose the best system, that is, the system with the smallest value of r_i:

$$r_l = \min_{l=1,\ldots,N} r_i = \min_{l=1,\ldots,N} E\{f(X^i)\}. \tag{6.5.2}$$

(We are minimizing r_i; the alternative problem of maximizing r_i can be considered as well.)

Iglehart [30] presents a method based on the following scheme. Two positive numbers P^* and δ^* are specified. Then with probability P^* the system with the smallest (largest) r_i is selected whenever that value of r_i is separated by at least δ^* from the other r_j's. Two procedures have been considered in Ref. 30 for this problem. The first procedure is sequential and the second is two-stage. Both procedures involve the use of normal approximations and require large samples in terms of the number of cycles of the regenerative processes simulated.

We consider here another adaptive approach suggested by Rubinstein [61]. Our method is based on an iterative procedure that selects the best system with probability 1.

We start solving the problem (6.5.2) by considering the following linear programming problem:

$$\min_p W(p) = \min_p \sum_{i=1}^{N} E\{f(X^i)\}p_i, \tag{6.5.3}$$

subject to

$$\sum_{i=1}^{N} p_i = 1, \qquad p_i \geq 0, i = 1, \ldots, N. \tag{6.5.4}$$

If there exists a unique solution of (6.5.2), then the problem (6.5.3)–(6.5.4) is equivalent to (6.5.2) and its solution is given by a vector p^* with a single nonzero component:

$$p^* = \{\underbrace{0, \ldots, 0, 1}_{l}, 0, \ldots, 0\}. \tag{6.5.5}$$

The algorithm for solving the problem (6.5.3)–(6.5.4) is based on a step-by-step correction of the probability vector $p[n]$, where n denotes the step number. There exists a mechanism, provided by (6.5.9) below, which ensures that $p_i[n] \geq \varepsilon[n]$, $i = 1, \ldots, N$, where $\{\varepsilon[n]\}_{n=0}^{\infty}$ is a monotone decreasing sequence of positive numbers, subject to (6.5.13) and (6.5.14) below. On the nth step the ith system, $i \in \{1, \ldots, N\}$, is chosen by simulating the distribution $p[n-1]$. We denote this event by $X[n] = X^i$. One cycle of the process $\{X^i(t) : t \geq 0\}$ is carried out. Denote by $\nu^i[n]$, $i = 1, \ldots, n$, the total number of renewal cycles made by the ith system up to and including the nth step. We check whether or not the inequality $\nu^k[n-1] \geq n\varepsilon[n]$, $k \in \{1, \ldots, i-1, i+1, \ldots, N\}$ is satisfied for all systems. If for some indices $k_1, \ldots, k_s \in \{1, \ldots, i-1, i+1, \ldots, N\}$, this inequality does not hold then one additional cycle is carried out for each system $k_1, \ldots, k_s$, so that ultimately

$$\nu^k[n] \geq n\varepsilon[n], \qquad k = 1, \ldots, N. \tag{6.5.6}$$

We record

$$\tau_n^k = T^k_{\nu^k[n]} - T^k_{\nu^k[n]-1}, \qquad k = i, k_1, \ldots, k_s,$$

the lengths of the cycles performed, and for each k calculate

$$Y_n^k = \int_{T^k_{\nu^k[n]-1}}^{T^k_{\nu^k[n]}} f(X^k(t))\, dt, \qquad k = i, k_1, \ldots, k_s \tag{6.5.7}$$

if the process $\{X^i(t) : t \geq 0\}$ is continuous-time.

In the case of a discrete-time process the integral should be replaced by the corresponding sum over the $\nu^k[n]$th cycle. Set also

$$Y_n^k = \tau_n^k = 0, \qquad \text{if } k \notin \{i, k_1, \dots, k_s\}. \tag{6.5.8}$$

We construct a new distribution $p[n]$ by the following recurrence formula:

$$p[n] = \pi_{S_{\varepsilon[n]}}\{p[n-1] - \gamma[n] B(n|i)\}. \tag{6.5.9}$$

Here S_ε is a simplex in R^N:

$$S_\varepsilon = \left\{ p = (p_1, \dots, p_N) \middle| \sum_{k=1}^{N} p_k = 1, 0 \le \varepsilon \le p_k \le 1 \right\},$$

π_{S_ε} is the projection operator onto the simplex S_ε, such that, for any $x \in R^N$,

$$\|z - \pi_{S_\varepsilon}(z)\| = \min_{y \in S_\varepsilon} \|z - y\|,$$

and $B(\cdot|\cdot)$ is a vector $\{B_1(\cdot|\cdot), \dots, B_N(\cdot|\cdot)\}$, where

$$B_k(n|i) = \delta_{ik} p_k^{-1}[n-1] r_k[n] \tag{6.5.10}$$

$$r_k[n] = \frac{Y^k[n]}{\tau^k[n]} \tag{6.5.11}$$

$$Y^k[n] = Y^k[n-1] + Y_n^k, \qquad \tau^k[n] = \tau^k[n-1] + \tau_n^k, \qquad k = 1, \dots, N \tag{6.5.12}$$

$$\delta_{ik} = \begin{cases} 1, & \text{if} \quad i = k \\ 0, & \text{if} \quad i \ne k \end{cases}.$$

The initial values of $p[0] \in S_{\varepsilon[0]}$, $Y[0] = (Y^1[0], \dots, Y^N[0])$, and $\tau[0] = (\tau^1[0], \dots, \tau^N[0])$ can be chosen arbitrarily, for example, $Y^k(0) = 0$, $\tau^k(0) = 0$, $k = 1, \dots, N$. The sequences $\{\gamma[n]\}_{n=1}^{\infty}$ and $\{\varepsilon[n]\}_{n=0}^{\infty}$ must be chosen so that the following conditions are satisfied:

$$\gamma[n] \downarrow 0, \qquad \varepsilon \in [n] \downarrow 0 \tag{6.5.13}$$

$$\sum_{n=1}^{\infty} \gamma[n] = \infty \tag{6.5.14}$$

$$\sum_{n=1}^{\infty} \frac{\gamma^2[n]}{\varepsilon[n-1]} < \infty \tag{6.5.15}$$

$$\sum_{n=1}^{\infty} \frac{\gamma[n]}{\varepsilon[n]\sqrt{n}} < \infty. \tag{6.5.16}$$

Remark 1 In order to satisfy conditions (6.5.13) through (6.5.16) take, for example,

$$\gamma[n] \sim n^{-1}, \qquad \varepsilon[n] \sim n^{-0.4}.$$

Remark 2 We assume that $\tau_n^k \geq \tau_0 > 0$, $k = 1, \ldots, m$, $n = 1, 2, \ldots,$ that is, that the cycle is taken into account only if it is of some minimal length (which can be considered as the sensitivity threshold of the measuring instrument).

Remark 3 The r.v.'s $Y^k[n]$ and $\tau^k[n]$, $k = 1, \ldots, N$, $n \geq 1$, defined in (6.5.12), store the information obtained up to and including the nth step. We should also note that, for each k fixed, only $\nu^k[n]$ summands in both $Y^k[n]$ and $\tau^k[n]$ are nonzero.

Theorem 6.5.1 If the values of the function f are uniformly bounded by some constant D and if there exists the unique optimal solution p^* of the problem (6.5.3)–(6.5.4), then for any initial distribution $p[0] \in S_{\varepsilon[0]}$ the sequence $\{p[n]\}_{n=1}^{\infty}$, generated by the algorithm (6.5.9)–(6.5.14), converges to p^* with probability 1.

Corollary The theorem remains valid if we assume that the values of the function f cannot be observed directly, but are measured with a random noise. In other words,

$$f(X^i) = E_\xi\{Q(X^i, \xi)\}, \qquad i = 1, \ldots, N,$$

where ξ is a random vector with an unknown time-independent probability distribution function. In this case we can consider another random process:

$$\{U^i(t) : t \geq 0\} = \{(X^i(t), \xi)\}; \qquad i = 1, \ldots, N.$$

If $\{X^i(t) : t \geq 0\}$ is regenerative, then $\{U^i(t) : t \geq 0\}$ is also regenerative and the values of Q are uniquely defined for each value of the steady-state r.v. U^i of the process $\{U^i(t) : t \geq 0\}$, and

$$E\{f(X^i)\} = E\{E_\xi\{Q(X^i, \xi)\}\} = E\{Q(U^i)\}, \qquad i = 1, \ldots, N.$$

Proof of the Theorem Before proving the theorem, let us introduce some notation. Let

$$t_i[n] = r_i[n] - r_i, \qquad t[n] = \max_i |t_i[n]| \tag{6.5.17}$$

where $r_i = E\{f(X^i)\}$, $i = 1, \ldots, N$, and $n = 1, 2, \ldots.$

On the nth step the state of the algorithm can be described by a $4N$-dimensional vector $\Xi[n] = (p[n], \tau[n] = (\tau^1[n], \ldots, \tau^N[n]), Y[n] = (Y^1[n], \ldots, Y^N[n]), \nu[n] = (\nu^1[n], \ldots, \nu^N[n]))$. We first prove the following lemma.

Lemma. For any $\Xi[0]$ such that $p[0] \in S_{\varepsilon[0]}$, and $\tau[0] > 0$,

$$\sum_{n=1}^{\infty} \gamma[n] E\{|t_i[n]||\Xi[0]\} < \infty, \qquad i = 1, \ldots, N. \tag{6.5.18}$$

Proof Without loss of generality set $i = 1$ and define

$$Z_n^1 = Y_n^1 - r_1\tau_n^1, \qquad n = 1, 2, \ldots.$$

If a cycle of the regenerative process $\{X^1(t) : t \geq 0\}$ was not carried out on the nth step, then $Z_n^1 = 0$. For all n's such that a cycle of $\{X^1(t) : t \geq 0\}$ was performed on the nth step, the Z_n^1 are i.i.d. r.v.'s with $E(Z_n^1) = 0$ and variance $\sigma_{Z^1}^2$. Define also

$$Z^1[n] = Z^1[n-1] + Z_n^1, \qquad n = 1, 2, \ldots, \quad Z^1[0] = Y^1[0] - r_1\tau^1[0].$$

Then by the Cauchy-Schwarz inequality,

$$\begin{aligned} E\{|t_1[n]||\Xi[0]\} &= E\left\{\left|\frac{Z^1[n]}{\tau^1[n]}\right| |\Xi[0]\right\} \\ &\leq E^{1/2}\{Z^1[n]^2|\Xi[0]\} \cdot E^{1/2}\{(\tau^1[n])^{-2}|\Xi[0]\} \\ &\leq \left((Z^1[0])^2 + n\sigma_{Z^1}^2\right)^{1/2} \cdot E^{1/2}\{(\tau^1[0] + \tau_0\nu^1[n])^{-2}|\Xi[0]\}, \end{aligned}$$

where τ_0 was defined in Remark 2. Since by (6.5.6) $\nu^1[n] \geq n\varepsilon[n]$, we have

$$E\{|t_1[n]||\Xi[0]\} \leq \left((Z^1[0])^2 + n\sigma_{Z^1}^2\right)^{1/2}(\tau^1[0] + \tau_0 n\varepsilon[n])^{-1}.$$

Thus for n large enough

$$E\{|t_1[n]||\Xi[0]\} \leq A_1\varepsilon^{-1}[n]n^{-1/2}, \tag{6.5.19}$$

where $A_1 = A_1(\Xi[0])$. Inequality (6.5.19) and condition (6.5.16) imply the convergence of the series (6.5.18). Q.E.D.

Corollary For any state $\Xi[n]$ of the algorithm on the nth step,

$$\sum_{m=n+1}^{\infty} \gamma[m] E\{t[m]|\Xi[n]\} < \infty.$$

Now we can prove our theorem.

Consider the vector $p^*[n] \in S_{\varepsilon[n]}$, such that

$$p_i^*[n] = \begin{cases} \varepsilon[n], & i \neq l \\ 1-(N-1)\varepsilon[n], & i = l, \end{cases} \tag{6.5.20}$$

where l is defined by (6.5.2) and is unique by the condition of Theorem (6.5.1). We have:

$$r(p^*[n]) = \min_{p \in S_{\varepsilon[n]}} r(p). \tag{6.5.21}$$

By the algorithm (6.5.9)–(6.5.11) and the properties of $\pi_{\varepsilon[n]}$ and (6.5.20), we have

$$\begin{aligned} \|p[n]-p^*[n]\|^2 \leq{}& \|p[n-1]-p^*[n-1]\|^2 + \|p^*[n-1]-p^*[n]\|^2 \\ &+ 2\langle p[n-1]-p^*[n-1], p^*[n-1]-p^*[n]\rangle \\ &+ \gamma^2[n](r^i[n])^2(p_i[n-1])^{-2} \\ &- 2\gamma[n](p_i[n-1]-p_i^*[n])r_i[n](p_i[n-1])^{-1} \\ \leq{}& \|p[n-1]-p^*[n-1]\|^2 + 3N(\varepsilon[n-1]-\varepsilon[n]) \\ &- 2\gamma[n](p_i[n-1]-p_i^*[n])r_i[n](p_i[n-1])^{-1} \\ &+ D^2\gamma^2[n](p_i[n-1])^{-2}, \end{aligned} \tag{6.5.22}$$

where $\langle \cdot, \cdot \rangle$ denotes the inner product. Taking the conditional expectation of both sides of (6.5.22) with $\Xi[n-1]$ fixed, we obtain

$$\begin{aligned} &E\{\|p[n]-p^*[n]\|^2 | \Xi[n-1]\} \\ &\leq \|p[n-1]-p^*[n-1]\|^2 + 3N(\varepsilon[n-1]-\varepsilon[n]) \\ &\quad + D^2N\gamma^2[n]\varepsilon^{-1}[n-1] - 2\gamma[n]\sum_{i=1}^{N}(p_i[n-1]-p_i^*[n])E\{r_i[n]|\Xi[n-1]\} \\ &\leq \|p[n-1]-p^*[n-1]\|^2 \\ &\quad + 3N(\varepsilon[n-1]-\varepsilon[n]) + D^2N\gamma^2[n]\varepsilon^{-1}[n-1] \\ &\quad - 2\gamma[n]\sum_{i=1}^{N}(p_i[n-1]-p_i^*[n-1])E\{r_i[n]|\Xi[n-1]\} \\ &\quad + 2\gamma[n]\left|\sum_{i=1}^{N}(p_i^*[n-1]-p_i^*[n])E\{r_i[n]|\Xi[n-1]\}\right| \\ &\leq \|p[n-1]-p^*[n-1]\|^2 \\ &\quad + L(\varepsilon[n-1]-\varepsilon[n]) + D^2N\gamma^2[n]\varepsilon^{-1}[n-1] \\ &\quad - 2\gamma[n]\sum_{i=1}^{N}(p_i[n-1]-p_i^*[n-1])E\{r_i[n]|\Xi[n-1]\}, \end{aligned} \tag{6.5.23}$$

where $L = 3N + 2ND \max_n \gamma[n]$. From (6.5.23) it follows that

$$\begin{aligned} E\{\| p[n] - p^*[n]\|^2 | \Xi[n-1]\} &\le \| p[n-1] - p^*[n-1]\|^2 + L(\varepsilon[n-1] - \varepsilon[n]) \\ &+ D^2 N \gamma^2[n] \varepsilon^{-1}[n-1] \\ &- 2\gamma[n] \sum_{i=1}^{N} (p_i[n-1] - p_i^*[n-1]) r_i \\ &- 2\gamma[n] \sum_{i=1}^{N} (p_i[n-1] \\ &- p_i^*[n-1]) E\{t_i[n] | \Xi[n-1]\} \\ &\le \| p[n-1] - p^*[n-1]\|^2 \\ &+ L(\varepsilon[n-1] - \varepsilon[n]) + D^2 N \gamma^2[n] \varepsilon^{-1}[n-1] \\ &- 2\gamma[n] \sum_{i=1}^{N} (p_i[n-1] \\ &- p_i^*[n-1]) E\{t_i[n] | \Xi[n-1]\}, \end{aligned} \tag{6.5.24}$$

since by (6.5.21)

$$\sum_{i=1}^{N} (p_i[n-1] - p_i^*[n-1]) r_i = r(p[n-1]) - r(p^*[n-1]) > 0.$$

Therefore

$$\begin{aligned} E\{\| p[n] - p^*[n]\|^2 | \Xi[n-1]\} &\le \| p[n-1] - p^*[n-1]\|^2 + L(\varepsilon[n-1] - \varepsilon[n]) \\ &+ D^2 N \gamma^2[n] \varepsilon^{-1}[n-1] \\ &+ 2N\gamma[n] E\{t[n] | \Xi[n-1]\}. \end{aligned} \tag{6.5.25}$$

Denote

$$\begin{aligned} v[n] = \| p[n] - p^*[n]\|^2 + L\varepsilon[n] + D^2 N \sum_{m=n+1}^{\infty} \gamma^2[m] \varepsilon^{-1}[m-1] \\ + 2N \sum_{m=n+1}^{\infty} \gamma[m] E\{t[m] | \Xi[n]\}. \end{aligned} \tag{6.5.26}$$

The first sum in (6.5.26) exists by (6.5.15) and the second by the corollary from the lemma. Taking the conditional expectation of both sides in (6.5.26), we obtain

$$E\{v[n]|\Xi[n-1]\} = E\{\|p[n]-p^*[n]\|^2|\Xi[n-1]\} + L\varepsilon[n] + D^2N\sum_{m=n+1}^{\infty}\gamma^2[m]\varepsilon^{-1}[m-1] + 2N\sum_{m=n+1}^{\infty}\gamma[m]E\{E\{t[m]|\Xi[n]\}|\Xi[n-1]\}$$
$$= E\{\|p[n]-p^*[n]\|^2|\Xi[n-1]\} + L\varepsilon[n] + D^2N\sum_{m=n+1}^{\infty}\gamma^2[m]\varepsilon^{-1}[m-1] + 2N\sum_{m=n+1}^{\infty}\gamma[m]E\{t[m]|\Xi[n-1]\}. \quad (6.5.27)$$

The last equality in (6.5.27) is justified by the fact that $\Xi[n]$ is a Markov chain taking values in R^{4N}[5]. Using (6.5.25),

$$E\{v[n]|\Xi[n-1]\} \le \|p[n-1]-p^*[n-1]\|^2 + L\varepsilon[n-1] + D^2N\sum_{m=n}^{\infty}\gamma^2[m]\varepsilon^{-1}[m-1] + 2N\sum_{m=n}^{\infty}\gamma[m]E\{t[m]|\Xi[n-1]\} = v[n-1]. \quad (6.5.28)$$

Thus $v[n]$ is a supermartingale [5] with respect to $\Xi[n]$, and $v[n]\to v$ a.s. as $n\to\infty$.

On the other hand,

$$u[n] = 2N\sum_{m=n+1}^{\infty}\gamma[m]E\{t[m]|\Xi[n]\} \quad (6.5.29)$$

is also a supermartingale, since

$$E\{u[n]|\Xi[n-1]\} = \sum_{m=n+1}^{\infty}\gamma[n]E\{t[m]|\Xi[n-1]\} \le u[n-1]. \quad (6.5.30)$$

Therefore $u[n]\to u$ a.s. as $n\to\infty$ and thus $\|p[n]-p^*[n]\|\to v-u$ a.s.

Taking the unconditional (i.e., conditioned by $\Xi[0]$) expectation of both sides of the first inequality in (6.5.24), using (6.5.25), and summing up from

$n = 1$ to $n = n_1$, we obtain

$$\sum_{n=1}^{n_1} E\{\| p[n] - p^*[n] \|^2\} \le \sum_{n=1}^{n_1} E\{\| p[n-1] - p^*[n-1] \|^2\} + L(\varepsilon[0] - \varepsilon[n_1]) + D^2 N \sum_{n=1}^{n_1} \gamma^2[n] \varepsilon^{-1}[n-1] - 2 \sum_{n=1}^{n_1} \gamma[n] E\{W(p[n-1]) - W(p^*[n-1])\} + 2N \sum_{n=1}^{n_1} \gamma[n] E\{t[n]\}. \tag{6.5.31}$$

As $n_1 \to \infty$ the last sum converges according to the lemma. Therefore

$$\sum_{n=1}^{\infty} \gamma[n+1] E\{r(p[n]) - r(p^*[n])\} < \infty. \tag{6.5.32}$$

By the Fatou lemma

$$\sum_{n=1}^{\infty} \gamma[n+1] \{r(p[n]) - r(p^*[n])\} < \infty \text{ a.s.} \tag{6.5.33}$$

From (6.5.33) and (6.5.14) follows the existence of a subsequence n_k such that

$$\| p[n_k] - p^*[n_k] \|^2 \to 0 \text{ a.s.} \qquad \text{as } n_k \to \infty.$$

Therefore $v - u = 0$ a.s. and $\| p[n] - p^*[n] \| \to 0$ a.s. as $n \to \infty$.

On the other hand, $p^*[n] \to p^*$, and so

$$p[n] \to p^* \text{ a.s.} \qquad \text{as } n \to \infty. \qquad \text{Q.E.D.}$$

Example Search for an optimal policy in a Markov decision process in the absence of a priori information.

Consider a system of I states, $S_1, \ldots, S_I$. At every stage $n = 1, 2, \ldots,$ one of M possible decisions $D_1, \ldots, D_M$ must be made. Denote by $S[n]$ and $D[n]$ the state and decision made in stage n, respectively. If $S[n] = S_i$ and $D[n] = D_k$, then the system moves at the next stage, $n + 1$, into the state S_j with an *a priori unknown probability*

$$\pi_{ij}^k = \Pr\{S[n+1] = S_j | S[n] = S_i, D[n] = D_k\}.$$

This transition, if it occurs, is followed by a random reward (or penalty) c_{ij}^k

with an *a priori unknown expectation*. The expected payoff at stage S_i, after the decision D_k is made, is given by

$$\phi_i^k = \sum_{j=1}^{I} \pi_{ij}^k c_{ij}^k.$$

A *policy* is a vector of indices $P=(k_1,\ldots,k_I)$, which determines what decision should be made at each state: for every $i=1,\ldots,I$, k_i is an integer lying between 1 and M, and at state S_i decision D_{k_i} should be made.

Suppose that some fixed policy $P=(k_1,\ldots,k_I)$ is maintained. The system then constitutes a Markov chain with transition probabilities

$$\Pr\{S[n+1]=S_j \mid S[n]=S_i\} = \pi_{ij}^{k_i}.$$

Henceforth it is assumed that for every policy P, the corresponding Markov chain is ergodic. Denote by $\pi_1^{(P)},\ldots,\pi_I^{(P)}$ the steady-state probabilities of this chain, that is,

$$\pi_i^{(P)} = \lim_{n\to\infty} \Pr\{S[n]=S_i\}, \qquad i=1,\ldots,I.$$

The problem is to find a policy P for which the expected payoff,

$$r^{(P)} = \sum_{i=1}^{I} \pi_i^{(P)} \phi_i^{k_i},$$

is minimal.

There are $N=M^I$ possible policies. For each policy $P_m=(k_1^m,\ldots,k_I^m)$, $m=1,\ldots,N$, let $r_m=r^{(P_m)}$. The problem is therefore to choose the policy with the smallest value of r_m.

The regenerative process $\{X^m(t):t\geq 0\}$, corresponding to the policy P_m, is the Markov chain whose states are $S_1,\ldots,S_I$ and whose transition probabilities are $\pi_{ij}^{k_i}$, $i,j=1,\ldots,I$. The regeneration times β_n^m, $n=0,1,2,\ldots$, for this policy are the times of visiting a certain fixed state, say S_1.

Since the algorithm (6.5.9)–(6.5.16) does not require any a priori information about the regenerative processes $\{X^m(t):t\geq 0\}$, $m=1,\ldots,N$, or about the values of $r_1,\ldots,r_N$, it can be applied for finding the optimal policy for the Markov decision process described above.

6.6 THE REGENERATIVE METHOD FOR CONSTRAINED OPTIMIZATION PROBLEMS [62]

In this section we consider an algorithm for solving a linear programming problem, whose coefficients present some unknown characteristics of regenerative processes.

Let us consider the following linear programming problem:

$$\min_p r_0(p) = \min_p \sum_{i=1}^{N} E\{f_0(X^i)\}p_i, \tag{6.6.1}$$

subject to

$$r_j(p) = \sum_{i=1}^{N} E\{f_j(X^i)\}p_i \le 0, \qquad j = 1, \ldots, M \tag{6.6.2}$$

$$p = (p_1, \ldots, p_N), \qquad p_i \ge 0, \ \sum_{i=1}^{N} p_i = 1. \tag{6.6.3}$$

Here X^i, $i = 1, 2, \ldots, N$, are the steady-state r.v.'s of the regenerative processes $\{X^i(t) : t \ge 0\}$, $i = 1, 2, \ldots, N$; the functions f_j, $j = 0, 1, \ldots, M$, are real measurable bounded functions defined on the ranges of these processes. $E\{f_0(X^i)\}$ can be viewed as a performance index of the ith system, $i = 1, \ldots, N$.

We assume that the values $E\{f_j(X^i)\}$, $i = 1, \ldots, N$, $j = 0, 1, \ldots, M$, are unknown a priori; therefore the standard simplex method for solving this linear programming problem cannot be applied. Our solution for this problem is based on the penalty function given below and the regenerative approach studied in the previous sections. Before we start solving this problem let us note that, if we drop (6.6.2) in the linear programming (LP) problem (6.6.1)–(6.6.3), then the problem (6.6.1)–(6.6.3) is identical to the problem (6.5.3)–(6.5.4), which is of course the same as the problem (6.5.2). The problem (6.5.3)–(6.5.4) is referred to as an unconstrained problem (UC) and is therefore a particular case of the constrained LP problem (6.6.1)–(6.6.3).

We start solving the problem (6.6.1)–(6.6.3) by introducing the following penalty function:

$$L(p, \mu) = \sum_{i=1}^{N} E\{f_0(X^i)\}p_i + \sum_{j=1}^{M} \frac{\mu_j}{2}\left(\left[\sum_{i=1}^{N} E\{f_j(X^i)\}p_i\right]^+\right)^2, \tag{6.6.4}$$

where $\mu_j > 0$, $j = 1, \ldots, M$. The operator $[\cdot]^+$ is defined by

$$[Z]^+ = \begin{cases} Z, & \text{for} \quad Z > 0 \\ 0, & \text{for} \quad Z \le 0 \end{cases}. \tag{6.6.5}$$

Now instead of the original LP problem, the following problem is solved:

$$\inf_p L(p, \mu_j[n]), \quad n \to \infty \tag{6.6.6}$$

where p satisfies (6.6.3) and the sequences $\{\mu_j[n]\}_{n=1}^{\infty}$, $j = 1, \ldots, N$, satisfy the following conditions:

$$\begin{aligned} &\mu_j[n] > 0 \quad \mu'[n] \le \mu_j[n] \le \mu''[n] \\ &\mu'[n]\uparrow\infty, \quad \mu[n] = (\mu_1[n],\ldots,\mu_M[n]). \end{aligned} \tag{6.6.7}$$

Now we propose an adaptive algorithm that converges with probability one to the optimal solution of the LP problem (6.6.1)–(6.6.3).

The algorithm is similar to the algorithm (6.5.8)–(6.5.16) and is based on a step-by-step correction of the probability vector p$[n]$, where n denotes the step number. As in the algorithm (6.5.8)–(6.5.16) there exists a mechanism, provided by (6.6.12) below, that ensures that $p_i[n] \ge \varepsilon[n]$, $i = 1,\ldots,N$, where $\{\varepsilon[n]\}_{n=0}^{\infty}$ is a monotone decreasing sequence of positive numbers, subject to (6.6.16) through (6.6.21) below. On the nth step the ith system, $i \in \{1,\ldots,N\}$, is chosen by simulating the distribution $p[n-1]$. We denote this event by $X[n] = X^i$. One cycle of the process $\{X^i(t): t \ge 0\}$ is carried out. Denote by $\nu^i[n]$, $i = 1,\ldots,n$, the total number of cycles made by the ith system up to and including the nth step. We check whether or not the inequality $\nu^k[n-1] \ge n\varepsilon[n]$, $k \in \{1,\ldots,i-1,i+1,\ldots,N\}$, is satisfied for all systems. If for some indices $k_1,\ldots,k_s \in \{1,\ldots,i-1,i+1,\ldots,N\}$ this inequality does not hold, then one additional cycle is carried out for each system $k_1,\ldots,k_s$, so that ultimately

$$\nu^k[n] \ge n\varepsilon[n], \qquad k = 1,\ldots,N, \tag{6.6.8}$$

holds.

We record also

$$\tau_n^k = T^k_{\nu^k[n]} - T^k_{\nu^k[n]-1}, \qquad k = i, k_1,\ldots,k_s, \tag{6.6.9}$$

the lengths of the cycles performed, and for each k calculate $M+1$ numbers

$$Y_n^{kj} = \int_{T_\nu^k k_{[n]-1}}^{T_\nu^k k_{[n]}} f_j(X^k(t))\,dt, \qquad k = 1, k_1,\ldots,k_s, \quad j = 0,1,\ldots,M, \tag{6.6.10}$$

if the process $\{X^k(t): t \ge 0\}$ is continuous-time.

In the case of discrete-parameter processes the integral should be replaced by the corresponding sum over the $(\nu^k[n])$th cycle. Set also

$$Y_n^{kj} = \tau_n^k = 0, \qquad \text{if } k \ne i, k_1,\ldots,k_s, j = 0,1,\ldots,M. \tag{6.6.11}$$

The new distribution $p[n]$ is updated according to the following recurrence formula:

$$p[n] = \pi_{S_\varepsilon[n]}\{p[n-1] - \gamma[n]\,B(n|i)\}. \tag{6.6.12}$$

Here S_ε is a simplex in R^N:

$$S_\varepsilon = \left\{ p = (p_1, \ldots, p_N) : \sum_{k=1}^{N} p_k = 1, 0 \le \varepsilon \le p_k \le 1 \right\},$$

π_{S_ε} is the projection operator onto the simplex S_ε, such that for any $Z \in R^N$,

$$\|Z - \pi_{S_\varepsilon}(Z)\| = \min_{y \in S_\varepsilon} \|Z - y\|,$$

and $B(\cdot|\cdot)$ is a vector $(B_1(\cdot|\cdot), \ldots, B_N(\cdot|\cdot))$, where

$$B_i(n|i) = \frac{g(n|i)}{p_i[n-1]} \tag{6.6.13}$$

$$B_{k \ne i}(n|i) = -\frac{g(n|i)}{(N-1)p_i[n-1]} \tag{6.6.14}$$

$$g(n|i) = r_{i0}[n] + \sum_{j=1}^{M} \mu_j[n] \left[\sum_{k=1}^{N} r_{kj}[n-1] p_k[n-1] \right]^+ r_{ij}[n] \tag{6.6.15}$$

$$r_{kj}[n] = \frac{Y^{kj}[n]}{\tau^k[n]}, \qquad Y^{kj}[n] = Y^{kj}[n-1] + Y_n^{kj},$$

$$\tau^k[n] = \tau^k[n-1] + \tau_n^k, \qquad k = 1, \ldots, N, j = 0, 1, \ldots, M. \tag{6.6.16}$$

The initial values of $p[0] \in S_{\varepsilon[0]}$, $Y[0]$, and $\tau\,[0]$ can be chosen arbitrarily. In the above, the sequences

$$\{\gamma[n]\}_{n=1}^{\infty}, \{\varepsilon[n]\}_{n=0}^{\infty}, \{\mu'[n]\}_{n=1}^{\infty}, \text{ and } \{\mu''[n]\}_{n=1}^{\infty}$$

must be chosen in such a way that the following conditions are satisfied:

$$\gamma[n]\downarrow 0, \qquad \varepsilon[n]\downarrow 0 \tag{6.6.17}$$

$$\sum_{n=1}^{\infty} \gamma[n] = \infty \tag{6.6.18}$$

$$\sum_{n=1}^{\infty} \left(\gamma[n]\mu''[n]^2 \varepsilon^{-1}[n-1]\right) < \infty \tag{6.6.19}$$

$$\sum_{n=1}^{\infty} n^{-1/2} \gamma[n] \mu''[n] \varepsilon^{-1}[n] < \infty \tag{6.6.20}$$

$$\sum_{n=1}^{\infty} \gamma[n](\mu'[n])^{-1} < \infty \tag{6.6.21}$$

$$\sum_{n=1}^{\infty} \gamma[n]\varepsilon[n] < \infty \tag{6.6.22}$$

Remark 1 In order to satisfy conditions (6.6.17) through (6.6.22) we can take, for example,

$$\gamma[n] \sim n^{-1}, \qquad \varepsilon[n] \sim n^{-0.2},$$
$$\mu'[n] \sim \mu''[n] \sim n^{0.2}.$$

Remark 2 We assume that $\tau_n^i \geq \tau_0 > 0$, $i = 1, \ldots, N$, $n = 1, 2, \ldots,$ that is, a cycle will be taken into account if it is of some minimal length (which can be considered as the sensitivity threshold of the measuring instrument).

Remark 3 The r.v.'s $Y^{kj}[n]$ and $\tau^k[n]$, $k = 1, \ldots, N, j = 0, 1, \ldots, M$, $n \geq 1$, defined in (6.6.16), accumulate the information obtained up to and including the $n$pth step. It is worth noting that, for each fixed k, only $\nu^k[n]$ summands in both $Y^{kj}[n]$ and $\tau^k[n]$ are nonzero.

Now we formulate a theorem, which is proven in Rubinstein and Karnovsky [62].

Theorem 6.6.1 If the values of the functions f_j, $j = 0, 1, \ldots, M$, are uniformly bounded by some constant D and if there exists the unique optimal solution p^* of the LP problem, then for any initial distribution $p[0] \in S_{\varepsilon[0]}$ the sequence $\{p[n]\}_{n=1}^{\infty}$ generated by the algorithm (6.6.7)–(6.6.22) converges with probability 1 to p^*.

Corollary 1 Since the UC problem (6.5.3)–(6.5.4) is a special case of the LP problem (6.6.1)–(6.6.3), the algorithm (6.6.7)–(6.6.22) solves the UC problem as well.

Corollary 2 The theorem remains valid if we assume that the values of the functions f_j cannot be observed directly, but can be measured with a random noise. In other words,

$$f_j(X^i) = E_{\xi}\{Q_j(X^i, \xi)\}, \qquad i = 1, \ldots, N, \quad j = 0, \ldots, M,$$

where ξ is a random vector with an unknown time-independent probability distribution function. In this case we can consider another random process:

$$\{f_j(X^i)\} = E\{E_{\xi}\{Q_j(X^i, \xi)\}\} = E\{Q_j(U^i)\}$$
$$i = 1, \ldots, N, j = 0, 1, \ldots, M.$$

6.7 VARIANCE REDUCTION TECHNIQUES

In Chapter 4 we studied several variance reduction techniques—namely: correlated and stratified sampling, antithetic and control variates—for estimating integrals, the mean waiting time in the $GI/G/1$ queueing system, and the expected completion time in networks. Here we deal further with variance reduction techniques for estimating some output parameters of the steady-state distribution of regeneration processes. To understand how expensive simulations can be, consider estimating, via simulation $E[W]$, the expected stationary waiting time in an $M/M/1$ queue. Usually, we would not simulate an $M/M/1$ queue since analytic results are available. However, despite its simplicity it can be very expensive to estimate $E[W]$. It is therefore a good candidate for testing simulation methodologies. Let the traffic intensity $\rho < 1$; then $\overline{W}_N$, the average of the first N waiting times, has an asymptotically normal distribution with mean $E[W]$ and variance σ^2/N. Therefore a confidence interval for $E[W]$ may be constructed.

A major problem in any simulation is how long to run it. One possibility is to run the simulation until the half length of a prescribed confidence interval. Table 6.7.1 lists the run lengths needed for the $M/M/1$ queue to have a half length of 0.10 $E(W)$. It follows from this table that as ρ

Table 6.7.1 Samples Sizes for the $M/M/1$ Queue Required

ρ	$E(W)$	σ^2	N
0.10	0.111	0.375	8,200
0.20	0.250	1.39	6,020
0.30	0.429	3.96	5,830
0.40	0.667	10.6	6,430
0.50	1.00	290	7,850
0.60	1.50	88.5	10,600
0.70	2.33	335	16,700
0.80	4.00	1,976	33,400
0.90	9.00	35,901	119,000
0.95	19.0	607,600	455,000
0.99	99.0	$3.95 \times 10^8 1.09 \times 10^7$	

Source: Ref. 24.

Note: N = Number of customers that must be simulated for a 90% confidence interval for $E[W]$ to have a half length of 0.1 $E[W]$,

$$\mu = 1,\ \rho = \lambda/\mu,\ E[W] = \lambda/\mu(\mu - \lambda).$$

increases beyond 0.3 the required run lengths increase rapidly, and for large values of ρ simulation is no longer a practical method.

In the following two sections we consider control variates and common random numbers (correlated sampling) techniques for variance reduction while simulating stochastic processes, and we give some practical recommendations for their application. The results of these sections are based on Heidelberger [24], Heidelberger and Iglehart [23], and Lavenberg, Moeller, and Sauer [45], and are reproduced mostly from them.

6.7.1 Control Variates

The method of control variates has already been described in Sections 4.3.3 and 4.3.12, and is only reviewed briefly here.

Let $\{X_n, n \geq 0\}$ be a sequence of i.i.d. random variables with unknown mean $r = E(X_n)$. We are interested in estimating r via simulation. Let $\sigma_x^2 = \sigma^2(X_n)$ be the variance of X_n. We can estimate r by

$$\bar{X}_N = \frac{\sum_{n=1}^{N} X_n}{N}$$

and then form a confidence interval by using the c.l.t.:

$$\frac{\sqrt{N}\left(\bar{X}_N - r\right)}{\sigma_x} \Rightarrow N(0,1) \qquad \text{as } N \to \infty.$$

Suppose now that we have another sequence of random variables $\{C_n, n \geq 0\}$, called control variates, such that C_n's are i.i.d., that X_n and C_n are correlated (usually achieved by simulating X_n and C_n with the same stream of random numbers) and that $r_c = E(C_n)$ is known. Let β be some constant and set

$$Z_n(\beta) = X_n - \beta(C_n - r_c). \tag{6.7.1}$$

Then $\{Z_n(\beta), n \geq 0\}$ are i.i.d. with mean r and some variance denoted by $\sigma^2(\beta)$. Let

$$\bar{Z}_N(\beta) = \frac{\sum_{n=1}^{N} Z_n(\beta)}{N};$$

then by the strong law of large numbers $\bar{Z}_N(\beta) \to r$ a.s. as $N \to \infty$ and, by the c.l.t.,

$$\frac{\sqrt{N}\left(\bar{Z}_N(\beta) - r\right)}{\sigma(\beta)} \Rightarrow N(0,1) \qquad \text{as } N \to \infty. \tag{6.7.2}$$

It can be readily shown (see also Section 4.4.3) that $\beta = \beta^*$, which minimizes the variance $\sigma^2(\beta)$, is equal to

$$\beta^* = \frac{\operatorname{cov}(X_n, C_n)}{\sigma^2(C_n)}$$

and that

$$\sigma^2(\beta^*) = \left(1 - \rho^2(X_n, C_n)\right)\sigma_x^2. \tag{6.7.3}$$

Formula (6.6.1) can be extended easily to the case of more than one control variate. Indeed, let $\mathbf{C} = (C_1, \ldots, C_Q)$ be a vector of Q control variates, let $\mathbf{r}_c = (r_1, \ldots, r_Q)$ be the known mean vector corresponding to $\mathbf{C}$, and let $\boldsymbol{\beta}$ be any vector. Then

$$Z_n(\boldsymbol{\beta}) = X_n - \boldsymbol{\beta}'(\mathbf{C}_n - \mathbf{r}_c) \tag{6.7.4}$$

is an unbiased estimator of r.

Another type of control variate $\mathbf{C} = (C_1, \ldots, C_Q)$ is one for which the vector $E(\mathbf{C})$ is unknown but its components $E(C_q)$, $q = 1, \ldots, Q$, are equal to r. In this case

$$Y(\boldsymbol{\beta}) = \beta_0 X + \sum_{i=1}^{Q} \beta_i C_i, \tag{6.7.5}$$

with $\sum_{i=0}^{Q} \beta_i = 1$, is again an unbiased estimator of r.

We now consider two examples of application control variates for which formulas (6.7.4) and (6.7.5) are applied and variance reduction is achieved. The first example deals with (6.7.5); the second with (6.7.4).

Example 1 Let $\{X_n, n \geq 0\}$ be an irreducible, aperiodic positive recurrent Markov chain with state space $I = \{0, 1, 2, \ldots,\}$ and transition matrix $P = \{p_{ij}, i, j \in I\}$. It is known from Section 6.2 that $X_n \Rightarrow X$ as $n \to \infty$($\Rightarrow$ denotes weak convergence), where X is the steady-state random variable having the stationary distribution $\pi = \{\pi_i : i \in I\}$ and π can be found from the solution of the system of linear equations $\pi = \pi P$.

Let $f: I \to R$ be a real-valued function on I and define

$$r = E\{f(X)\} = \pi f = \sum_{i \in I} \pi_i f(i).$$

Here $\pi f = \sum_{i \in I} \pi_i f(i)$ is the inner product* of π and f. We are interested in estimating r. If the matrix P is unknown or the state space I is large (i.e., it is difficult to solve $\pi = \pi P$), it may become necessary to estimate $r = \pi f$

*For simplicity we use this form rather than the more conventional $\langle \pi, f \rangle$.

via simulation. This can be done as follows (see also (6.2.1) through (6.2.4)):

Pick some state in I, say 0, and set $T_0 = 0$. Define

$$T_m = \inf\{n > T_{m-1} : X_n = 0\}, \qquad m \geq 0.$$

We say that a regeneration occurs at time T_m and the time between T_m and T_{m+1}, that is, $\tau_m = T_{m+1} - T_m$, is referred to as the length of the m cycle. Let k be some positive integer and let $r_\nu = \pi f_\nu = E\{f_\nu(x)\}$, $\nu = 0, 1, \ldots, k$. For each $m \geq 0$ and $\nu = 0, 1, \ldots, k$, define $Y_m(\nu)$ by

$$Y_m(\nu) = \sum_{n=T_m}^{T_{m+1}-1} f_\nu(X_n).$$

It follows from Proposition 6.2.2 that, if $\pi|f_\nu| < \infty$, then

$$r_\nu = \pi f_\nu = \frac{E(Y_m(\nu))}{E(\tau_m)}. \tag{6.7.6}$$

Let $Z_m(\nu) = Y_m(\nu) - r_\nu \tau_m$. By (6.7.6) we have for each $\nu = 0, 1, \ldots, k$ and each $m \geq 0$

$$E(Z_m(\nu)) = 0 \tag{6.7.7}$$

Define

$$\hat{r}_\nu(M) = \frac{\sum_{m=1}^{M} Y_m(\nu)}{\sum_{m=1}^{M} \tau_m} \tag{6.7.8}$$

and

$$\hat{X}_\nu(N) = \frac{\sum_{n=0}^{N} f_\nu(X_n)}{N+1} \tag{6.7.9}$$

for each $\nu = 0, 1, \ldots, k$.

Then $\hat{r}_\nu(M) \to r_\nu$ a.s. as $M \to \infty$ and $\hat{X}_\nu(N) \to r_\nu$ a.s. as $N \to \infty$. Observe that $r_\nu(M)$ is an estimator for r_ν based on M cycles of the process and $\hat{X}_\nu(N)$ is an estimation for r_ν based on N transitions of the process. Because $\{Z_m(\nu) : m \geq 0\}$ are i.i.d., it is readily possible to prove the following two c.l.t.'s:

$$\frac{\sqrt{M}\,(\hat{r}_\nu(M) - r_\nu)}{\sigma/E(\tau_1)} \Rightarrow N(0,1) \qquad \text{as } M \to \infty \tag{6.7.10}$$

$$\frac{\sqrt{N}\,(\hat{X}_\nu(N) - r_\nu)}{\sigma/E(\tau_1)^{1/2}} \Rightarrow N(0,1) \qquad \text{as } N \to \infty \tag{6.7.11}$$

Proposition 6.7.1 Let Σ_k be a $(k+1) \times (k+1)$-dimensional covariance matrix of $Z_m(\nu)$'s, whose (i,j)th entry is $\sigma_{ij} = E[Z_m(i)Z_m(j)]$. If $E(|f_\nu(x)|) < \infty$ for each $\nu = 0, 1, \ldots, k$, then

$$\left[\sqrt{M}\,(r_0(M) - r_0), \ldots, \sqrt{M}\,(\hat{r}_k(M) - r_k)\right] \Rightarrow N\left(0, \frac{\Sigma_k}{E^2(\tau_1)}\right) \tag{6.7.12}$$

$$\left[\sqrt{N}\,(\hat{X}_0(N) - r_0), \ldots, \sqrt{N}\,(\hat{X}_k(N) - r_k)\right] \Rightarrow N\left(0, \frac{\Sigma_k}{E(\tau_1)}\right). \tag{6.7.13}$$

The proof of this proposition is given in Ref. 24.

Now let $\boldsymbol{\beta}$ be a $(k+1)$-dimensional row vector of real numbers whose νth entry is $\beta(\nu)$. Let $\mathbf{r}, \hat{\mathbf{r}}(M)$, and $\hat{\mathbf{x}}(N)$ denote $(k+1)$-dimensional column vectors whose νth entries are r_ν, $\hat{r}_\nu(M)$, and $\hat{x}_\nu(N)$, respectively. A simple application of the continuous mapping theorem (Theorem (5.1) of Billingsley [1]) yields the following.

Proposition 6.7.2 Let $\sigma_k^2(\boldsymbol{\beta}) = \boldsymbol{\beta}\Sigma_k\boldsymbol{\beta}' = \Sigma_{i=0}^k \Sigma_{j=0}^k \beta(i)\sigma_{ij}\beta(j)$. Under the hypotheses of Proposition 6.7.1,

$$\frac{\sqrt{M}\,(\boldsymbol{\beta}\hat{\mathbf{r}}(M) - \boldsymbol{\beta}\mathbf{r})}{\sigma_k(\boldsymbol{\beta})/E(\tau_1)} \Rightarrow N(0,1) \qquad \text{as } M \to \infty, \tag{6.7.14}$$

and

$$\frac{\sqrt{N}\,(\boldsymbol{\beta}\hat{\mathbf{x}}(N) - \boldsymbol{\beta}\mathbf{r})}{\sigma_k(\boldsymbol{\beta})/E(\tau_1)^{1/2}} \Rightarrow N(0,1) \qquad \text{as } N \to \infty, \tag{6.7.15}$$

where $\boldsymbol{\beta}\mathbf{r} = \Sigma_{\nu=0}^k \beta(\nu) r_\nu$ is the inner product of $\boldsymbol{\beta}$ and $\mathbf{r}$.

In order to form confidence intervals for the r_ν's (or for linear combinations of the r_ν's) it is necessary to know the σ_{ij}'s as well as $E(\tau_1)$. These constants are usually unknown and must be estimated. In addition $\boldsymbol{\beta}$ may be a fixed, but unknown, vector so it too must be estimated. The following proposition, the proof of which is also given in Ref. 24, tells us that we may replace these quantities in Proposition 6.7.2 by any sequence of strongly consistent estimators preserving the asymptotic normality.

Proposition 6.7.3 Suppose that $\bar{\tau}_1(M) \to E(\tau_1)$ a.s., that $\hat{\sigma}_{ij}(M) \to \sigma_{ij}$ a.s. for each i and j, and that $\hat{\beta}(i, M) \to \beta(i)$ a.s. for each i. Let $\hat{\Sigma}_k(M)$ be the matrix whose (i,j)th entry is $\hat{\sigma}_{ij}(M)$, let $\hat{\boldsymbol{\beta}}(M)$ be the vector whose ith component is $\hat{\beta}(i, M)$, and let $\hat{\sigma}_k(\hat{\beta}, M) = \hat{\boldsymbol{\beta}}(M)\hat{\Sigma}_k(M)\hat{\boldsymbol{\beta}}'(M)$. Then

$$\frac{\sqrt{M}\left(\hat{\boldsymbol{\beta}}(M)\hat{\mathbf{r}}(M) - \hat{\boldsymbol{\beta}}(M)\mathbf{r}\right)}{\hat{\sigma}_k(\hat{\boldsymbol{\beta}}, M)/\bar{\tau}_1(M)} \Rightarrow N(0,1) \qquad \text{as } M \to \infty. \quad (6.7.16)$$

We turn now to the problem of choosing the functions f_ν with a view to achieving variance reductions.

Heidelberger [24–26] suggested several ways of choosing f_ν, $\nu = 0, \ldots, k$. We consider only one of them [24].

Let

$$f_\nu = P^\nu f, \; \nu = 0, 1, \ldots, k, \tag{6.7.17}$$

where P^ν is the ν step matrix function of the process. It is shown in Ref. 24 that in this case, that is, when $f_\nu = P^\nu f$, all $r_\nu = \pi f_\nu$, $\nu = 0, 1, \ldots, k$, are equal to r, and if $E\{|f(x)|\} < \infty$, then $\pi f = \pi(Pf)$. Since $r_\nu = r$, $\nu = 0, 1, \ldots, k$, it is obvious that

$$\hat{r}_\nu(M) = \frac{\sum_{m=1}^{M} Y_m(\nu)}{\sum_{m=1}^{M} \tau_m} \xrightarrow{\text{a.s.}} \frac{E(Y_1(\nu))}{E(\tau_1)} = \pi f_\nu = r. \tag{6.7.18}$$

Therefore each $\hat{r}_\nu(M)$, $\nu = 0, 1, \ldots, k$, is a strongly consistent estimator for r, and we can use one of them for this purpose. However, better results can be achieved by using all of them simultaneously, for instance, using (6.7.5), which can be written as

$$\hat{r}_\beta(M) = \sum_{\nu=0}^{k} \beta(\nu)\hat{r}_\nu(M), \tag{6.7.19}$$

where $\sum_{\nu=0}^{k}\beta(\nu) = 1$. Variance reduction can be achieved if we choose the $\beta(\nu)$'s so as to minimize the asymptotic variance $\sigma_k^2(\boldsymbol{\beta})$ of $\hat{r}_\beta(M)$. Mathematically it can be written as

$$\text{minimize } \sigma_k^2(\beta) = \boldsymbol{\beta}\Sigma\boldsymbol{\beta}' \tag{6.7.20}$$

$$\text{subject to } \sum_{\nu=0}^{k} \beta(\nu) = 1. \tag{6.7.21}$$

The solution of this problem, which can be obtained using Lagrange multipliers, is

$$\beta^* = \frac{e\Sigma^{-1}}{e\Sigma^{-1}e^t} \tag{6.7.22}$$

$$\sigma_k^2(\beta^*) = \frac{1}{e\Sigma^{-1}e^t}, \tag{6.7.23}$$

where e denotes the $(k+1)$-dimensional row vector each of whose components is 1, and where t is the transpose operation. Formulas (6.7.10) and (6.7.11) can be now rewritten as

$$\frac{\sqrt{M}\,\big(\hat{r}_{\beta^*}(M)-r\big)}{\sigma_k(\boldsymbol{\beta}^*)/E(\tau_1)} \Rightarrow N(0,1) \tag{6.7.24}$$

$$\frac{\sqrt{N}\,\big(\hat{X}_{\beta^*}(N)-r\big)}{\sigma_k(\beta^*)/E(\tau_1)^{1/2}} \Rightarrow N(0,1), \tag{6.7.25}$$

where $\hat{X}_\beta(N)=\sum_{\nu=0}^{k}\beta(\nu)\hat{X}_\nu(N)$, and both

$$\hat{r}(M)\to r \text{ a.s.} \qquad \text{as } M\to\infty$$

and

$$\hat{X}_\beta(N)\to r \text{ a.s.} \qquad \text{as } N\to\infty.$$

Since the covariance matrix Σ is in general unknown, it is necessary to estimate it. If $\hat{\Sigma}(M)$ is any estimator such that $\hat{\Sigma}(M)\to\Sigma$ a.s. as $M\to\infty$, then it is clear that $\hat{\Sigma}^{-1}(M)\to\Sigma^{-1}$ a.s. as $M\to\infty$. Letting

$$r_{\beta^*}(M)=\sum_{\nu=0}^{k}\hat{\beta}^*(\nu,M)\hat{r}_\nu(M), \tag{6.7.26}$$

and applying Proposition 6.7.3, we have

$$\frac{\sqrt{M}\,\big(r_{\hat{\beta}^*}(M)-r\big)}{\hat{\sigma}_k\big(\hat{\boldsymbol{\beta}}_k^*,M\big)/\bar{\tau}_1(M)} \Rightarrow N(0,1), \tag{6.7.27}$$

where $\hat{\sigma}_k(\hat{\boldsymbol{\beta}}_k^*,M)\to\sigma_k(\boldsymbol{\beta}^*)$ a.s. as $M\to\infty$ and $\bar{\tau}_1(M)$ is any sequence of numbers such that $\bar{\tau}_1(M)\to E(\tau_1)$ a.s. as $M\to\infty$. A corresponding c.l.t. exists for the $\hat{X}_\nu(N)$'s as well.

This method is called the "method of multiple estimates" because it combines several different estimates of the same quantity.

In order to apply this method the functions f_ν must be computed (usually before the start of the simulation). For computation efficiency f_ν can be defined recursively by $f_0=f$ and $f_\nu=Pf_{\nu-1}$ for $\nu\geq 1$. This saves having to compute the ν step transition function P^ν, a potentially large computational economy. If the state space is finite and the transition matrix is sparse, the work involved in calculating f_ν for a few values of ν may not be too heavy.

We note that to form the estimates $\hat{x}_\nu(N)$ (or $\hat{r}_\nu(M)$) we must evaluate $f_\nu(X_n)$ for each value of ν and each transition n. This tends to increase the amount of time needed for each transition simulated. However, if the variance reduction obtained is sufficiently large, the potential savings in the number of transitions that need to be simulated will more than offset the extra work per transition. We also note that additional work must be done at the end of each cycle to update the estimates of the covariance matrix Σ_k (using no variance reducing technique, we need only update σ_0^2). It is shown (see [24]) that $\sigma_k^2(\boldsymbol{\beta}^*)\to 0$ as $k\to\infty$. For many types of Markov chains we can expect substantial variance reductions even when k is relatively small (say 2 or 3). For countable I we have

$$f_k(i)=\sum_{j=0}^{\infty} P_{ij}^k f(j)=E\left[f(X_{n+k})\mid X_n=i\right]. \tag{6.7.28}$$

Thus if the Markov chain makes transitions only to "neighboring" states and if $f(j)$ is close to $f(i)$ for j close to i, it can be seen from (6.7.28) that, for small k, $f_k(i)$ and $f(i)$ should be nearly the same. This means that $\hat{x}_k(N)$ and $\hat{x}_0(N)$ will be highly correlated, a condition that generally results in good variance reduction. Many queueing networks exhibit this special type of structure.

Ideally, we would like to be able to have the "optimal" value of k in the sense that, for a given computer budget, we would like to pick the value that yields the narrowest confidence intervals for r (part of the budget must be allocated to calculation of the f_ν's). To perform such an optimization we would have to know $\sigma_\nu^2(\boldsymbol{\beta}^*)$ for each $\nu\geq 0$. These quantities are generally unknown, and even to estimate them would require calculating the f_ν's and then simulating the Markov process for an additional number of cycles. The disadvantage of such a procedure is that the cost of computation of f_ν may be higher than the gain achieved through variance reduction. Generally speaking, the success of this technique depends on our ability to compute and store efficiently the functions f_ν.

The method of multiple estimates can be extended to certain types of continuous-time processes such as continuous-time Markov chains and semi-Markov processes (see [24]).

To find out the efficiencies of this method Heidelberger [24] considered the following four examples: the queue length process in a finite capacity $M/M/1$ queue, the queue length process in the repair problem with spares, and the waiting time processes in both $M/M/1$ and $M/M/2$ queues. These processes were chosen because analytic results are readily available, thereby making a comparison between analytic and simulation results possible. Despite their simplicity, these processes are by no means "easy" to simulate, in particular the heavily loaded queues, which require very large run lengths to get good simulation estimates. The simulation results, which are also presented in Ref. 24, show that for all four examples substantial variance reduction was obtained. However, as this method entails additional computations both before and during the course of simulation, we would recommend using it only when it is computationally advantageous to do so. In the case of Markov chain it is likely that the method will be most effective if the transition matrix of the process is sparse, in which case the preliminary calculations can be carried out with relative ease. It is for this type of process that the method is recommended.

Example 2 We consider now another example of variance reduction, taken from Ref. 45. Before starting this example we need more mathematical background on the regenerative method.

Let X again be the steady-state vector of the regenerative process $\{x(t): t \geq 0\}$, let f and g be given real-valued measurable functions, and suppose we want to estimate

$$\mu = \frac{E\{f(X)\}}{E\{g(X)\}}. \tag{6.7.29}$$

It follows from Proposition 6.2.2 that, if $E\{|f(X)|\} < \infty$ and $E\{|g(x)|\} < \infty$, then

$$\mu = \frac{E(Y_i)}{E(Z_i)} = \frac{E(Y)}{E(Z)}, \qquad i = 1, 2, \ldots, \tag{6.7.30}$$

where

$$Y_i = \int_{T_i}^{T_{i+1}} f\{X(t)\}\, dt$$

and

$$Z_i = \int_{T_i}^{T_{i+1}} g\{X(t)\}\, dt$$

are dependent random variables defined with respect to a single cycle $\tau_i = T_{i+1} - T_i$. In the particular case where $g = 1$, we have $Z_i = \tau_i$ and (6.7.30) becomes (6.2.4). The classical point estimator for μ obtained from M cycles is

$$\hat{\mu} = \frac{\sum_{i=1}^{M} Y_i}{\sum_{i=1}^{M} Z_i}, \qquad i = 1, 2, \ldots. \tag{6.7.31}$$

and for sufficiently large M

$$\frac{\sqrt{M}\,(\hat{\mu} - \mu)}{\sigma} \Rightarrow N(0, 1), \tag{6.7.32}$$

where

$$\sigma^2 = \frac{\operatorname{Var}(Y_i - \mu Z_i)}{E(Z_i)}, \qquad i = 1, 2, \ldots.$$

Furthermore, if we replace σ with its estimator $\hat{\sigma}$ such that

$$\hat{\sigma}^2 = \frac{(1/(M-1)) \sum_{i=1}^{M} (Y_i - \mu Z_i)^2}{\left((1/M) \sum_{i=1}^{M} Z_i\right)^2}, \tag{6.7.33}$$

the c.l.t (6.7.32) will also hold; therefore a confidence interval for μ can be obtained.

Assume now that we have Q pairs of dependent random variables $\{Y^{(q)}, Z^{(q)}\}$, $q = 1, \ldots, Q$, defined with respect to a single cycle. Denote

$$\mu_q = \frac{E(Y^{(q)})}{E(Z^{(q)})}. \tag{6.7.34}$$

Assume also that μ_q, $q = 1, \ldots, Q$, is known, but that the expected values of the pairs $\{Y^{(q)}, Z^{(q)}\}$ are unknown. In order to apply control variates in this case the sequence of i.i.d. pairs of random vectors

$$\mathbf{R}_n = \left\{(Y_n, Z_n), \left(Y_n^{(1)}, Z_n^{(1)}\right), \ldots, \left(Y_n^{(Q)}, Z_n^{(Q)}\right)\right\}, \qquad n = 1, \ldots, M \tag{6.7.35}$$

is collected, and then the Q-dimensional vector of control variates $\mathbf{C} = (C_1, \ldots, C_Q)$ is defined as

$$C_q = \frac{\sum_{n=1}^{M} Y_n^{(q)}}{\sum_{n=1}^{M} Z_n^{(q)}}, \qquad q = 1, \ldots, Q. \tag{6.7.36}$$

Now, by analogy with (6.7.4), for any vector $\boldsymbol{\beta}$ a point estimator for μ using these control variates is

$$\hat{\mu}(\boldsymbol{\beta}) = \hat{\mu} - \boldsymbol{\beta}'(\mathbf{C} - \boldsymbol{\mu}_c), \tag{6.7.37}$$

where $\boldsymbol{\mu}_c = (\mu_1, \ldots, \mu_Q)$. Note that because $\hat{\mu}$ and C_q, $q = 1, \ldots, Q$, are biased estimators, respectively, for μ and μ_q, $q = 1, \ldots, Q$, the estimator $\hat{\mu}(\beta)$ is also biased, which differentiates it from the unbiased estimator for $Z_n(\boldsymbol{\beta})$ in (6.7.4). However, $\hat{\mu}(\boldsymbol{\beta})$ is a strongly consistent estimator of μ and, for M sufficiently large,

$$\frac{\sqrt{M}[\hat{\mu}(\beta) - \mu]}{\sigma(\beta)} \Rightarrow N(0, 1), \tag{6.7.38}$$

where

$$\sigma^2(\boldsymbol{\beta}) = \operatorname{var}\left[\frac{Y - \mu Z}{E[Z]} - \sum_{q=1}^{Q} \frac{\beta_q\left(Y^{(q)} - \mu_q Z^{(q)}\right)}{E\left[Z^{(q)}\right]}\right]. \tag{6.7.39}$$

The value of $\boldsymbol{\beta}$ that minimizes $\sigma^2(\boldsymbol{\beta})$ is (see (4.3.30))

$$\boldsymbol{\beta}^* = \Sigma^{-1}\boldsymbol{\sigma}, \tag{6.7.40}$$

where the matrix Σ and $\boldsymbol{\sigma}$ have elements

$$(\Sigma)_{qp} = \operatorname{cov}\left[\frac{Y - \mu Z}{E[Z]}, \frac{Y^{(p)} - \mu_q Z^{(p)}}{E\left[Z^{(p)}\right]}\right] \tag{6.7.41}$$

and

$$(\boldsymbol{\sigma})_q = \operatorname{cov}\left[\frac{Y - \mu Z}{E[Z]}, \frac{Y^{(q)} - \mu_q Z^{(q)}}{E\left[Z^{(q)}\right]}\right]. \tag{6.7.42}$$

The resulting minimum value of $\sigma^2(\boldsymbol{\beta})$ is

$$\sigma^2(\boldsymbol{\beta}^*) = (1 - R^2)\sigma^2, \tag{6.7.43}$$

where

$$R^2 = \frac{\boldsymbol{\sigma}'\Sigma^{-1}\boldsymbol{\sigma}}{\operatorname{var}[\hat{\mu}]}. \tag{6.7.44}$$

Finally, for M sufficiently large

$$\frac{\sqrt{M}\left(\hat{\mu}(\hat{\boldsymbol{\beta}}^*) - \mu\right)}{\hat{\sigma}(\boldsymbol{\beta}^*)} \Rightarrow N(0, 1), \tag{6.7.45}$$

where $\hat{\boldsymbol{\beta}}^*$ is an estimator of $\boldsymbol{\beta}^*$ and $\hat{\sigma}^2(\boldsymbol{\beta}^*)$ is an estimator of $\sigma^2(\boldsymbol{\beta}^*)$. As M increases $\hat{\sigma}^2(\boldsymbol{\beta}^*)/\hat{\sigma}^2$ approaches $1 - R^2$, and therefore variance reduction can be achieved.

Now we start with the example given in Ref. 45. Consider a $GI/G/1$ queue with i.i.d interarrival times A_i and i.i.d. service times S_i. Let μ_2 be the mean interarrival time. Assume that the traffic intensity $\rho = \mu_2/\mu_1 < 1$; this means that the queueing time $\{W_i, i \geq 0\}$, which is defined by $W_i = (W_{i-1} + S_{i-1} - A_i)^+$, $i \geq 1$, and $W_0 = 0$, is a regenerative process with regenerative points $\{T_k, k = 1, 2, \dots\}$, where T_k is the serial number of the kth customer that arrives to find the system empty and $T_1 = 1$ (consult Section 6.4.1).

The steady-state waiting time $E(W) = \mu$ can be estimated by

$$\hat{\mu} = \frac{\sum_{i=1}^{M} Y_i}{\sum_{i=1}^{M} Z_i}$$

where

$$Y_i = \sum_{j=T_i}^{T_{i+1}-1} W_j$$

and $Z_i = T_{i+1} - T_i$. Define

$$Y^{(1)} = \sum_{j=T_i}^{T_{i+1}-1} A_j,$$

the duration of the ith busy cycle (busy period plus idle time), and

$$Y^{(2)} = \sum_{j=T_i}^{T_{i+1}-1} S_j,$$

the duration of the th busy period. It is known [45] that $E(Y^{(1)}) = \mu_1 E(Z)$ and $E(Y^{(2)}) = \mu_2 E(Z)$, where $E(Z) = E(Z_i)$, $i = 1, 2, \ldots$.

The following vector of control variates $\mathbf{C} = (C_1, C_2)$ with components (see also (6.7.36))

$$C_q = \frac{\sum_{n=1}^{M} Y_n^{(q)}}{\sum_{n=1}^{M} Z_n} \qquad q = 1, 2,$$

is considered in Ref. 45 and the point estimator $\hat{\mu}(\boldsymbol{\beta})$ given in (6.7.37) is adapted for the parameter $\mu = E(W)$. It is shown numerically in Ref. 45 that substantial variance reduction is obtained by simulating the $GI/G/1$ queue and some other queueing models, while using these control variates.

6.7.2 Common Random Numbers in Comparing Stochastic Systems [23]

In this section we show how the method of common random numbers may be used in simulation of discrete and continuous Markov chains for variance reductions.

Suppose we have two irreducible, aperiodic, positive recurrent Markov chains in discrete time and we wish to construct a confidence interval for $r_1 - r_2 = E\{f_1(X^1)\} - E\{f_2(X^2)\}$ by simulating the two processes. Here X^i, $i = 1, 2$, is the steady-state r.v. of the regenerative process $\mathbf{X}^i = \{X_t^i : t \geq 0\}$ and the f_i are given real-valued functions defined on the state space I_i of process $\mathbf{X}^i$.

Let us consider the following two point estimates of r_i:

$$\hat{r}_n^i = \frac{(1/n) \sum_{k=1}^{n} Y_k^i}{(1/n) \sum_{k=1}^{n} \tau_k^i} \tag{6.7.46}$$

and

$$\tilde{r}_N^i = \frac{1}{N} \sum_{k=0}^{N-1} f_i(X_k^i), \tag{6.7.47}$$

where n is the number of simulated cycles, N is the number of steps, and it is assumed without loss of generality that $T_0^i = 0$, $X_0^i = 0$, $i = 1, 2$. The two c.l.t.'s are the following:

$$\frac{n^{1/2}\left[\hat{r}_n^i - r^i\right]}{\sigma_i / E_0\{\tau_1^i\}} \Rightarrow N(0,1) \tag{6.7.48}$$

$$\frac{N^{1/2}\left[\tilde{r}_N^i - r^i\right]}{\sigma_i^{1/2} / E_0\{\tau_1^i\}} \Rightarrow N(0,1) \tag{6.7.49}$$

as n and $N \to \infty$.

To construct a confidence interval for $r^1 - r^2$ we can simulate the two processes $\mathbf{X}^1$ and $\mathbf{X}^2$ independently and apply the bivariate c.l.t.

$$N^{1/2}\left[\tilde{\mathbf{r}}_N - \mathbf{r}\right] \Rightarrow N(\mathbf{0}, \mathbf{A}), \tag{6.7.50}$$

where $\tilde{\mathbf{r}}_N = (\tilde{r}_N^1, \tilde{r}_N^2)$, $\mathbf{r} = (r^1, r^2)$, $N(\mathbf{0}, \mathbf{A})$ is a two-dimensional normal vector with mean vector $\mathbf{0} = (0,0)$ and covariance matrix

$$\mathbf{A} = \begin{bmatrix} \dfrac{\sigma_1^2}{E_0\{\tau_1^1\}} & 0 \\ 0 & \dfrac{\sigma_2^2}{E_0\{\tau_1^2\}} \end{bmatrix}.$$

It can be readily shown (see [23]) that

$$\frac{N^{1/2}\left[(\tilde{r}_N^1 - \tilde{r}_N^2) - (r_1 - r_2)\right]}{\sigma} \Rightarrow N(0,1), \tag{6.7.51}$$

where

$$\sigma^2 = \frac{\sigma_1^2}{E_0\{\tau_1^1\}} + \frac{\sigma_2^2}{E_0\{\tau_1^2\}}.$$

A c.l.t. similar to (6.7.51), but based on simulating m cycles, can also be obtained to construct a confidence interval for $r_1 - r_2$.

Now we turn our attention to the problem of using common random numbers while generating sample paths for $\mathbf{X}^1$ and $\mathbf{X}^2$. Our goal in using common random numbers is to produce a shorter confidence interval for $r_1 - r_2$ for the same length of simulation run. In other words, we seek a c.l.t. similar to (6.7.51) but with a smaller value of σ. To accomplish this we generate the bivariate M.C. $\mathbf{X} = \{X_n : n \geq 0\}$, where $X_n = (X_n^1, X_n^2)$. At each jump of the process $\mathbf{X}$ the same random number is used to generate the jumps of the two marginal chains $\mathbf{X}^1$ and $\mathbf{X}^2$. The marginals of the process $\mathbf{X}$ are seen to have the same distributions as the original chains $\mathbf{X}^1$ and $\mathbf{X}^2$; however, the marginal chains are now dependent. The state space of the chain $\mathbf{X}$ is denoted by F which is a (possibly proper) subset of $I_1 \times I_2$. We assume here that the chain $\mathbf{X}$ is also irreducible, aperiodic, and positive-recurrent. (These conditions are not automatic but usually hold for practical simulations.) Furthermore, we assume for convenience that $(0,0) \in F$ and use that state to form regenerative cycles. Note that $X_n \Rightarrow X$ as $n \to \infty$, and the marginal distributions of X are the same as those of X^1 and X^2, namely, $\{\pi_j(i) : j \in I\}$ for $i = 1, 2$. For any real-valued function $f: F \to R$ satisfying $E\{|f(X)|\} < \infty$, the regenerative method can be applied to $\mathbf{X}$ to estimate $E\{f(X)\}$. Let $X_0 = (0,0)$, $T_0 = 0$, and define the mth entrance to state $(0,0)$ by $\mathbf{X}$ to be

$$T_{m+1} = \inf\{n > T_m : X_n = (0,0)\}, \qquad m \geq 0.$$

Also, let $\tau_m = T_{m+1} - T_m$, $m \geq 0$, be the length of the mth cycle and

$$Y'_m(i) = \sum_{n=T_m}^{T_{m+1}-1} f_i(X_n), \qquad m \geq 0.$$

Set $Z'_m(i) = Y'_m(i) - r_i \tau_m$. Since the ratio formula (6.2.4) still holds for the process $\mathbf{X}$, $E_{(0,0)}\{Z'_m(i)\} = 0$ for $i = 1, 2$. Let

$$\sigma_{ij} = E_{(0,0)}\{Z'_1(i) Z'_1(j)\}, \qquad i, j = 1, 2,$$

which we assume is finite and nonzero. Since the vectors $\mathbf{Z}'_m = (Z'_m(1), Z'_m(2))$ are i.i.d., the standard c.l.t. yields

$$n^{1/2} \sum_{m=1}^{n} \mathbf{Z}'_m \Rightarrow N(\mathbf{0}, \mathbf{\Sigma}), \tag{6.7.52}$$

where $\mathbf{\Sigma} = \{\sigma_{ij}\}$. By analogy with (6.7.49) and (6.7.50) it can be shown (see [23]) that

$$N^{1/2}[\tilde{r}_N - \mathbf{r}] \Rightarrow N(\mathbf{0}, \mathbf{B}), \tag{6.7.53}$$

and

$$\frac{N^{1/2}[(\tilde{r}_N^1 - \tilde{r}_N^2) - (r_1 - r_2)]}{v} \Rightarrow N(0,1). \tag{6.7.54}$$

Here $\mathbf{B} = \sigma_{ij} / E_{(0,0)}\{\tau_1\}$ and

$$v^2 = \frac{(\sigma_{11} + \sigma_{22} - 2\sigma_{12})}{E_{(0,0)}\{\tau_1\}}.$$

A c.l.t. similar to (6.7.54), but in terms of n regenerative cycles, can also be obtained. Now consider the marginals of (6.7.53) in conjunction with (6.7.49). Since the marginals of the chain **X** have the same stochastic structure as the chains $\mathbf{X}^1$ and $\mathbf{X}^2$ considered separately, these two c.l.t.'s must be identical. Hence

$$\frac{\sigma_1^2}{E_0\{\tau_1^i\}} = \frac{\sigma_{ii}}{E_{(0,0)}\{\tau_1\}}. \tag{6.7.55}$$

Thus upon comparing the constant σ^2 in (6.7.51) and v^2 in (6.7.54), we conclude that $v^2 < \sigma^2$ if and only if $\sigma_{12} > 0$.

The measure of variance reduction we use is

$$R^2 = \frac{\sigma^2}{v^2}. \tag{6.7.56}$$

So, for example, if $R = 2$, then only half as many steps of the Markov chain **X** need be simulated to obtain a confidence interval of specified length for $r_1 - r_2$ as would be required when simulating $\mathbf{X}^1$ and $\mathbf{X}^2$ independently. In addition, of course, only one stream of random numbers need be generated. While we have worked here with discrete-time Markov chains, the same method can be used for continuous-time Markov chains, semi-Markov processes, and discrete-time Markov processes with a general state space.

The following definition and properties will be used in obtaining nonnegative correlation.

Definition 1 Random variables $Y = (Y_1, \ldots, Y_n)$ are said to be associated if $\operatorname{cov}\{f(Y), g(Y)\} \geq 0$ for all nondecreasing functions f and g for which $E\{f(Y)\}, E\{g(Y)\}$ and $E\{f(Y), g(Y)\}$ exist.

PROPERTY 1. Any subset of associated random variables are associated.

PROPERTY 2. If two sets of associated random variables are independent of one another, then their union is a set of associated random variables.

PROPERTY 3. The set consisting of a single random variable is associated.

PROPERTY 4. Nondecreasing functions of associated random variables are associated.

A class of processes for which nonnegative correlation can be guaranteed is stochastically monotone Markov chains (s.m.m.c.). In the following definition let i be a fixed index.

Definition 2 Let $\mathbf{X}^i = \{X_n^i, n \geq 0\}$ be a real-valued Markov process with initial distribution $P_i(x) = P\{X_0^i \leq x\}$ and transition function $P_i(x, A) = P\{X_{n+1}(i) \in A | X_n(i) = x\}$ (for measurable sets A). $\mathbf{X}^i$ is said to be an s.m.m.c. if, for every y, $P_i(x, (-\infty, y])$ is a nonincreasing function of x.

Define the inverse distribution functions $P_i^{-1}(\cdot)$ and $P_i^{-1}(x, \cdot)$ by

$$P_i^{-1}(u) = \inf\{y : P_i(y) \geq u\} \tag{6.7.57}$$

$$P_i^{-1}(x, u) = \inf\{y : P_i(x, (-\infty, y]) \geq u\}. \tag{6.7.58}$$

Henceforth we assume that the sample paths of $\mathbf{X}^i$ are generated on the computer, using the inverse transformation scheme

$$X_0^i = P_i^{-1}(U_0) \tag{6.7.59}$$

$$X_n^i = P_i^{-1}(X_{n-1}^i, U_n), \qquad n \geq 1, \tag{6.7.60}$$

where $\{U_n, n \geq 0\}$ is a sequence of random numbers.

Notice that, if $\mathbf{X}^i$ is an s.m.m.c., then $P_i^{-1}(x, u)$ is an increasing function in both arguments. This fact enables us to show that for each $n \geq 0$ $\{X_0^1, \ldots, X_n^1, X_0^2, \ldots, X_n^2\}$ are associated.

Theorem 6.7.1 If $\mathbf{X}^1$ and $\mathbf{X}^2$ are both s.m.m.c.'s with sample paths generated by (6.7.59) and (6.7.60), then, for each $n \geq 0$, $\{X_0^1, \ldots, X_n^1, X_0^2, \ldots, X_n^2\}$ are associated random variables.

Proof The proof is by induction. For $n = 0$ Property 3 implies that $\{U_0\}$ is associated and since $P_i^{-1}(U_0)$ is a nondecreasing function of U_0 for each i, yields that $\{X_0^1, X_0^2\}$ are associated. Assume now that $\{X_0^1, \ldots, X_n^1, X_0^2, \ldots, X_n^2\}$ are associated. Since U_{n+1} is independent of this set, $\{X_0^1, \ldots, X_n^1, X_0^2, \ldots, X_n^2, U_{n+1}\}$ are associated by Property 4. The map that takes these random variables into $\{X_0^1, \ldots, X_n^1, X_{n+1}^1, X_0^2, \ldots, X_n^2, X_{n+1}^2\}$ is nondecreasing because $\mathbf{X}^1$ and $\mathbf{X}^2$ are both s.m.m.c.'s. Property 4 then yields the final result. Q.E.D.

The following theorem, whose proof is found in Ref. 23, shows that, when simulating s.m.m.c.'s using common random numbers, a reduction in variance is achieved.

Theorem 6.7.2 Let $\mathbf{X}^1$ and $\mathbf{X}^2$ both be s.m.m.c.'s with sample paths generated by (6.7.59) and (6.7.60). Let f_1 and f_2 be nondecreasing functions.

If

$$E\{\tau_m^2\} < \infty, \tag{6.7.61}$$

$$E\left\{\left(\sum_{n=T_m}^{T_{m+1}-1} |f_i(X_n^i)|\right)^2\right\} < \infty, \qquad \text{for } i = 1,2, \tag{6.7.62}$$

then $\sigma_{12} \geq 0$.

The efficiency of common random numbers in variance reduction was checked for different output parameters of regenerative processes and substantial variance reduction was achieved only for some particular cases. The effect of variance reduction decreases with increasing complexity of the processes being simulated. The method is effective only where the expected cycle length is sufficiently short. If preliminary simulation runs indicate that the expected cycle length is excessive, it is suggested that independent simulations be performed.

EXERCISES

1 For the data given in Fig. 6.4.1 construct a 90% confidence interval using the classical estimator

$$r = \frac{\sum_{i=1}^{n} Y_i}{\sum_{i=1}^{n} \tau_i},$$

where n is the number of cycles.

2 Prove by induction that, if $f_\nu = P^\nu f$, where P^ν is the νth step transition matrix, then $r_\nu = \pi_\nu$ is equal to $r = \pi f$. Here π is the steady-state distribution of P. From Heidelberger [24].

3 Prove that, if $\pi|f| < \infty$, then $\pi f = \pi(Pf)$. From Heidelberger [24].

4 Prove that the solution of the problem (6.7.20)–(6.7.21) is (6.7.22)–(6.7.23).

5 Consider the following system of linear equations:

$$Y = \alpha PY + f,$$

where P is an $(n \times n)$ ergodic Markov chain with stationary distribution $\pi = \pi P$, $\alpha < \infty$. Prove that

$$\langle \pi, Y \rangle = \frac{r}{1-\alpha}$$

where $r = \langle \pi, f \rangle$.

6 *Inventory Model*. Consider a situation in which a commodity is stocked in order to satisfy some demand. An inventory (s, S) policy is characterized by two postive numbers s and S with $S > s$. If the available stock quantity is greater than s, do not order. If the amount of inventory on hand plus on order is less than s, order to bring the quantity of the s stock to S. Let X_j denote the level of inventory on hand plus on order in the period i after ordering. Let d_j denote the demand in period j; then the stock values

$$X_{j+1} = \begin{cases} X_j - d_i, & \text{if } d_j \leq X_{j-s} \\ S, & \text{otherwise} \end{cases}$$

define a Markov chain with state space $I = \{s, s+1, \ldots, S-1, S\}$, where it is assumed that $s \leq X_0 \leq S$. As a numerical example let $s = 2$, $S = 5$, and $\left\{p(d_j = 0) = \frac{2}{8}, p(d_j = 1) = \frac{1}{8}, P(d_j = 2) = \frac{2}{8}, \text{and } P(d_j = 3) = \frac{3}{8}\right\}$. Then the transition matrix is

$$P = \begin{vmatrix} \frac{2}{8} & 0 & 0 & \frac{6}{8} \\ \frac{1}{8} & \frac{2}{8} & 0 & \frac{5}{8} \\ \frac{2}{8} & \frac{1}{8} & \frac{2}{8} & \frac{3}{8} \\ \frac{3}{8} & \frac{2}{8} & \frac{1}{8} & \frac{2}{8} \end{vmatrix}.$$

(a) Find the stationary probabilities π_i, $i \in I$, analyically and by simulation the Markov chain, making a run of 1000 cycles.

(b) Describe a program to simulate the regenerative process $\{X(n) : n > 0\}$ including a flow diagram, a listing of the program, and the random number generator.

7 *M/M/1 Queue*. Run this queueing model for 2000 cycles. From the simulated data:

(a) Fill out a table similar to the Table 6.4.1, taking the same parameters, that is, $\lambda = 5$, $\mu = 10$, and the 90% confidence interval.

(b) Describe your random number generator, a flow diagram, and a listing of your program.

8 *Repairman model with spares*. Select the same parameters as in Section 6.4.2, that is assume $n = 10$, $m = 5$, $s = 4$, $\mu = 2$, $\lambda = 5$, and choose the 95% confidence level. Run the model for 500 cycles and, from the simulated data:

(a) Fill out a table similar to Table 6.4.2.

(b) Describe your random number generator, a flow diagram of your program, and a listing of your program.

REFERENCES

1 Bilingsley, P., *Convergence of Probability Measures*, Wiley, New York, 1968.

2 Carson, J. S., Variance reduction techniques for simulated queuing processes, Ph.D. thesis, Department of Industrial Engineering, University of Wisconsin, Madison, Wisconsin, 1978.

3 Carson, J. S. and A. M. Law, Conservation equations and variance reduction in queuing simulations. Technical Report 77-25, Department of Industrial Engineering, University of Wisconsin, Madison, Wisconsin, 1977.

4 Cox, D. R. and W. L. Smith, *Queues*, Methuen, London, 1961.

5 Činlar, E., *Introduction to Stochastic Processes*, Prentice-Hall, Englewood Cliffs, New Jersey, 1975.

6 Crane, M. A. and D. L. Iglehart, Simulating stable stochastic systems, I: General multi-server queues, *J. Assoc. Comp. Mach.*, **21**, 1974, 103–113.

7 Crane, M. A. and D. L. Iglehart, Simulating stable stochastic systems, II: Markov chains, *J. Assoc. Comp. Mach.*, **21**, 1974, 114–123.

8 Crane, M. A. and D. L. Iglehart, Simulating stable stochastic systems, III: Regenerative processes and discrete-event simulations, *Oper. Res.*, **23**, 1975, 33–45.

9 Crane, M. A. and D. L. Iglehart, Simulating stable stochastic systems, IV: Approximation techniques, *Manage. Sci.*, **21**, 1975, 1215–1224.

10 Crane, M. A. and A. J. Lemoine, *An Introduction to the Regenerative Method for Simulation Analysis*, Springer-Verlag, New York, 1977.

11 Esary, J. D., F. Proschan, and D. W. Walkup, Association of random variables with application, *Ann. Math. Stat.*, **38**, 1967, 1466–1474.

12 Fiacco, A. V. and G. P. McCormick, *Nonlinear Programming: Sequential Unconstrained Minimization Techniques*, Wiley, New York, 1968.

13 Fishman, G. S., *Concepts and Methods in Discrete Digital Simulation*, Wiley, New York, 1973.

14 Fishman, G. S., Statistical analysis for queueing simulations, *Manage. Sci.*, **20**, 1973, 363–369.

15 Fishman, G. S., Estimation in multiserver queueing simulations, *Oper. Res.*, **22**, 1974, 72–78.

16 Fishman, G. S., Achieving specific accuracy in simulation output analysis, *Comm. Assoc. Comp. Mach.*, **29**, 1977, 310–315.

17 Gass, S. I., *Linear Programming Methods and Applications*, 3rd ed., McGraw-Hill, New York, 1969.

18 Gaver, D. P. and G. S. Shedler, Control variable methods in the simulation of a model of a multiprogrammed computer system, *Nav. Res. Logist. Quart.*, **18**, 1971, 435–450.

19 Gaver, D. P. and G. L. Thompson, *Programming and Probability Models in Operations Research*, Brooks/Cole, Monterey, California, 1973.

20 Gunther, F. L., The almost regenerative method for stochastic system simulations, Technical Report ORC 75-21, Operations Research Center, University of California, Berkeley, California, 1975.

21 Heidelberger, P., Variance reduction techniques for the simulation of Markov processes, Ph.D. thesis, Department of Operations Research, Stanford University, Stanford, California, 1978.

22 Heidelberger, P. and M. Meketon, Bias reduction in regenerative simulation, Research Report RC 8397, IBM Corporation, Yorktown Heights, New York, 1980.

23 Heidelberger, P. and D. L. Iglehart, Comparing stochastic systems using regenerative simulation and common random numbers, *Adv. Appl. Prob.*, **11**, 1979, 804–819.

24 Heidelberger, P., Variance reduction techniques for the simulation of Markov processes, I: Multiple estimates, *IBM J. Res. Develop.* (to appear).

25 Heidelberger, P., Variance reduction techniques for the simulation of Markov processes, II: Matrix iterative methods, *Acta Inform.*, **13**, 1980, 21–37.

26 Heidelberger, P., A variance reduction technique that increases the regeneration frequency, in *Current Issues in Computer Simulation*, Academic, New York, 1979, pp. 257–269.

27 Hordijk, A., D. L. Iglehart, and R. Schassberger, Discrete time methods for simulating continuous time Markov chains, *Adv. Appl. Prob.*, **8**, 1976, 772–788.

28 Iglehart, D. L., Simulating stochastic systems, V: Comparison of ratio estimators, *Nav. Res. Logist. Quart.*, **22**, 1975, 553–565.

29 Iglehart, D. L., Simulating stable stochastic systems, VI: Quantile estimation, *J. Assoc. Comp. Mach.*, **23**, 1976, 347–360.

30 Iglehart, D. L., Simulating stable stochastic systems, VII: Selecting best system, in *Algorithmic Methods in Probability*, Vol. 7, edited by M. Neuts, North-Holland, Amsterdam, 1977, 37–50.

31 Iglehart, D. L., Regenerative simulation for extreme values, Technical Report 43, Department of Operations Research, Stanford University, Stanford, California, 1977.

32 Iglehart, D. L. and P. A. W. Lewis, Variance reduction for regenerative simulations, I: Internal control and stratified sampling for queues, Technical Report 86-22, Control Analysis Corporation, Palo Alto, California, 1976.

33 Iglehart, D. L. and G. S. Shedler, Regenerative simulation of response times in network of queues, *J. Assoc. Comp. Mach.*, **25**, 1978, 449–460.

34 Iglehart, D. L. and G. S. Shedler, Simulation of response times in finite-capacity open networks of queues, *Oper. Res.*, **26**, 896–914.

35 Iglehart, D. L. and G. S. Shedler, Regenerative simulation of response times in networks of queues, II: Multiple job types, Research Report RJ 2256, IBM Corporation, San Jose, California, 1978.

36 Iglehart, D. L., The regenerative method for simulation analysis, in *Current Trends in Programming Methodology*, Vol III, *Software Engineering*, edited by K. M. Chandy and R. T. Yeh, Prentice-Hall, Englewood Cliffs, New Jersey, 1978.

37 Iglehart, D. L. and G. S. Shedler, Regenerative simulation of response times in networks of queuees, Springer-Verlag, New York, 1980.

38 Iglehart, D. L., Regenerative simulation. Forthcoming.

39 Kabak, I. W., Stopping rules for queueing simulations, *Oper. Res.*, **16**, 1968, 431–437.

40 Karlin, S. and H. M. Taylor, *A First Course in Stochastic Processes*, 2nd ed., Academic, New York, 1975.

41 Kiefer, J. and J. Wolfowitz, On the theory of queues with many servers, *Trans. Amer. Math. Soc.*, **78**, 1955, 1–18.

42 Knuth, D. E., *The Art of Computer Programming*, Vol. 2, *Seminumerical Algorithms*, Addison-Wesley, Reading, Massachusetts, 1969.

43 Lavenberg, S. S., Efficient estimation of work rates in closed queueing networks, in *Proceedings in Computational Statistics*, Physica Verlag, Vienna, 1974, pp. 353–362.

44 Lavenberg, S. S., Regenerative simulation of queueing networks, Research Report RC 7087, IBM Corporation, Yorktown Heights, New York, 1978.

45 Lavenberg, S. S., T. L. Moeller, and C. H. Sauer, Concominant control variables applied to the regenerative simulation of queueing systems, *Oper. Res.*, **21**, 1979, 134–160.

46 Lavenberg, S. S., T. L. Moeller, and P. D. Welch, Control variables applied to the simulation of queueing models of computer systems, in *Computer Performance*, North Holland, Amsterdam, 1977, pp. 459–467.

47 Lavenberg, S. S., T. L. Moeller, and P. D. Welch, Statistical results on multiple control variables with application to variance reduction in queueing network simulation, Research Report RC 7423, IBM Corporation, 1978.

48 Lavenberg, S. S. and C. H. Sauer, Sequential stopping rules for the regenerative method of simulation, *IBM J. Res. Develop.*, **21**, 1977, 545–558.

49 Lavenberg, S. S. and G. S. Shedler, Derivation of confidence intervals for work rate estimators in a closed queueing network, *Soc. Indust. Appl. Math. J. Comp.*, **4**, 1975, 108–124.

50 Lavenberg, S. S. and D. R. Slutz, Introduction to regenerative simulation, *IBM J. Res. Develop.*, **19**, 1975, 458–462.

51 Lavenberg, S. S. and D. R. Slutz, Regenerative simulation of an automated tape library, *IBM J. Res. Develop.*, **19**, 1975, 463–475.

52 Law, A. M., Efficient estimators for simulated queueing systems, Technical Report ORC 74-7, Operations Research Center, University of California, Berkeley, California, 1974.

53 Law, A. M., Efficient estimators for simulated queueing systems, *Manage. Sci.*, **22**, 1975, 30–41.

54 Law, A. M., Confidence intervals in discrete event simulation: A comparison of replication and batch means, Technical Report 76-13, Department of Industrial Engineering, University of Wisconsin, Madison, Wisconsin, 1976.

55 Law, A. M. and J. S. Carson, A sequential procedure for determining the length of a steady-state simulation, Technical Report 77-12, Department of Industrial Engineering, University of Wisconsin, Madison, Wisconsin, 1977.

56 Law, A. M. and W. D. Kelton, Confidence intervals for steady-state simulations, II: A survey of sequential procedures, Technical Report 78-6, Department of Industrial Engineering, University of Wisconsin, Madison, Wisconsin, 1978.

57 Lewis, P. A. W., A. S. Goodman, and J. M. Miller, A pseudo-random number generator for the System/360, *IBM Syst. J.*, **8**, 1969, 199–200.

58 Mitchell, B., Variance reduction by antithetic variates in $GI/G/1$ queueing simulations, *Oper. Res.*, **21**, 1973, 988–997.

59 Poliak, D. G., Some methods of efficient simulation for queueing system, *Eng. Cybern.* (NY), **1**, 1970, 75–85.

60 Robinson, D. W., Determinants of run lengths in simulation of stable stochastic systems, Technical Report 86-21, Control Analysis Corporation, Palo Alto, California, 1976.

61 Rubinstein, Y. R., Selecting the best stable stochastic system, *Stochastic Processes Appl.*, **10**, 1980, 75–851.

62 Rubinstein, Y. R. and A. Karnovsky, The regenerative method for constrained optimization problems, in *OR'79*, edited by K. B. Haley, North-Holland, Amsterdam, 1979, 931–949.

63 Seila, A. F., Quantile estimation methods in discrete event simulations of stochastic systems, Technical Report 76-12, Curriculum in Operations Research and Systems Analysis, University of North Carolina, Chapel Hill, North Carolina, 1976.

64 Varga, R. S., *Matrix Iterative Analysis*, Prentice-Hall, Englewood Cliffs, New Jersey, 1962.

CHAPTER 7

Monte Carlo Optimization

Optimization is the science of selecting the best of many possible decisions in a complex real-life environment. The subject of this chapter is Monte Carlo optimization, a subject playing an important role in finding extrema—that is, minima or maxima of complicated nonconvex real-valued functions. We show how Monte Carlo methods can be successfully applied while solving complex optimization problems where the convex optimization methods (see Avriel [2]) fail. Before proceeding to the rest of the chapter, however, we explain what we mean by *local and global extrema* for unconstrained optimization.

Consider a real-valued function g with domain D in R^n. The function g is said to have a *local maximum* at point $x^* \in D$ if there exists a real number $\delta > 0$ such that $g(x) \leq g(x^*)$ for all $x \in D$ satisfying $\|x - x^*\| < \delta$. We define a *local minimum* in a similar way, but in the sense that inequality $g(x) \leq g(x^*)$ is reversed. If the inequality $g(x) \leq g(x^*)$ is replaced by a strict inequality

$$g(x) < g(x^*), \qquad x \in D, x \neq x^*,$$

we have a *strict* local maximum; and if the sense of the inequality $g(x) < g(x^*)$ is reversed, we have a strict local minimum. We say that the function g has a *global* (*absolute*) *maximum* (strict global maximum) at $x^* \in D$ if $g(x) \leq g(x^*)$, $[g(x) < g(x^*)]$ holds for every $x \in D$. A similar definition holds for a *global minimum* (strict global minimum). A global maximum at x^* implies that $g(x)$ takes on its greatest value $g(x^*)$ at that point no matter where else we may search in the set D. A local maximum, on the other hand, only guarantees that the value of $g(x)$ is a maximum with respect to other points nearby, specifically in a δ-region about x^*.

Thus a function may have many local maxima, each with a different value of $g(x)$, say, $g(x_j^0)$, $j=1,\ldots,k$. The global maximum can always be chosen from among these local maxima by comparing their values and choosing one such that

$$g(x^*) \geq g(x_j^0), \qquad j=1,\ldots,k,$$

where

$$x^* \in \{x_j^0, j=1,\ldots,k\}.$$

It is clear that every global maximum (minimum) is also a local maximum (minimum); however, the converse of this statement is, in general, not true. If $g(x)$ is a convex function in R^n and $D \subset R^n$ is a convex set then every local minimum of g at $x \in D$ also a global minimum of g over D [2].

7.1 RANDOM SEARCH ALGORITHMS

Consider the following *deterministic optimization problem*:

$$\max_{x \in D \subset R^n} g(x) = g(x^*) = g^*, \tag{7.1.1}$$

where $g(x)$ is a real-valued bounded function defined on a closed bounded domain $D \subset R^n$. It is assumed that g achieves its maximum value at a unique point x^*. The function $g(x)$ may have many local maxima in D but only one global maximum.

When $g(x)$ and D have some attractive properties, for instance, $g(x)$ is a differentiable concave function and D is a convex region, then, as previously mentioned, a local maximum is also a global maximum and problem (7.1.1) can be solved explicitly by mathematical programming methods (see Avriel [2]). If the problem cannot be solved explicitly, then numerical methods, in particular Monte Carlo methods, can be applied. For better understanding of the subsequent text we describe an iterative gradient algorithm, assuming for simplicity that the set $D = R^n$.

According to the gradient algorithm, we approximate the point x^* step by step. If on the ith iteration ($i=1,2,\ldots$) we have reached point x_i, then the next point x_{i+1} is chosen as

$$x_{i+1} = x_i + \alpha_i \nabla g(x_i), \qquad \alpha_i > 0 \tag{7.1.2}$$

where

$$\nabla g(x) = \left\{ \frac{\partial g(x_1)}{\partial x_1}, \ldots, \frac{\partial g(x_n)}{\partial x_n} \right\}$$

is the gradient of $g(x)$, where $\partial g(x_k)/\partial x_k$, $k = 1, \ldots, n$, are the partial derivatives, and where $\alpha_i > 0$ is the step parameter.

If the function $g(x)$ is not differentiable or if the analytic expression of $g(x)$ is not given explicitly (only the values of $g(x)$ can be observed at each point $x \in D$), then the finite difference gradient algorithm

$$x_{i+1} = x_i + \alpha_i \hat{\nabla} g(x_i) \tag{7.1.3}$$

can be applied. In (7.1.3)

$$\begin{aligned}\hat{\nabla} g(x) &= \left\{ \frac{\hat{\partial} g(x_1)}{\partial x_1}, \ldots, \frac{\hat{\partial} g(x_n)}{\partial x_n} \right\} \\ &= \left\{ \frac{g(x_1 + \beta_1, x_2, \ldots, x_n) - g(x_1 - \beta_1, x_2, \ldots, x_n)}{2\beta_1}, \ldots, \right. \\ &\quad \left. \times \frac{g(x_1, \ldots, x_n + \beta_n) - g(x_1, \ldots, x_n - \beta_n)}{2\beta_n} \right\}\end{aligned}$$

is the finite difference estimate of the gradient $\nabla g(x_i)$.

Under some rather mild conditions (see Avriel [2]) on $g(x)$ and α_i, the algorithm (7.1.3) converges to the local extremum x^*. In the case where either $g(x)$ or the region D is nonconvex, the classical numerical optimization methods fail. However, Monte Carlo methods, in particular random search algorithms, can be applied.

If we assume, for instance, that $g(x)$ is a multiextremal function, then procedures (7.1.2) and (7.1.3) converge only to one of the local extrema, subject to choice of the initial point x_0 from which the algorithms (7.1.2) and (7.1.3) start.

We consider several random search algorithms capable of finding the extremum x^* for complex nonconvex functions.

The random search algorithms have been described in many papers and books (see Ermolyev [9], Katkovnik [17], Rastrigin [28], and Rubinstein [31–36]), and successfully implemented for various complex optimization problems. We now consider several random search algorithms.

Random Search Double Trial Algorithm (Algorithm RS-1)

$$x_{i+1} = x_i + \frac{\alpha_i}{2\beta_i}\left[g(x_i + \beta_i \Xi_i) - g(x_i - \beta_i \Xi_i) \right] \Xi_i, \qquad \alpha_i > 0, \beta_i > 0. \tag{7.1.4}$$

According to this algorithm, at the ith iteration we generate a random vector Ξ_i continuously distributed on the n-dimensional unit sphere, calculate the increment (see Fig. 7.1.1)

$$\Delta g_{\pm}(\Xi_i) = g(x_i + \beta_i \Xi_i) - g(x_i - \beta_i \Xi_i), \tag{7.1.5}$$

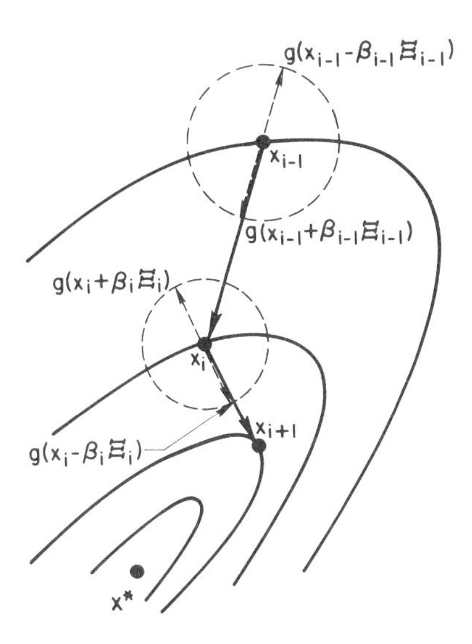

Fig. 7.1.1 Graphical representation of the double trials random search algorithm RS-1.

and choose the next point according to (7.1.4). It is not difficult to see that this algorithm generalizes the gradient algorithm (7.1.3). Only in the particular case where Ξ_i is taken in the direction of the gradient do procedures (7.1.3) and (7.1.4) coincide.

Nonlinear Tactic Random Search Algorithm (Algorithm RS-2)

$$x_{i+1} = x_i + \frac{\alpha_i}{\beta_i} Y_i \operatorname{Sign} Y_i \Xi_i, \qquad \alpha_i > 0, \beta_i > 0, \tag{7.1.6}$$

where

$$Y_i = g(x_i + \beta_i \Xi_i) - g(x_i) \tag{7.1.7}$$

$$\operatorname{Sign} Y_i \begin{cases} 1, & \text{if } Y_i > 0 \\ 0, & \text{if } Y_i \geq 0. \end{cases}$$

According to this algorithm, we perform a trial step in the random direction Ξ_i and check the Sign Y_i. If $Y_i > 0$, then $x_{i+1} = x_i + (\alpha_i/\beta_i) Y_i \Xi_i$. If $Y_i \leq 0$, then $x_{i+1} = x_i$ and no iteration is made.

Linear Tactic Random Search Algorithm (Algorithm RS-3)

This algorithm contains the following steps:

1 $i \leftarrow 0$, generate Ξ_0.

2 Calculate the increment

$$Y_i = g(x_i + \beta_i \Xi_i) - g(x_i).$$

3 If $Y_i < 0$, go to step 6.

4

$$x_{i+1} = x_i + \frac{\alpha_i}{\beta_i} Y_i \Xi_i, \qquad \alpha_i > 0, \beta_i > 0. \tag{7.1.8}$$

5 Go to step 7.
6 $x_{i+1} \leftarrow x_i$, $i \leftarrow i+1$; generate Ξ_i.
7 Go to step 2.

Thus if $Y_i > 0$, we perform as many iterations as possible in the initial chosen random direction $x_i + \beta_i \Xi_{i-1}$; if $Y_i \leq 0$, we generate a random vector Ξ_i and perform only one iteration according to the nonlinear tactic random search algorithm RS-2.

It is not difficult to see that search in the same direction versus choice of a new direction is subject to the shape of $g(x)$. The flatter the gradient lines, the more iterations will be performed according to step 4 and correspondingly the fewer iterations according to step 6. In the particular case where $g(x)$ is a linear function, all iterations will be performed according to step 4 in the direction of the vector $x_0 + \alpha + \alpha_0 \beta_0^{-1} Y_0 \Xi_0$, where Ξ_0 is the first random vector such that $Y_0 > 0$ and no iteration will be performed according to step 6. This is the reason why this algorithm is called a linear tactic random search algorithm.

Optimum Trial Random Search Algorithm (Algorithm RS-4)

This algorithm comprises the following steps:

1 Choose $N > 1$ independent random points $x_i + \beta_i \Xi_{ik}$ on the sphere $\{x_i + \beta_i \Xi_i\}$, where Ξ_i is a random vector continuously distributed on the unit sphere with realizations Ξ_{ik}, $k = 1, \ldots, N$.

2 Consider the sequence of increments

$$Y_{ik} = g(x_i + \beta_i \Xi_{ik}) - g(x_i), \qquad k = 1, \ldots, N. \tag{7.1.9}$$

3 Set

$$Y_{iN}^0 = \max(Y_{i1}, \ldots, Y_{iN}) \tag{7.1.10}$$

and let Ξ_{iN}^0 denote the direction that has produced this maximum.

4 The point x_{i+1} is chosen according to the following iterative procedure:

$$x_{i+1} = x_i + \alpha_i \beta_i^{-1} Y_{iN}^0 \Xi_{iN}^0, \qquad \alpha_i > 0, \beta_i > 0. \tag{7.1.11}$$

Thus the next point x_{i+1} is chosen in the direction Ξ_{iN}^0 of the greatest increase Y_{iN}^0 of the function $g(x)$, that is, the vector Ξ_{iN}^0 corresponds to the trial optimal among those available.

Statistical Gradient Random Search Algorithm (Algorithm RS-5)

This algorithm can be described as follows.

1 Choose $N > 1$ independent random points $x_i + \beta_i \Xi_{ik}$ on the sphere $\{x_i + \beta_i \Xi_i\}$, where Ξ_i is a random vector continuously distributed on the unit sphere with realizations Ξ_{ik}, $k = 1, \ldots, N$.

2 Calculate the sequence of increments

$$Y_{ik} = g(x_i + \beta_i \Xi_{ik}) - g(x_i), \qquad k = 1, \ldots, N. \tag{7.1.12}$$

3 Set

$$V_{i\Xi} = \frac{1}{N} \sum_{k=1}^{N} Y_{ik} \Xi_{ik}. \tag{7.1.13}$$

4 The point x_{i+1} is chosen according to

$$x_{i+1} = x_i + \alpha_i \beta_i^{-1} V_{i\Xi}, \qquad \alpha_i > 0, \beta_i > 0. \tag{7.1.14}$$

Thus given x_i, the next point x_{i+1} is chosen in the direction $V_{i\Xi}$, which is a result of averaging the sample $\Xi_{i1}, \ldots, \Xi_{iN}$ weighted with their corresponding increments Y_{ik} (7.1.12). In the particular case where $N = n$ and

$$\Xi_{ik} = e_{ki} = \underbrace{0, \ldots, 0}_{k}, 1, 0, \ldots, 0, \qquad k = 1, \ldots, n,$$

we obtain the following finite difference gradient algorithm:

$$x_{i+1} = x_i + \alpha_i \tilde{\nabla} g(x_i) \tag{7.1.15}$$

where

$$\tilde{\nabla} g(x) = \left\{ \frac{\tilde{\partial} g(x)}{\partial x_1}, \ldots, \frac{\tilde{\partial} g(x)}{\partial x_n} \right\}$$

$$= \left\{ \frac{g(x_1 + \beta_1, x_2, \ldots, x_n) - g(x)}{\beta_1}, \ldots, \frac{g(x_1, \ldots, x_n + \beta_n) - g(x)}{\beta_n} \right\}.$$

It is not difficult to prove that for a linear function the direction of $V_{i\Xi}$, on the average, coincides with that of the gradient of $g(x)$. This is the reason why the algorithm is called "statistical gradient algorithm."

Consider the following *stochastic optimization problem*.

$$\max_{x \in D \subset R^n} E[\phi(x, W)] = \max_{x \in D \subset R^n} g(x) = g(x^*) = g^*. \tag{7.1.16}$$

Here $\phi(x, W)$ is a function of two variables, x and W, x^* is the optimal point of $g(x)$, which is assumed to be unique, and W is an r.v. with unknown p.d.f. $f_W(w)$. We assume that at each point $x \in D$ only the individual realization of $\phi(x, W)$ can be observed.

It is clear that, if the p.d.f. $f_W(w)$ is unknown, problem (7.1.16) cannot be solved analytically. However, numerical methods can be applied.

One widely used numerical method for solving (7.1.16) is the *stochastic approximation method*. This method was originated by Robbins and Monro [30], who suggested a procedure for finding a root of a regression function measured with a noise. Kiefer and Wolfowitz [19] considered a procedure for finding x^* in the optimization problem (7.1.16) where $x \in R^1$. The procedures of Robbins-Monro and Kiefer-Wolfowitz were generalized by Dvoretzky [8]. Hundreds of papers and many books have been written in the past 15 years about stochastic approximation, their convergence, and their applications. The reader is referred to Wilde [44] and Wasan [43].

We consider the following algorithm:

$$x_{i+1} = x_i + \alpha_i \hat{\nabla} \phi(x_i, W_i), \tag{7.1.17}$$

where

$$\hat{\nabla}(\phi(x, W)) = \left\{ \frac{\hat{\partial}\phi(x_1, W_1)}{\partial x_1}, \ldots, \frac{\hat{\partial}\phi(x_n, W_n)}{\partial x_n} \right\}$$

$$= \left\{ \frac{\phi(x_1 + \beta_1, x_2, \ldots, x_n, W_{11}) - \phi(x_1 - \beta_1, x_2, \ldots, x_n, W_{12})}{2\beta_1}, \ldots, \right.$$

$$\left. \frac{\phi(x_1, x_2, \ldots, x_n + \beta_n, W_{n1}) - \phi(x_1, x_2, \ldots, x_n - \beta_n, W_{n2})}{2\beta_n} \right\}$$

is the estimate of the gradient $\hat{\nabla} g(x)$.

It is readily seen that in the absence of noise, that is, when $W = 0$, $\hat{\nabla}\phi(x, W) = \hat{\nabla} g(x)$ and (7.1.17) coincides with (7.1.3). In addition, if the realizations of the noise are independent and $E(W) = 0$, then $\hat{\nabla}\phi(x, W)$ is an unbiased estimator of $\hat{\nabla} g(x)$.

Proof of convergence of algorithm (7.1.17) to x^*, subject to some conditions on the sequences $\{\alpha_i\}_{i=1}^{\infty}$, $\{\beta_i\}_{i=1}^{\infty}$ and the function $\phi(x, W)$, can be found, for instance, in Dvoretzky [8], Gladyshev [13], and Wasan [43].

It is not difficult to understand that the random search algorithm can also be used for solving problem (7.1.16). For instance, by analogy with (7.1.17) the random search double trial algorithm (Algorithm RS-1) can be written as

$$x_{i+1} = x_i + \frac{\alpha_i}{2\beta_i}\left[\phi(x_i + \beta_i \Xi_i, W_{i1}) - \phi(x_i - \beta_i \Xi_i, W_{i2})\right]\Xi_i. \tag{7.1.18}$$

We can see that, for the same reasons as the random search algorithm (7.1.4) extends the gradient algorithm (7.1.3), the random search algorithm (7.1.18) extends the stochastic approximation algorithm (7.1.17).

Proof of convergence of (7.1.18) to x^* can be found in Rubinstein [31]. In analogy with (7.1.18) we can adopt any of the random search algorithms RS-2 through RS-5 for solving problem (7.1.16).

7.2 EFFICIENCY OF THE RANDOM SEARCH ALGORITHMS

The random search algorithms can be compared according to different criteria. Usually, they are compared according to their local and integral properties [28, 29].

Local properties are associated with a single iteration of the random search algorithm, integral properties-with many iterations. Comparing different algorithms according to integral properties we usually define:

1 The initial condition from which search starts.

2 A set of test functions (linear, quadratic, parabolic, multiextremal, etc.) for which the extremum is sought.

3 Some criteria that must be achieved during optimization. The following criteria can be used. Find an index k corresponding to the best algorithm among S algorithms available, such that:

(a)

$$\min_{s=1,\ldots,S} E(\|x_i^{(s)} - x^*\|) = E(|x_i^{(k)} - x^*|),$$

where the number of iteration i is given.

(b)

$$\max_{s=1,\ldots,S} \Pr(\|x_i^{(s)} - x^*\| \le \varepsilon_x) = P_r(\|x_i^{(k)} - x^*\| \le \varepsilon_x),$$

where i and ε_x are given.

(c)

$$\max_{s=1,\ldots,S} \Pr(|g(x_i^{(s)}) - g(x^*)| \le \varepsilon_g) = \Pr(|g(x_i^{(k)}) - g(x^*)| \le \varepsilon_g),$$

where i and ε_g are given.

(d)

$$\min_{s=1,\ldots,S} i^{(s)} = i^{(k)},$$

subject to

$$E(x_i^{(s)}) \in [R_x = \{x : \|x - x^*\| \le \varepsilon_x\}]$$

or

$$E(x_i^{(s)}) \in [R_g = \{x : \|g(x) - g(x^*)\| \le \varepsilon_g\}].$$

(e)

$$\min_{s=1,\ldots,S} i^{(s)} = i^{(k)},$$

subject to

$$E(x_i^{(s)}) \in [R_x = \{x : \|x - x^*\| < \varepsilon_x\}]$$

$$\operatorname{var} x_i^{(s)} \le c, \qquad c > 0.$$

It is readily seen that the first three problems are associated with finding the best algorithm when the number of iteration i is given; the last two involve finding the best algorithm that hits, at the minimum number of iterations, a given region R_x or R_q containing the extremum point x^*. In Section 7.3 we consider some local and integral properties of Algorithm RS-4.

Generally, the problem of comparison of different algorithms according to their integral properties is difficult to solve. Some attempts to overcome this difficulty have been made by Rastrigin [28]. Another interesting problem is how to find the optimal combination of algorithms, each of which is capable of finding the extremum of $g(x)$. This problem is solved in Rubinstein [33] and uses Bellman's principle of optimality.

Now we consider some local properties of the random search algorithms, assuming that some point x_i has been reached, and that we are allowed to make only a single step (iteration). Let $x_{i+1}^{(s)}$, $s=1,\ldots,S$, be the point (the state of the system) after this single iteration. Let us define the efficiency of the random search algorithms as

$$C_n = \frac{E\big(\tilde{\Delta} x_i^{(s)}\big)}{E\big(N_i^{(s)}\big)}, \tag{7.2.1}$$

where

$$\tilde{\Delta} x_i^{(s)} = \frac{\langle x_{i+1}^{(s)} - x_i, x_i - x^* \rangle}{\| x_i - x^* \|},$$

that is, where $\tilde{\Delta} x_i^{(s)}$ is the projection of the vector $x_{i+1}^{(s)} - x_i$ on the direction of the vector $x_i - x^*$, and $N_i^{(s)}$ is the number of observations (measurements) of $g(x)$ required for the algorithms in the ith step. For simplicity we consider only the case where $g(x)$ is approximately a linear function, which is the same as to assume that in Taylor expansion

$$g(x_{i+1}) = g(x_i + \Delta x_i) = g(x_i) + \langle \Delta x_i, \nabla g(x_i) \rangle + o(\Delta x_i). \tag{7.2.2}$$

Therefore at each iteration made by the random search algorithms, we approximate $g(x)$ linearly on the interval Δx_i. It is proven in [32] that, for a rather wide class of functions optimized by random search algorithms under the conditions

$$\sum_{i=1}^{\infty} \alpha_i = \infty, \qquad \sum_{i=1}^{\infty} \alpha_i^2 \beta_i^{-2} < \infty,$$

there exists a number I, sufficiently large and such that for $i \geq I$ a linear approximation of $g(x)$, that is, (7.2.2), is valid.

Substituting (7.2.2) in any of the four random search Algorithms RS-1, RS-2, RS-4, and RS-5 (see, respectively, (7.1.4), (7.1.6), (7.1.11), and

(7.1.14)), we readily obtain

$$x_{i+1}^{(s)} = x_i + \alpha_i^{(s)} \nabla g(x_i) \cos \varphi_i^{(s)} + o(\Delta x_i^{(s)}) \tag{7.2.3}$$

where

$$\cos \varphi_i^{(s)} = \frac{\langle \nabla g(x_i), \Xi_i^{(s)} \rangle}{\| \nabla g(x_i) \|} \tag{7.2.4}$$

and $s = 1, 2, 4, 5$ corresponds to RS-1, RS-2, RS-4, and RS-5. The distribution of $\varphi_i^{(s)}$ depends on the specific algorithm and on the distribution of the random vector $\Xi_i^{(s)}$. Let us assume without loss of generality that $\alpha^{(s)} = 1$. Then taking into account that for a linear function $g(x)$ the direction of the vector $x^* - x_i$ coincides with the direction of the gradient $\nabla g(x_i)$, we can express the efficiency C_n (see (7.2.1)) as

$$C_n = \frac{E(\cos \varphi_i^{(s)})}{E(N_i^{(s)})}. \tag{7.2.5}$$

We consider here only the efficiencies of the random search Algorithms RS-1, and RS-4, assuming that the vector Ξ is uniformly distributed on the surface of the unit n-dimensional sphere.

(a) The Double Trial Random Search Algorithm RS-1 It follows from (7.1.4), (7.2.3), and (7.2.4) that

$$\cos \varphi_i^{(1)} = \frac{\langle \nabla g(x_i), \Xi_i^{(1)} \rangle}{\| \nabla g(x_i) \|}, \qquad -\frac{\pi}{2} \leq \varphi_i^{(1)} \leq \frac{\pi}{2},$$

where $\varphi_i^{(1)}$ is a random angle between the vector $\Xi_i^{(1)}$ uniformly distributed on the n-dimensional sphere and the vector $\nabla g(x_i)$. We assume here that the direction of the gradient corresponds to $\varphi_i^{(1)} = 0$. Furthermore, it follows from (7.2.5) that the distribution of $\varphi_i^{(1)}$ does not depend on i; therefore the index i can be omitted. We also omit for convenience index (1) in $\varphi_i^{(1)}$. It is shown in the Appendix that φ has a p.d.f.*

$$h_n(\varphi) = B_n \sin^{n-2} \varphi, \qquad -\frac{\pi}{2} \leq \varphi \leq \frac{\pi}{2}, \tag{7.2.6}$$

where

$$B_n = \frac{\Gamma(n/2)}{\sqrt{\pi}\, \Gamma((n-1)/2)} \tag{7.2.7}$$

*We use for convenience $-\frac{\pi}{2} \leq \varphi \leq \frac{\pi}{2}$ rather than $0 \leq \varphi \leq \pi$ (see Appendix).

Since for Algorithm RS-1 we need two observations of $g(x)$ at points $g(x+\beta\Xi)$ and $g(x-\beta\Xi)$, respectively, the efficiency C_n (see (7.2.5)) is

$$C_n^{(1)} = \frac{E(\cos\varphi)}{2}. \tag{7.2.8}$$

The expected value and the variance of cos φ are, respectively,

$$E(\cos\varphi) = \int_{-\pi/2}^{\pi/2} \cos\varphi h_n(\varphi)\, d\varphi = 2B_n \int_0^{\pi/2} \cos\varphi \sin^{n-2}\varphi\, d\varphi = \frac{2B_n}{n-1} \tag{7.2.9}$$

$$\begin{aligned} \operatorname{var}(\cos\varphi) &= E(\cos^2\varphi) - [E(\cos\varphi)]^2 \\ &= 2B_n \int_0^{\pi/2} \cos^2\varphi \sin^{n-2}\varphi\, d\varphi - \left(\frac{2B_n}{n-1}\right)^2 = \frac{1}{n} - \left(\frac{2B_n}{n-1}\right)^2. \end{aligned} \tag{7.2.10}$$

Substituting (7.2.9) in (7.2.8), we obtain

$$C_n = \frac{B_n}{n-1} \tag{7.2.11}$$

and the following relationships can also be easily verified:

$$C_{n+1} = \frac{1}{2\pi n C_n}$$

$$C_{n+2} = \frac{n}{n+1} C_n.$$

Table 7.2.1 and Fig. 7.2.1 represent the efficiency C_n and $\operatorname{var}(\cos\varphi) = \sigma^2$ as

Table 7.2.1 The Efficiency and σ^2 as Functions of n for Algorithm RS-1

n	C_n	σ^2	$\frac{C_n}{\sigma}$
2	0.3184	0.5995	0.4112
3	0.25	0.416	0.3876
4	0.2125	0.314	0.3792
5	0.1875	0.26	0.3677
6	0.1702	0.221	0.3602
7	0.1556	0.1957	0.3518
8	0.1452	0.166	0.3564
9	0.1367	0.1401	0.3652
10	0.1294	0.1344	0.3529
11	0.123	0.2268	0.3538

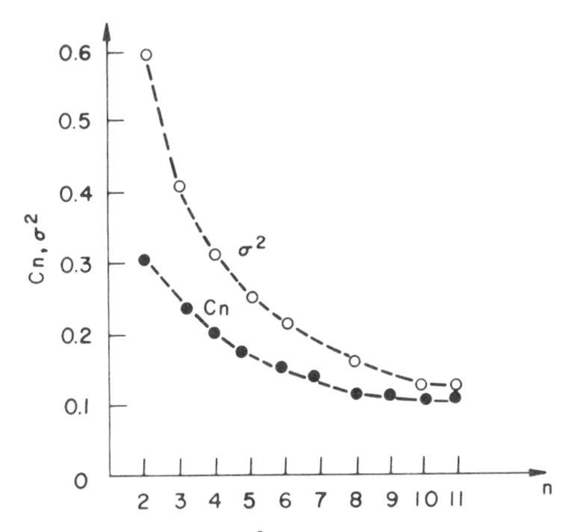

Fig. 7.2.1 The efficiency and σ^2 as functions of n for Algorithm RS-1.

a function of space size n, from which it follows that, as n increases, both the efficiency and the variance decrease. When $n \to \infty$, $E(\cos\varphi) \to 0$ and $C_n \to 0$, that is, the random search Algorithm RS-1 becomes inefficient.

(b) The Optimum Trials Random Search Algorithm RS-4 It follows from (7.1.11), (7.2.3), and (7.2.4) that

$$\cos\varphi_i^{(4)} = \frac{\langle \nabla g(x_i), \Xi_i^{(4)} \rangle}{\|\nabla g(x_i)\|}, \qquad 0 \le \varphi_i^{(4)} \le 2\pi,$$

where $\cos\varphi_i^{(4)} = \max(\cos\varphi_{i1}, \ldots, \cos\varphi_{iN})$. Since the distribution of $\varphi_i^{(4)}$ does not depend on the step number i, we can again omit the index i. We also omit for convenience index (4) in $\varphi_i^{(4)}$. To find the efficiency of Algorithm RS-4 let us find the distribution of $V = \cos\varphi$, where φ is distributed (compare with (7.2.6) and (7.2.7))

$$\hat{h}_n(\varphi) = \frac{B_n}{2}|\sin^{n-2}\varphi|, \qquad 0 \le \varphi \le 2\pi$$

and

$$B_n = \frac{\Gamma(n/2)}{\sqrt{\pi}\,\Gamma((n-1)/2)}.$$

By the transformation method (see Section 3.5.2) we obtain

$$P_n(v) = B_n(1-v^2)^{\frac{n-3}{2}}, \qquad -1 \le v \le 1. \tag{7.2.12}$$

The c.d.f. and p.d.f. of $V_N^0 = \max(V_1, \ldots, V_N)$ are, respectively,

$$F_n(v_N^0) = [F_n(v)]^N \tag{7.2.13}$$

and

$$P_n(v_N^0) = N[F_n(v)]^{N-1} P_n(v). \tag{7.2.14}$$

The expected value and the variance of V_N^0 are, respectively,

$$E(V_N^0) = \int_{-1}^{1} v_N^0 P_n(v_N^0)\, dv_N^0 \tag{7.2.15}$$

and

$$\operatorname{var}(V_N^0) = E\left[(V_N^0)^2\right] - \left[E(V_N^0)\right]^2. \tag{7.2.16}$$

For $n = 3$ we have

$$P_3(v_N^0) = \frac{N}{2^N}(1 + v)^{N-1} \tag{7.2.17}$$

$$E(V_N^0) = \frac{N-1}{N+1} \tag{7.2.18}$$

$$\operatorname{var}(V_N^0) = \frac{4N}{(N-1)(N+1)^2}. \tag{7.2.19}$$

It follows from (7.2.5) that the efficiency of Algorithm RS-4 is

$$C_n = \frac{E(V_N^0)}{N}. \tag{7.2.20}$$

For $n = 3$ we obtain

$$C_3 = \frac{(N-1)}{(N+1)N}. \tag{7.2.21}$$

The optimal value of C_3 equals $\frac{1}{6}$ and is achieved when N is equal to 2 or 3.
Generally, it is difficult to find C_n and $\operatorname{var}(V_N^0)$ for $n > 3$. Table 7.2.2 and Fig. 7.2.2 represent simulation results for C_n and $\operatorname{var}(V_N^0)$ as a function of n for the optimal number of trials N^* on the base of 100 runs. It is interesting to note that the optimal $N^* = 2$ and does not depend on n.

Comparing Algorithms RS-1 and RS-4 for a linear function, we conclude that RS-1 is more efficient than RS-4 for all $n > 1$. The variance associated with Algorithm RS-4 for the optimal $N^* = 2$ is always less than that associated with RS-1. The intuitive explanation for it can be given as follows. Taking two random trials according to Algorithm RS-1, we always

Table 7.2.2 The Efficiency and the var(V_N^0) as Functions of n for Algorithm RS-4

n	C_n	var(V_N^0)	N^*	$\frac{C_n}{\sigma}$
3	0.198	0.236	2	4075
4	0.159	0.171	2	3845
5	0.137	0.134	2	3743
6	0.121	0.110	2	3647
7	0.109	0.053	2	3575
9	0.092	0.070	2	3478
11	0.081	0.050	2	3622

Note: The sample size is equal to 100.

find a feasible random direction toward the extremum, which is generally not true for Algorithm RS-4. Indeed, the probability of finding such a direction (success) in N independent trials is equal $P(N) = 1 - (1-p)^N$. Here p is the probability of success in a single trial. Taking into account that for a linear function $p = \frac{1}{2}$, we obtain, for the optimal $N^* = 2$, $P(N^* = 2) = \frac{3}{4}$, that is, the probability of a success in Algorithm RS-4 is equal to $\frac{3}{4}$. Defining the efficiency as $C_n^{(s)}/\sigma^{(s)}$, where $\sigma^{(s)} = [\text{var}(\cos\varphi^{(s)})]^{1/2}$, we see from Tables 7.2.1 and 7.2.2 that both Algorithms RS-1 and RS-4 have approximately the same efficiency.

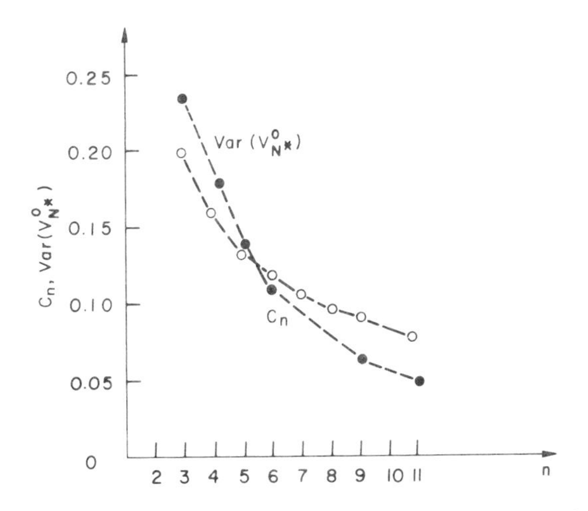

Fig. 7.2.2 The efficiency and the var($V_{N^*}^0$) as functions of n for Algorithm RS-4 (the sample size is equal to 100).

7.3 LOCAL AND INTEGRAL PROPERTIES OF THE OPTIMUM TRIAL RANDOM SEARCH ALGORITHM RS-4

This section is based on Ref. 35.

7.3.1 Local Properties of the Algorithm

The term "local properties" refers here to convergence of the vector $x_{i+1} - x_i$ to the direction of greatest increase of the function $g(x)$, as the number of trials m tends to infinity.

Assume that $g(x)$ is a continuous function and

$$\phi(x, W) = g(x) + W, \qquad x \in D \subset R^n, \tag{7.3.1}$$

that is, each measurement of the function $g(x)$ is accompanied by additive noise W, and assume that the vector Ξ is continuously distributed on the unit sphere with a density $f(\Xi)$. Let B be the set on the surface of the unit sphere defined by the condition $f(\Xi) > 0$ and let $\bar{B}$ be the closure of B. Let us also assume that the maximum

$$\max_{\Xi \in \bar{B}} g(x + \beta\Xi) = g(x + \beta\Xi^0), \qquad \Xi^0 \in \bar{B} \tag{7.3.2}$$

occurs at the unique point $x + \beta\Xi^0$.

We are concerned with the asymptotic behavior of the sequence of optimum-trial directions $\{\Xi_m^0\}_{m=1}^{\infty}$ defined by

$$\phi(x + \beta\Xi_m^0) = \max_{1 \le k \le m} \phi(x + \beta\Xi_k). \tag{7.3.3}$$

Theorem 7.3.1 Vector Ξ^0 is almost surely (a.s.) the only limiting vector of the sequence $\{\Xi_m^0\}_{m=1}^{\infty}$ if and only if the noise W satisfies the following property: For a.s. any sequence $\{W_k\}_{k=1}^{\infty}$ of W's realizations and for any $c > 0$, there exists a natural number K_c (which depends on the sequence) such that

$$W_k < \bar{W}_k + c, \qquad K_c \le k < \infty, \tag{7.3.4}$$

where

$$\bar{W}_k = \max_{1 \le j \le k-1} W_j. \tag{7.3.5}$$

Proof **(1)** *Suffiency* Let us prove that for every $\delta > 0$, the δ-neighborhood $S(\Xi^0, \delta)$ of the point Ξ^0 contains almost all optimum-trial directions Ξ_m^0, when m is sufficiently large. The proof is by contradiction. Assume that there exists $\delta > 0$ such that the following holds: There is a positive probability that a realization $\{\Xi_m\}_1^\infty$ contains a subsequence $\{\Xi_{m_k}\}_{k=1}^\infty$ such that

$$g(x+\beta\Xi_{m_k}) + W_{m_k} > g(x+\beta\Xi_j) + W_j, \qquad 1 \le j \le m_k - 1 \quad (7.3.6)$$

and at the same time

$$\Xi_{m_k} \notin S(\Xi^0, \delta).$$

Continuity of $g(x)$ implies that we can choose $\eta > 0$ and $\delta_1 < \delta$, such that

$$\inf_{\Xi \in \bar{B} \cap S(\Xi^0, \delta_1)} g(x+\beta\Xi) > \sup_{\Xi \in \bar{B}/S(\Xi^0, \delta)} g(x+\beta\Xi) + 2\eta. \quad (7.3.7)$$

(a) *The case of unbounded noise* Assume the sequence $\{W_{m_k}\}_{k=1}^\infty$ is unbounded and satisfies

$$W_{m_k} < \overline{W}_{m_k} + \eta, \qquad m_k > K_\eta \quad (7.3.8)$$

Denote by $\bar{m}_k$ the number of the trial in which the maximum $\overline{W}_{m_k}$ is achieved, that is, $W_{\bar{m}_k} = \overline{W}_{m_k}$ and $\bar{m}_k < m_k$ hold. The sequence of indices $\{\bar{m}_k\}_{k=1}^\infty$ is a.s. unbounded, because $\{W_{m_k}\}_{k=1}^\infty$ is a.s. unbounded. Therefore the event

$$\Xi_{\bar{m}_{k0}} \in \bar{B} \cap S(\Xi^0, \delta_1)$$

will a.s. occur for some $\bar{m}_{k_0} > K_\eta$, since at each trial there is a constant nonzero probability of its occurrence. Comparing the results obtained in trials $\bar{m}_{k_i}$ and m_{k_0}, it follows from (7.3.7) and (7.3.8) that

$$g(x+\beta\Xi_{\bar{m}_{k0}}) + W_{\bar{m}_{k0}} > g(x+\beta\Xi_{m_{k0}}) + W_{m_{k0}} + \eta,$$

which contradicts (7.3.6). Q.E.D.

(b) *The case of bounded noise.* If $\sup W = W_{\max} < \infty$, then the sequence $\{W_m\}_{m=1}^\infty$ a.s. contains an infinite subsequence $\{W_{m_i}\}_{i=1}^\infty$ such that

$$W_{\max} - \eta < W_{m_i} \le W_{\max}, \qquad i = 1, 2, \ldots$$

On the other hand, there exists a.s. a particular subscript m'_{i0} such that

$$\Xi_{m'_{i0}} \in \bar{B} \cap S(\Xi^0, \delta_1).$$

Thus for any $m > m'_{i0}$ satisfying $\Xi_m \notin S(\Xi^0, \delta)$,

$$g(x+\beta\Xi_{m_{i0}}) + W_{m_{i0}} > g(x+\beta\Xi_m) + W_{\max} + \eta$$
$$> g(x+g\Xi_m) + W_m,$$

which contradicts (7.3.6). Q.E.D.

(2) *Necessity* Assume that the set $\bar{C}$ of sequences $\{W_k\}_{k=1}^\infty$ not satisfying the theorem's condition has a probability $P(\bar{C}) > 0$ For each sequence from $\bar{C}$ there exists a number $c > 0$ and a subsequence $\{W_{k_j}\}_{j=1}^{\infty}$ such that

$$W_{k_j} \geq \overline{W}_{k_j} + c. \tag{7.3.9}$$

Our task now is to prove that with probability $P(\bar{C})$ the vector Ξ^0 is the only limiting vector of the sequence $\{\Xi_m^0\}_{m=1}^{\infty}$. What we actually prove is a somewhat stronger statement: namely, that the set of limiting vectors contains the set

$$V_c = \bar{B} \cap \{\Xi | g(x + \beta\Xi^0) - c < g(x + \beta\Xi) \leq g(x + \beta\Xi^0)\}. \tag{7.3.10}$$

To prove this statement it suffices to show that for any $y \in \bar{B}$ and any $\delta > 0$ the sequence $\{\Xi_m^0\}_{m=1}^{\infty}$ will visit the neighborhood $S(y, \delta)$ infinitely often. Indeed, for any trial there exists a constant positive probability of entering the set $S(y, \delta) \cap V_c$. This implies that the subsequence of trials $\{k_j\}$ satisfying (7.3.9) a.s. contains a new subsequence $\{k_{j_l}\}$ such that $\Xi_{k_{j_l}} \in S(y, \delta) \cap V_c$ holds. The vectors $\Xi_{k_{j_l}}$ will be optimum-trial directions, since for any i, $1 \leq i \leq k_{j_l} - 1$,

$$\begin{aligned} g(x + \beta\Xi_{k_{j_l}}) + W_{k_{j_l}} &> g(x + \beta\Xi_i) + W_{k_{j_l}} \\ &\geq g(x + \beta\Xi_i) + W_i. \end{aligned}$$

Q.E.D.

Remark In the case without noise ($W = 0$ a.s.) we can explicitly calculate the number of trials required to enter a prescribed δ-neighborhood $S(\Xi^0, \delta)$ of the point Ξ^0 with a prescribed probability p.

Define

$$\alpha = \left(\int_{\bar{B}} f(\Xi)\, d\Xi\right)^{-1} \int_{\bar{B} \cap S(\Xi^0, \delta)} f(\Xi)\, d\Xi,$$

that is α is the probability of visiting $S(\Xi^0, \delta)$ at each single trial. The probability of visiting $S(\Xi^0, \delta)$ at least once by making m trials is equal to

$$p_m = 1 - (1 - \alpha)^m. \tag{7.3.11}$$

Thus if we want $p_m \geq p$, it suffices to produce

$$m \geq \frac{\ln(1 - p)}{\ln(1 - \alpha)} \tag{7.3.12}$$

trials.

In the case where $p = 1 - \alpha$,

$$m \geq \frac{\ln \alpha}{\ln(1 - \alpha)}. \tag{7.3.13}$$

Table 7.3.1 shows some values of m as a function of α.

Table 7.3.1 Dependence of m on α

α	0.500	0.200	0.100	0.050	0.020	0.010	0.005	0.002	0.001
m	1	8	22	58	194	458	1057	3104	6903

7.3.2 Integral Properties of the Algorithm

The term "integral properties" refers to convergence of Algorithm RS-4 to the point of extremum x^*.

Theorem 7.3.2 Suppose that $g(x)$ has bounded second derivatives. Let

$$E(\|\xi_i\|^2 | x_0, x_1, \ldots, x_i) \leq h_i^2 < \infty \tag{7.3.14}$$

for

$$\|x_j\| \leq B < \infty, j = 0, 1, \ldots, i, \text{ where } \xi_i = \beta_i^{-1} Y_{iN}^0 \Xi_{iN}^0.$$

Let the normalizing factor γ_i satisfy the condition

$$0 < \gamma_i(\tau_i \|x_i\| + h_i) < \infty, \tag{7.3.15}$$

where

$$\tau_i = 1, \qquad \text{if } \|V_i \beta_i\| > 0$$

and

$$\tau_i = 0, \qquad \text{if } \|V_i \beta_i\| = 0$$

(V_i is defined in (7.3.18)), and let a_i and β_i be such that

$$a_i \geq 0, \qquad \beta_i \geq 0, \qquad \sum_{i=1}^{\infty} a_i \beta_i < \infty, \qquad \sum_{i=1}^{\infty} \beta_i^2 < \infty, \qquad \sum_{i=1}^{\infty} a_i = \infty; \tag{7.3.16}$$

then the optimal trial random search algorithm

$$x_{i+1} = \pi(x_i - a_i \gamma_i \xi_i) \tag{7.3.17}$$

converges a.s. to x^*. Here $\pi(\cdot)$ denotes the projection operator on D (i.e., for every $x \in R^n$, $\pi(x) \in D$ and $\|x - \pi(x)\| = \min_{y \in D} \|x - y\|$.

Proof Since $g(x)$ has bounded second derivatives, it is readily shown that

$$E(\xi_i | x_i) = C_i \nabla g(x_i) + \beta_i V_i, \tag{7.3.18}$$

where C_i and the vector V_i have bounded components, that is, $C_i < \infty$, $\|V_i\| < \infty$. Further, convergence of (7.3.17) to x^* follows from Ref. 10, Theorem 1.

7.4 MONTE CARLO METHOD FOR GLOBAL OPTIMIZATION

(a) Deterministic Optimization Problem The problem of finding the global extremum of $g(x)$ (see (7.1.1)) has been approached in a number of different ways. The earliest methods were associated with the grid technique and the function was evaluated at equispaced points throughout D. We shall consider only Evtushenko's algorithm [11] in such a deterministic sense. Some other deterministic approaches for global optimization are given in Dixon [7], Shubert [37], and Strongin [39]. Evtushenko makes the following assumptions about the function and the objective:

1 The function satisfies the Lipscitz condition, that is,

$$|g(x_1) - g(x_2)| < L\|x_1 - x_2\|,$$

for any $x_1, x_2 \in D$, $L > 0$.

2 Each $x \in D_\varepsilon$, where

$$D_\varepsilon = \{x : |g(x) - g(x^*)| < \varepsilon\},$$

is accepted as an approximation for x^*.

Evtushenko's algorithm is as follows.

Algorithm Gl – 1

1 Evaluate the function at N equispaced points $x_1, \ldots, x_N$ throughout D and define

$$y_k = g(x_k), \qquad k = 1, \ldots, N.$$

2 Estimate g^* by

$$M_N = \max(y_1, \ldots, y_N).$$

The theoretical background to this approach is very simple. Let V_i be the sphere $\|x - x_i\| \leq r_i$ where

$$r_i = L^{-1}(g(x_i) - M_N + \varepsilon).$$

Then for any $x \in V_i$

$$g(x) \geq g(x_i) - Lr_i = M_N - \varepsilon.$$

Hence if the sphere V_i, $i = 1, \ldots, N$, covers the whole set D, then M_N cannot differ from g^* by more than ε, and the problem is solved.

In the simplest case where D is an interval, $a \leq x \leq b$, Evtushenko proposed the following procedure:

$$x_1 = a + \frac{\varepsilon}{L}, \qquad M_1 = g(x_1)$$

$$x_{k+1} = x_K + \frac{2\varepsilon + g(x_k) - M_k}{L}$$

$$M_N = \max(g(x_N), M_{N-1}).$$

The number of function evaluations required to solve the problem is greatest in the case of a monotonically increasing function, namely

$$N = \frac{L(b-a)}{2\varepsilon}.$$

Most algorithms for global optimization contain random elements and are related to the Monte Carlo method. We consider some such algorithms.

Brooks [4] suggested, for solving problem (7.1.1), the following "pure" random search algorithm.

Algorithm Gl-2

1 Generate $X_1, \ldots, X_N$ from any p.d.f. $f_X(x)$ such that $f_X(x) > 0$, when $X \in D$.

2 Find $Y_k = g(X_k)$, $k = 1, \ldots, N$.

3 Estimate g^* by

$$M_N = \max(Y_1, \ldots, Y_N).$$

This algorithm was also discussed in Ref. 36. Our nomenclature follows that reference, and our discussion is based on it.

Let p be the probability measure defined on B, the Borel σ-field of D, so that (D, B, P) is a probability space.

Let $g^{-1}(a,b) = \{x \in D : a < g(x) \leq b\}$, and let $F(y) = P\{Y_i \leq y\}$; then

$$F(y) = P\{g(X_i) \leq y\} = P\{g^{-1}(-\infty, y)\}$$

and $Y_1, \ldots, Y_N$ are independent identically distributed (i.i.d.) random variables (r.v.'s) on R^1 with a cumulative probability distribution function (c.d.f.) F.

Proposition 7.4.1 **Suppose P assigns a positive probability to every neighborhood of x^*, and suppose g is continuous at x^*, then**

$$\lim_{N\to\infty} M_N = g^* \quad \text{a.s.} \tag{7.4.1}$$

Proof It is clear that $F(g^*) = 1$ and for each $\delta > 0$ we have $1 - F(g^* - \delta) = P\{g^* - \delta < g(X_i) \leq g^*\} > 0$ by our assumption. Let $A_N(\delta)$ be the event $\{M_N \leq g^* - \delta\}$; then $P\{A_N(\delta)\} = F^N(g^* - \delta)$ and $\sum_{N=1}^{\infty} P\{A_N(\delta)\} = F(g^* - \delta)/(1 - F(g^* - \delta)) < \infty$. By the Borel-Cantelli lemma $P\{M_N \leq g^* - \delta \text{ infinitely often}\} = 0$ for all $\delta > 0$ and thus (7.4.1) follows. Q.E.D.

The choice of P, and consequently the resulting F, depends on our prior knowledge of x^*. If it is known that a certain region is more likely to include x^*, then it would be more efficient to assign a higher probability to that region. If nothing is known a priori about x^*, a uniform distribution over D can be assumed.

In guaranteeing (7.4.1) the exact choice of P is immaterial. However, the rate of convergence is determined by the properties of F. For example, by a theorem of Gnedenko [14], if there exists a constant $\alpha > 0$ such that

$$\lim_{\delta \downarrow 0} \frac{1 - F(g^* - c\delta)}{1 - F(g^* - \delta)} = c^{\alpha}, \qquad \forall c > 0 \tag{7.4.2}$$

then

$$\lim_{N \uparrow \infty} P\left\{ \frac{M_N - g^*}{a_N} \leq x \right\} = \exp\{-|x|^{\alpha}\}, \qquad x \leq 0, \tag{7.4.3}$$

with a_N determined by $F(g^* - a_N) = (N-1)/N$. Some more properties of M_N are listed below.

1 Geometric distribution Let N_δ be the first N for which $M_N > g^* - \delta$. Then N_δ is a *geometric* r.v., that is,

$$P\{N_\delta = k\} = F^{k-1}(g^* - \delta)[1 - F(g^* - \delta)], \qquad k = 1, 2, \ldots . \tag{7.4.4}$$

Consequently, it is well known that

$$EN_\delta = \frac{1}{1 - F(g^* - \delta)} \equiv \eta_\delta$$

and

$$P\{N_\delta \leq k\} = 1 - F^k(g^* - \delta) \equiv P_{\delta, k}.$$

It is clear that $\eta_\delta \to \infty$ as $\delta \downarrow 0$ and thus $P_{\delta, [\eta_\delta]} = 1 - (1 - 1/\eta_\delta)^{[\eta_\delta]} \to 1 - e^{-1} = 0.63$ (here $[\eta]$ is the integer part of η). Hence $\eta_\delta = EN_\delta$ is approximately a 63% confidence bound for N_δ, the number of trials necessary to make $M_N > g^* - \delta$ ($\delta > 0$ small). Let $\alpha = 1 - F(g^* - \delta)$; then $P_{\delta,k} = 1 - (1-\alpha)^k$. For every given pair (α, β) the smallest k for which $P_{\delta,k} \geq \beta$ is $k(\alpha, \beta) = \ln(1-\beta)/\ln(1-\alpha)$, and Table 7.3.1 with $k(\alpha, 1-\alpha) \equiv m$ can be used again.

2 Lack of memory It is well known that (7.4.4) implies

$$P\{N_\delta > k + m \mid N_\delta > m\} = P\{N_\delta > k\}. \tag{7.4.5}$$

In terms of N_M we thus have

$$P\{M_{k+m} \leq g^* - \delta \mid M_m \leq g^* - \delta\} = P\{M_k \leq g^* - \delta\},$$

because the events $\{N_\delta > k\}$ and $\{M_k \leq g^* - \delta\}$ are identical. It follows that, given m successive failures (to enter $\{y : y > g^* - \delta\}$), the conditional distribution of the number of trials necessary for the first success equals its unconditional distribution. In particular we have

$$E(N_\delta | M_m \leq g^* - \delta) = m + EN_\delta. \tag{7.4.6}$$

3 Poisson approximation If (7.4.2) or (7.4.3) hold, then $Z_{\delta, N}$, the number of Y_i, $i = 1, 2, \ldots, N$, for which $Y_i > g^* - \delta$, is asymptotically Poisson distributed. More precisely, for fixed N and $\delta > 0$, $Z_{\delta, N}$ is a binomial r.v. with parameters N and $p = 1 - F(g^* - \delta)$. When (7.4.2) holds by substituting $\delta = a_N$ in (7.4.2), we obtain $N[1 - F(g^* - \delta a_N)] \to c^\alpha$, which implies that $Z_{ca_N, N}$ converges in distribution to a Poisson r.v. with parameter c^α.

The problem of finding the global maximum of $g(x)$ can be reduced to that of finding the mode for association with $g(x)$ density function. Indeed, if $g(x) \geq 0$, $x \in D$, then $\psi(x) = c^{-1}g(x)$ where $c^{-1} = (\int g(x)\,dx)^{-1}$ is a density function, and the problems of finding the global maximum of $g(x)$ and finding the mode of $\psi(x)$ are equivalent. This can be solved by one of the methods mentioned in Refs. 41, 42, and 46.

If $g(x)$ is unrestricted in sign but bounded, that is, if $|g(x)| \leq k$, then $f_X(x) = c^{-1}(g(x) + k)$, where $c^{-1} = [\int (g(x) + k)\,dx]^{-1}$ is again a density function.

A natural extension of the "pure" random search algorithm Gl-2 is the so-called multistart algorithm [7], which is probably the one most frequently used in practice for global optimization. In this approach we use any iterative procedure (gradient, random search, etc.) for local optimization and run it from a number of different starting points x_{0j}, $j = 1, \ldots, N$. The set of all terminating points hopefully includes the global maximum x^*.

The multistart algorithm is as follows.

Algorithm Gl-3

1 Generate $X_{01}, \ldots, X_{0N}$ from any p.d.f. $f_{X_0}(x) > 0$, $x \in D$ (usually X_0 is chosen to be uniformly distributed over D).

2 Consider $X_{01}, \ldots, X_{0N}$ as the starting points, then apply N times a local optimization algorithm (gradient, random search, etc.) and find the local extrema $x_1^*, \ldots, x_N^*$ of $g(x)$ associated with $X_{01}, \ldots, X_{0N}$.

3. Estimate x^* by

$$\max(x_1^*, \ldots, x_N^*).$$

Let us define D_j as the set of starting points X_{0j} from which the algorithm will converge to j-th local maximum. We call D_j the *region of attraction* of the jth local maximum. Let us assume that the number of local maxima is finite, and let X_0 be uniformly distributed over D; then the probability of at least one X_{0j}, from a sequence of N points drawn at random over D, falling in the region of attraction of the global maximum D_j^*, equals

$$p = 1 - \left[1 - \frac{m(D_j^*)}{m(D)}\right]^N, \tag{7.4.7}$$

where $m(D)$ is the measure of D.

A more sophisticated approach to the global optimization problem was suggested by Chichinadze [5], who introduced a probability function $P(v)$ as the probability of $g(x) < v$, that is, if $m(V)$ is the measure of the level set

$$V = \{x : g(x) < v\},$$

then

$$P(v) = \frac{m(V)}{m(D)}. \tag{7.4.8}$$

The function $P(v)$ is, of course, not available, but if we calculate $g(x)$ at N points distributed at random over D, and count the number M of these points for which $g(x) < v$, then M/N approximates $P(v)$. It is not difficult to see that the global maximum corresponds to $P(v) = 1$ and the global minimum to $P(v) = 0$. To find the solution $P(v) = 1$, Chichinadze suggested approximating $P(v)$ by a linear combination of a set of given polynomial functions $P_i(v)$, $i = 1, \ldots, k$,

$$P(v) = \sum_{i=1}^{k} \lambda_i P_i(v). \tag{7.4.9}$$

The range of v was divided at the points v_j, $j = 1, \ldots, s$, and the optimal values of λ_i were determined by minimizing

$$\sum_{j=1}^{s} W_j \left(\frac{M_j}{N} - \sum_{i=1}^{k} \lambda_i P_i(v_j) \right)^2, \tag{7.4.10}$$

where M_j is the number of points for which $g(x) < v_j$, and $W_j > 0$, $j = 1, \ldots, s$. The root v^* of $P(v) = 1$ was then determined to obtain an estimate of the global maximum of $g(x)$.

Considerable attention has been paid in the multiextremal optimization to the random search algorithms. Gaviano [12] showed that if

$$x_{i+1} = x_i + \alpha_i \Xi_i \tag{7.4.11}$$

and

$$\alpha_i = \arg(\text{global max } g(x_i + \alpha_i \Xi_i)), \tag{7.4.12}$$

then

$$\lim_{i\to\infty} P\{g(x_i) - g(x^*) < \varepsilon\} = 1 \tag{7.4.13}$$

for every $\varepsilon > 0$. Here Ξ is a vector uniformly distributed on the surface of a unit n-dimensional sphere.

If D is a finite space and if a bound on the first derivative of $g(x)$ is known, then Evtushenko's [11] or Shubert's [37] one-dimensional global optimization techniques could be used to find the optimal α_i. However, for a general function, a global optimization along the lines of (7.4.12) is difficult to perform.

Matyas [22] proved the convergence to x^* of the following random search algorithm.

Algorithm Gl-4

1 Generate $Y_1, Y_2, \ldots,$ from an n-dimensional normal distribution with zero mean and covariance matrix Σ, that is $Y \sim N(0, \Sigma)$.

2 Select an initial point $x_1 \in D$.

3 Compute $g(x_1)$.

4 $i \leftarrow 1$.

5 If $x_1 + Y_i \in D$, go to step 8.

6 $x_i \leftarrow x_{i+1}$.

7 Go to step 10.

8 Compute $g(x_i + Y_i)$.

9

$$x_{i+1} = \begin{cases} x_i + Y_i, & \text{if } g(x_i + Y_i) \geq g(x_i) - \varepsilon, \text{ where } \varepsilon > 0 \\ x_i, & \text{otherwise.} \end{cases}$$

10 $i \leftarrow i + 1$.

11 Go to step 5.

According to this algorithm, a step is made from the point x_i in the direction Y_i only if $x_i + Y_i \in D$ and $g(x_i + Y_i) \geq g(x_i) - \varepsilon$.

The following procedure, based on cluster analysis, was introduced into global optimization by Becker and Lago [3].

Algorithm Gl-5

1 Select N points uniformly distributed in D.

2 Take $N_1 < N$ of these points with the greatest function values.

3 Apply a cluster analysis to these N_1 points, grouping them into discrete clusters; then find the boundaries of each cluster and define a new domain $D_1 \subset D$, which hopefully contains the global maximum.

4 Replace D by D_1 and perform steps 1 through 3 several times.

This is a heuristic algorithm and its ability to find the global maximum depends on the cluster analysis technique used in step 3 and on the parameters N and N_1. There exists a positive probability of missing the global maximum. However, in practice this technique is widely used for global optimization. More on cluster analysis for global optimization can be found in Gomulka [15], Price [27], and Törn [40].

(b) Stochastic Optimization Problem Consider the stochastic optimization problem (7.1.16), assuming that

$$g(x, W) = g(x) + W, \tag{7.4.14}$$

which means that $g(x)$ is measured with some error W. The following Monte Carlo algorithm, which is similar to Algorithm Gl-2, can be used for estimating g^* in (7.1.16).

Algorithm Gl-2′

1 Generate $X_1, \ldots, X_N$ from any probability distribution function (p.d.f.) $f_X(x)$, $(f_X(x) > 0, x \in D)$.

2 Find $Y_k = g(X_k, W_k) = g(X_k) + W_k$, $k = 1, \ldots, N$.

3 Estimate g^* by

$$M_N = \max(Y_1, \ldots, Y_N).$$

Let W_k be i.i.d. r.v.'s with a given c.d.f. H. We also assume that the W_k and the X_k are independent and that $W_* = \inf\{u : H(u) = 1\} \leq \infty$. The following proposition is proven in Ref. 36.

Proposition 7.4.2. Under the conditions of Proposition 7.4.1

$$\lim_{N\to\infty} M_N = g^* + W_* \quad \text{a.s.} \tag{7.4.15}$$

Proof: Let

$$E_N = \max_{1 \leq i \leq N} W_i.$$

We say that $\{E_N\}$ is stable if there exists a sequence of constants $\{\eta_N\}$ such that for all $\delta > 0$

$$\lim_{N\to\infty} P\{|E_N - \eta_N| > \delta\} = 0. \tag{7.4.16}$$

We consider three cases.

1 $W_* < \infty$, in which case our estimate for g^* is $M_N - W_*$, and we certainly have

$$\lim_{N\to\infty} (M_N - W_*) = g^* \quad \text{a.s.} \tag{7.4.17}$$

2 $W_* = \infty$, but $\{E_N\}$ is stable, in which case (7.4.16) implies

$$\lim_{N\to\infty} (M_N - \eta_N) = g^* \text{ in probability,} \tag{7.4.18}$$

and η_N is determined by $H(\eta_N) = (N-1)/N$. A necessary and sufficient condition for case 2 is [14]

$$\lim_{u\to\infty} \frac{1 - H(u+\delta)}{1 - H(u)} = 0, \qquad \forall\delta > 0. \tag{7.4.19}$$

We thus see that, if W_* and η_N are known, we still have convergent algorithms in (7.4.17) and (7.4.18).

3 $W_* = \infty$, but $\{E_N\}$ is not stable. Here we have by (7.4.15) $M_N \to \infty$ a.s. Q.E.D.

The following examples will demonstrate these ideas.

1 If the W_i are normally distributed with mean 0 and variance σ^2, then (7.4.19) holds and $\{E_N\}$ is stable with $\eta_N = \sigma(2 \log N)^{1/2}$.

2 Suppose that the W_i's have the generalized double exponential distribution, that is,

$$h(x) = \begin{cases} \frac{1}{2} e^{-|x|^\alpha}, & x \le 0 \\ 1 - \frac{1}{2} e^{-x^\alpha}, & x \ge 0. \end{cases}$$

Then by (7.4.19) $\{E_N\}$ is not stable for $\alpha \le 1$, but is stable for $\alpha > 1$ with $\eta_N = (\log(N/2))^{1/\alpha}$.

Algorithm Gl-3 can be also adapted for the stochastic optimization problem (7.1.16), rewriting step 2 as follows:

2 Consider $X_{01}, \ldots, X_{0N}$ as the starting points; then apply N times a local iterative procedure (stochastic approximation, random search, etc.) that is able to find the association local extrema $x_1^*, \ldots, x_N^*$ of $E[g(x, W)] = g(x)$.

(c) Constrained Optimization Consider the following constrained optimization problem:

$$\max_{x \subset D \in R^n} g_0(x), \tag{7.4.20}$$

subject to

$$g_k(x) \le 0, \qquad k = 1, \ldots, m. \tag{7.4.21}$$

We assume that the convex programming methods (see Avriel [2]) cannot be applied because the convexity assumptions do not hold either for the region $D = \{x : g_k(x) \le 0, k = 1, \ldots, m\}$ or for the function $g_0(x)$.

Let us consider two cases.

1 If the region $D=\{x: g_k(x)\le 0, k=1,\ldots,m\}$ is known, and we can readily generate r.v.'s at D, then Algorithms Gl-2 through Gl-5 can be directly applied for finding the global extremum of (7.4.20) and (7.4.21).

2 If the region $D=\{x: g_k(x)\le 0,\ k=1,\ldots,m\}$ is either unknown explicitly or is complex, but another region D_1 that contains D and has a simple shape is known, then we generate r.v.'s at D_1 and accept or reject them according to whether $X\in D$ or $X\in(D_1-D)$. Next we can apply again Algorithms Gl-2 through Gl-5.

7.5 A CLOSED FORM SOLUTION FOR GLOBAL OPTIMIZATION

This section is based on the results of Meerkov [23] and Pincus [25]. Both papers deal with the multiextremal optimization and use the classical Laplace formula for certain integrals. We follow Pincus [25].

Consider the optimization problem

$$\min_{x\in D\subset R^n} g(x)=g(x^*)=g^*,$$

where $g(x)$ is a continuous function, D is a closed bounded domain, and x^* is the unique optimum point. Pincus [25] proved the following theorem.

Theorem 7.5.1. Let $g(x)=g(x_1,\ldots,x_n)$ be a real-valued continuous function over a closed bounded domain $D\in R^n$. Further, assume there is a unique point $x^*\in D$ at which $\min_{x\in D} g(x)$ is attained (there are no restrictions on relative minima). Then the coordinates x_i^* of the minimization point are given by

$$x_i^* = \lim_{\lambda\to\infty} \frac{\int_D x_i \exp(-\lambda g(x))\,dx}{\int_D \exp(-\lambda g(x))\,dx}, \qquad i=1,\ldots,n. \tag{7.5.1}$$

In particular the theorem is valid when D is convex and the objective function g is strictly convex. The proof of the theorem is based on the Laplace formula, which for sufficiently large λ can be written as

$$\int_D x_i \exp(-\lambda g(x))\,dx \approx x_i^* \exp(-\lambda g(x^*)) \tag{7.5.2}$$

$$\int_D \exp(-\lambda g(x))\,dx \approx \exp(-\lambda g(x^*)). \tag{7.5.3}$$

We now outline a Monte Carlo method based on Metropolis et al. work [24] (see also [26]) for evaluating the coordinates of the minimization point

$x^* = (x_1^*, \ldots, x_n^*)$, that is, for approximating the ratio appearing on the right-hand side of (7.5.1). For fixed λ (7.5.1) can be written as

$$\frac{\int_D x_i \exp(-\lambda g(x))\,dx}{\int_D \exp(-\lambda g(x))\,dx}. \tag{7.5.4}$$

For large λ the major contribution to the integrals appearing in (7.5.1) comes from a small neighborhood of the minimizing point x^*. Metropolis' sampling procedure [24], described below, is based on simulating a Markov chain that spends, in the long run, most of the time visiting states near the minimizing point and is more efficient than a direct Monte Carlo, which estimates both the numerator and the denominator separately.

The idea of the method is to generate samples with density

$$f_X(x) = \frac{\exp(-\lambda g(x))}{\int_D \exp(-\lambda g(x))\,dx}, \qquad x \in D, \tag{7.5.5}$$

where the denominator of (7.5.5) is not known. This is done as follows.

Partition the region D into a finite number N of mutually disjoint subregions D_j and replace integrals over D by corresponding Riemann sums using the partition $\{D_j\}$. Fix a point $y^j = (y_1^j, \ldots, y_n^j) \in D_j$. Then construct an irreducible ergodic Markov chain $\{X_k\}$ with state space $\{y^1, \ldots, y^N\}$ and with transition probabilities p_{ij}, $1 \le i, j \le N$, satisfying $\pi_j = \Sigma_i \pi_i p_{ij}$, $j = 1, \ldots, N$, where $\pi_j = \exp 1[-\lambda g(y^j)]/\Sigma_{h=1}^N \exp 1[-\lambda g(y^h)]$; that is, $\{\pi_j\}$ is the invariant distribution for the Markov chain. It should be noted that, in the last expression for π, we have assumed for simplicity that all subregions D_j have equal volumes. Then using the strong law of large numbers for Markov chains, we have with probability 1

$$\frac{1}{m}\sum_{k=1}^m X_k \underset{m\to\infty}{\rightarrow} \frac{\left\{\sum_{j=1}^N y_1^j \exp\left[-\lambda g(y^j)\right], \ldots, \sum_{j=1}^N y_n^j \exp\left[-\lambda g(y^j)\right]\right\}}{\sum_{j=1}^N \exp\left[-\lambda g(y^j)\right]}$$

$$\approx \frac{\left\{\int_D x_1 \exp(-\lambda g(x))\,dx, \ldots, \int_D x_n \exp(-\lambda g(x))\,dx\right.}{\int_D \exp(-\lambda g(x))\,dx} \approx x^*. \tag{7.5.6}$$

The sampling error for each component X_k^i of the vector X_k is (see [26]) $E[m^{-1}(\sum_{k=1}^{m} X_k^i - \mu_i)^2] \le c/m$, where c is a positive number,

$$\mu_i = \frac{\sum_{j=1}^{j=N} y_i^j \exp\left[-\lambda F(y^j)\right]}{\sum_{j=1}^{j=N} \exp\left[-\lambda F(y^j)\right]}.$$

From Chebyshev's inequality we have

$$P\left[\left|m^{-1}\sum_{k=1}^{k=m} X_k^i - \mu_i\right| \ge \varepsilon\right] \le \frac{c}{\varepsilon^2 m}.$$

We now turn to the question of how Metropolis constructs a Markov chain with the required invariant distribution. He starts with a symmetric transition probability matrix $P^* = (p_{ij}^*)$, $1 \le i, j \le N$, that is, $p_{ij}^* = p_{ji}^*$, $p_{ij}^* > 0$, $\sum_{j=1}^{j=n} p_{ij}^* = 1$, the known ratios π_i/π_j, and defines the transition matrix of the Markov chain $\{X_k\}$ as follows:

$$p_{ij} = \begin{cases} \dfrac{p_{ij}^* \pi_j}{\pi_i}, & \text{if } \dfrac{\pi_j}{\pi_i} < 1,\ i \ne j, \\ p_{ij}^*, & \text{if } \dfrac{\pi_j}{\pi_i} \ge 1,\ i \ne j, \\ p_{ii}^* + \sum_{l:\pi_l < \pi_i} & p_{ij}\left(1 - \dfrac{\pi_l}{\pi_i}\right),\ i = j. \end{cases} \tag{7.5.7}$$

It is shown in Ref. 16 that a Markov chain with the above transition matrix has the invariant distribution $\{\pi_i\}$, that is, $\pi_i = \sum_j p_{ij}\pi_j$. A chain with such a transition matrix can be realized as follows. Given that the chain is in state y^i at time k, that is, $\{X_k = y^i\}$, the state at time $k+1$ is determined by choosing a new state according to the distribution $\{p_{ij}^*, j = 1, \dots, N\}$. If the state chosen is y^j, we calculate the ratio π_j/π_i. If $\pi_j/\pi_i \ge 1$, we accept y^j as the new state at time $k+1$; if $\pi_j/\pi_i < 1$, we take y^j as the state of the Markov chain at time $k+1$ with probability π_j/π_i and y^i as the new state at time $k+1$ with probability $1 - \pi_j/\pi_i$. It is also shown in Ref. 16 that this procedure leads to a Markov chain with transition matrix $P = (p_{ij})$.

It should be noted that (7.5.1) can be useful not only for finding the global optimum in a multiextremal problem, but also for solving nonlinear equations (see [20]) and some kinds of problems in statistical mechanics as well (see [16]).

7.6 OPTIMIZATION BY SMOOTHED FUNCTIONALS

Consider the following stochastic optimization problem (see (7.1.16))

$$\min_{x \in D \subset R^n} E_W[\phi(x, W)] = \min_{x \in D \subset R^n} g(x) = g(x^*) \quad (7.1.16')$$

where $\phi(x, W)$ is a stochastic function with unknown p.d.f. $p(x)$, D is a convex bounded domain, and x^* is the unique optimal point. We also assume that $g(x)$ is bounded for each $x \in D$ and $\text{var}_W[\phi(x, W)] < \infty$. For solving this problem let us introduce the following convolution function:

$$\hat{g}(x, \beta) = \int_{\infty}^{\infty} \hat{h}(v, \beta) g(x - v) \, dv = \int_{-\infty}^{\infty} \hat{h}((x - v), \beta) g(v) \, dv, \quad (7.6.1)$$

which is called a *smoothed functional* [18].

In order for $\hat{g}(x, \beta)$ to have nice smoothed properties, let us make some assumptions about the kernel $\hat{h}(v, \beta)$.

1 $\hat{h}(v, \beta) = (1/\beta^n) h(v/\beta) = (1/\beta^n) h(v_1/\beta, \ldots, v_n/\beta)$ is a piece-wise differentiable function with respect to v.
2 $\lim_{\beta \to 0} \hat{h}(v, \beta) = \delta(v)$, where $\delta(v)$ is Dirac's delta function.
3 $\lim_{\beta \to 0} \hat{g}(x, \beta) = g(x)$.
4 $\hat{h}(v, \beta)$ is a p.d.f., that is, $\hat{g}(x, \beta) = E_V[g(x - V)]$.

We assume that the original function $g(x)$ is not "well behaved." For instance, it can be a multiextremal function or have a fluctuating character (see Fig. 7.6.1).

We expect "better behavior" from the smoothed function $\hat{g}(x, \beta)$ than from the original one.

The idea of smoothed functionals is as follows: for a given function $g(x)$ construct a smoothed function $\hat{g}(x, \beta)$ and, operating only with $\hat{g}(x, \beta)$, find the extremum for $g(x)$. In other words, while operating only with

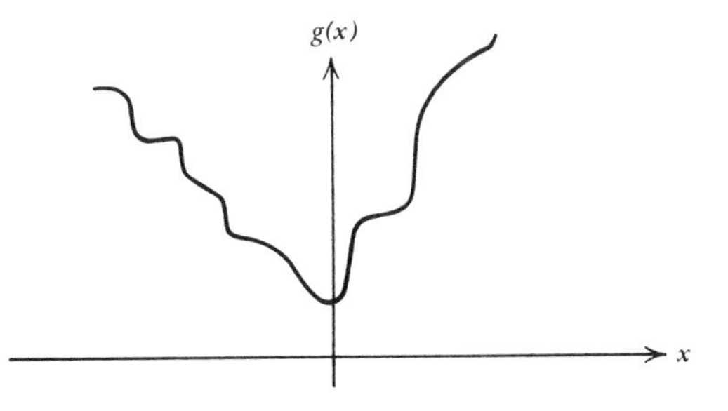

Fig. 7.6.1 A bed "behaved" function.

$\hat{g}(x, \beta)$, we want to avoid all fluctuation and local extrema of $g(x)$ and find x^*.

It is obvious that the effect of smoothing depends on the parameter β: for large β the effect of smoothing is large, and vice versa. When $\beta \to 0$ it follows from condition 2 that $\hat{g}(x, \beta) \to g(x)$ and that there is no smoothing.

It is intuitively clear that, to avoid fluctuations and local extrema, β has to be sufficiently large at the start of the optimization. However, on approaching the optimum we can reduce the effect of smoothing by letting β vanish, since at the extremum point x^* we want coincidence of both extrema, $g(x)$ and $\hat{g}(x, \beta)$. Accordingly, we speak of a set of smoothed functions $\hat{g}(x, \beta_s)$, $s = 1, 2, \ldots,$ while constructing an iterative procedure for finding x^*.

Before describing the iterative procedure for solving the problem (7.1.16′), we derive some attractive properties of $\hat{g}(x, \beta)$.

PROPERTY 1 If $g(x)$ is convex, then $\hat{g}(x, \beta)$ is also convex.

The proof of this property is straightforward. For $0 < \lambda < 1$

$$\lambda \hat{g}(x, \beta) + (1 - \lambda)\hat{g}(y, \beta) - \hat{g}(\lambda x + (1 - \lambda)y, \beta)$$

$$= \int \hat{h}(v, \beta)[\lambda g(x - v) + (1 - \lambda)g(y - v) - g(\lambda x + (1 - \lambda)y - v)]\, dv. \tag{7.6.2}$$

The convexity of $g(x)$ implies

$$\begin{aligned} g(\lambda x + (1 - \lambda)y - v) &= g(\lambda(x - v) + (1 - \lambda)(y - v)) \\ &\le \lambda g(x - v) + (1 - \lambda)g(y - v). \end{aligned} \tag{7.6.3}$$

Substituting (7.6.3) in (7.6.2) and taking into account that $\hat{h}(v, \beta) \ge 0$, we obtain the proof immediately.

PROPERTY 2 It is readily seen that the gradient of the smoothed function $\hat{g}(x, \beta)$ may be expressed as

$$\hat{g}_x(x, \beta) = \int_{-\infty}^{\infty} \hat{h}_x((x - v), \beta)g(v)\, dv = \int_{-\infty}^{\infty} \hat{h}_v(v, \beta)g(x - v)\, dv \tag{7.6.4}$$

and is called a *smoothed gradient*. Using the right-hand side of (7.6.4), together with condition (1), we obtain

$$g_x(x, \beta) = \frac{1}{\beta}\int_{-\infty}^{\infty} h_v(v)g(x - \beta v)\, dv, \tag{7.6.5}$$

where

$$h_v(v) = \left\{ \frac{\partial h(v)}{\partial v_1}, \ldots, \frac{\partial h(v)}{\partial v_n} \right\} \tag{7.6.6}$$

is the gradient of $h(v)$ and $\partial h(v)/\partial v_k$, $k = 1, \ldots, n$, are the partial derivatives.

It is important to note that, to find a gradient of the smoothed function $\hat{g}(x, \beta)$, we do not need to know the gradient of $g(x)$, which sometimes does not exist at all.

We consider also the following smoothed function:

$$\tilde{g}(x, \beta) = \int_{-\infty}^{\infty} \hat{h}(v, \beta)[g(x+v) + g(x-v)]\, dv. \tag{7.6.7}$$

By analogy with (7.6.4) and (7.6.5) we can obtain the smoothed gradient for $\tilde{g}(x, \beta)$:

$$\begin{aligned} \tilde{g}_x(x, \beta) &= \int_{-\infty}^{\infty} h_v(v, \beta)[g(x-v) - g(x+v)]\, dv \\ &= \frac{1}{\beta} \int_{-\infty}^{\infty} h_v(v)[g(x - \beta v) - g(x + \beta v)]\, dv \\ &= \frac{1}{\beta} E_V[g(x - \beta V) - g(x + \beta V)]. \end{aligned} \tag{7.6.8}$$

Now we give two examples of kernels $\hat{h}(v, \beta)$, which satisfy conditions 1 through 4, and find their smoothed gradients according to (7.6.8).

Example 1 Let $h(v)$ be an n-dimensional standard multinormal distribution

$$h(v) = \frac{1}{(2\pi)^{n/2}} \exp\left(-\tfrac{1}{2} \sum_{s=1}^{n} v_i^2 \right), \qquad -\infty < v_s < \infty, s = 1, \ldots, n. \tag{7.6.9}$$

Then the smoothed gradient of $g(x)$ is

$$\tilde{g}_x(x, \beta) = \frac{1}{\beta} \int_{-\infty}^{\infty} v h(v)[g(x + \beta v) - g(x - \beta v)]\, dv. \tag{7.6.10}$$

Example 2 Let

$$h(v) = \begin{cases} \dfrac{\Gamma(n/2)}{2\pi^{n/2}}, & \|v\| = 1 \\ 0, & \|v\| \neq 1 \end{cases} \tag{7.6.11}$$

that is, let the random vector v be uniformly distributed over the surface of the unit sphere. The smoothed gradient equals

$$\tilde{g}_x(x,\beta)=\frac{n}{\beta}\int_{\|v\|=1} vh(v)[g(x+\beta v)-g(x-\beta v)]\,dv. \quad (7.6.12)$$

Having $\tilde{g}_x(x,\beta)$ at our disposal, we can construct, for instance, an iterative gradient algorithm

$$x_{i+1}=\pi\{x_i-\alpha\tilde{g}_x(x_i,\beta_i)\}, \qquad \alpha>0 \quad (7.6.13)$$

and find the conditions under which x_i converges to x^* in the deterministic optimization problem $\min_{x\in D\subset R^n} g(x)=g(x^*)$, which is a particular case of (7.1.16′), with $p(w)$ being a Dirac δ function.

Here $\pi(\cdot)$ denotes the projection operation on D (i.e., for every $x\in R^n$, $\pi(x)\in D$ and $\|x-\pi(x)\|=\min_{y\in D}\|x-y\|$), and α is a step parameter.

Since $g(x)$ is not a "well behaved" function, calculation of the multiple integrals $\hat{g}_x(x,\beta)$ and $\tilde{g}(x,\beta)$ are usually not available in explicit form and numerical methods have to be used. One of them is, as we know, the Monte Carlo method. For instance, an estimator of $\tilde{g}_x(x,\beta)$ can be found by the sample-mean Monte Carlo method (see Section 4.2.2)

$$\hat{\xi}(x,\beta)=\frac{1}{N\beta}\sum_{j=1}^{N}\frac{h_v(V_j)}{f(V_j)}\left[g(x-\beta V_j)-g(x+\beta V_j)\right] \quad (7.6.14)$$

and is called *parametrical statistical gradient* (PSG) [18].

Here $f(v)$ is a p.d.f. from which a sample of length N is taken. Assuming that $f(v)\equiv h(v)$, we obtain, respectively, the PSG in examples 1 and 2, as

$$\hat{\xi}(x,\beta)=\frac{1}{N\beta}\sum_{j=1}^{N}V_j[g(x+\beta V_j)-g(x-\beta V_j)] \quad (7.6.15)$$

and

$$\hat{\xi}(x,\beta)=\frac{n}{N\beta}\sum_{j=1}^{N}V_j[g(x+\beta V_j)-g(x-\beta V_j)]. \quad (7.6.16)$$

The r.v.'s in (7.6.15) and (7.6.16) are generated from (7.6.9) and (7.6.11), respectively.

By analogy with (7.6.7) the smoothed gradient of $\phi(x,W)$ is

$$\begin{aligned}\hat{\phi}_x(x,\beta,W)&=\int_{-\infty}^{\infty}h_v(v,\beta)[\phi(x-v,W_1)-\phi(x+v,W_2)]\,dv\\&=\frac{1}{\beta}\int_{-\infty}^{\infty}h_v(v)\phi[(x-\beta v,W_1)-\phi(x+\beta v,W_2)]\,dv,\end{aligned}$$

(7.6.17)

and by analogy with (7.6.14) the sample-mean Monte Carlo estimator for the smoothed gradient of $\phi(x, W)$ is

$$\xi(x,\beta)=\frac{1}{N\beta}\sum_{j=1}^{N}\frac{h_v(V_j)}{f(V_j)}\left[\phi(x-\beta V_j, W_{1j})-\phi(x+\beta V_j, W_{2j})\right]. \tag{7.6.18}$$

Assuming $f(v)\equiv h(v)$ by analogy with (7.6.15) and (7.6.16), we have the PSG for Examples 1 and 2, respectively:

$$\xi(x,\beta)=\frac{1}{N\beta}\sum_{j=1}^{N}V_j\left[\phi(x+\beta V_j, W_{1j})-\phi(x-\beta V_j, W_{2j})\right] \tag{7.6.19}$$

$$\xi(x,\beta)=\frac{n}{N\beta}\sum_{j=1}^{N}V_j\left[\phi(x+\beta V_j, W_{1j})-\phi(x-\beta V_j, W_{2j})\right]. \tag{7.6.20}$$

From (7.6.18) through (7.6.20) it follows that the estimator $\xi(x,\beta)$ of the smoothed gradient $\hat{\phi}_x(x,\beta,W)$ is constructed on the basis of observations of $\phi(x, W)$ alone. Both the "artificial" random variable V and "natural" random variable W are averaged in these equations. Table 7.6.1 presents some smoothed gradients and their estimators.

Assuming that the r.v.'s V and W are mutually independent and taking the expectation of $\xi(x,\beta)$ with respect to W and V, we obtain

$$E(\xi(x,\beta))=\frac{1}{N}E\left(\sum_{j=1}^{N}\xi_j\right)=g_x(x,\beta), \tag{7.6.21}$$

where

$$\xi_j=\frac{1}{\beta}\frac{h_v(v_j)}{f(v_j)}\left[\phi(x+\beta V_j, W_{1j})-\phi(x-\beta V_j, W_{2j})\right]. \tag{7.6.22}$$

That is, the PSG $\xi(x,\beta)$ is an unbiased estimator for the smoothed gradient $\hat{\phi}_x(x,\beta,W)$. Assuming also the independence of W_j's, $j=1,\ldots,N$, we obtain the variance of the sth component of $\xi(x,\beta)$:

$$\operatorname{var}\xi_s(x,\beta)=\frac{1}{N^2}\sum_{j=1}^{N}\operatorname{var}\xi_{sj}=\frac{1}{N}\operatorname{var}\xi_s(x,\beta)=N^{-1}\beta^{-2}\sigma_s^2(x),$$
$$s=1,\ldots,n, \tag{7.6.23}$$

where

$$\sigma_s^2(x)=\operatorname{var}\left[\frac{h_{vs}(V)}{f(v)}\left[\phi(x-\beta V, W_1)-\phi(x+\beta V, W_2)\right)\right] \tag{7.6.24}$$

TABLE 7.6.1. Smoothed Gradients and their Estimators

$\hat{h}(v,\beta)$	$h(v)$	$\hat{\phi}_x(x,\beta,W)$	$\xi(x,\beta)$	Algorithm
$\frac{1}{2^n\beta^n}\prod_{s=1}^{n}[U(v_s-\beta) - U(v_s+\beta)]$	$\frac{1}{2^n}\prod_{s=1}^{n}[U(v_s-1) - U(v_s+1)]$	$\frac{1}{\beta}\sum_{s=1}^{n}[\phi(x+\beta l_s, W_1) - (x-\beta l_s, W_2)]l_s$	$\frac{1}{\beta}\sum_{s=1}^{n}[\phi(x+\beta l_s, W_1) - \phi(x-\beta l_s, W_2)]l_s$	Stochastic approximation [13]
$\frac{1}{(2\pi)^{n/2}\beta^n}\exp\{-\sum_{s=1}^{n}\frac{v_s^2}{2\beta^2}\}$, $-\infty < v_s < \infty$	$\frac{1}{(2\pi)^{n/2}}\exp\{\sum_{s=1}^{n} - \frac{v_s^2}{2}\}$, $-\infty < v_s < \infty$	$\frac{1}{\beta}E_V[\phi(x+\beta V, W_1) - \phi(x-\beta V, W_2)]V$	$\frac{1}{N\beta}\sum_{j=1}^{N}[\phi(x+\beta V_j, W_{1j}) - \phi(x-\beta V_j, W_{2j})]V_j$	Formula (7.6.19)
$\frac{1}{\beta^n}(1-\sum_{s=1}^{n}\frac{v_s^2}{\beta^2}) \times U(1-\sum_{s=1}^{n}\frac{v_s^2}{\beta^2})$, $0 \le v_s \le 1$	$(1-\sum_{s=1}^{n}v_s^2)U(1-\sum_{s=1}^{n}v_s^2)$, $0 \le v_s \le 1$	$\frac{1}{\beta}E_V[\phi(x+\beta V, W_1) - \phi(x-\beta V, W_2)]V$	$\frac{1}{N\beta}\sum_{j=1}^{N}[(x+\beta V_j, W_{1j}) - \phi(x-\beta V_j, W_{2j})]V_j$	Random search [28]

Note: Here

$$U(x) = \begin{cases} 1, & \text{if } x \ge 0 \\ 0, & \text{if } x < 0. \end{cases}$$

and

$$E\{\langle\xi(x,\beta),\xi(x,\beta)\rangle\} = \operatorname{var}\left[\sum_{s=1}^{n}\xi_s(x,\beta)\right] + \langle E[\xi(x,\beta)], E[\xi(x,\beta)]\rangle$$
$$\leq \sigma^2 N^{-1}\beta^{-2} + \langle\hat{g}_x(x,\beta),\hat{g}_x(x,\beta)\rangle. \tag{7.6.25}$$

Here $h_{vs}(V)$ is the sth coordinate of the vector $h_v(V)$, $\langle\ ,\ \rangle$ denotes the scalar product, and

$$\sigma^2 = n^2 \max_s \sup_{x\in D} \sigma_s^2(x). \tag{7.6.26}$$

Note that n^2 appears in (7.6.26) rather than n because of the covariance terms.

Taking into account that $g(x)$ is bounded for all $x \in D$ and $\operatorname{var}_W[\phi(x, W)] < \infty$, we can readily conclude that $\sigma_s^2(x) < \infty$ for all $x \in D$ and therefore $\sigma^2 < \infty$.

Now problem (7.1.16′) can be solved by the following algorithm:

$$x_{i+1} = \pi(x_i - \alpha\xi(x_i, \beta_i)). \tag{7.6.27}$$

Theorem 7.6.1 Assume that the iterative process is constructed in accordance with (7.6.27) and that for every $x \in D$ and for every i the following conditions are satisfied:

$$\langle(x - x^*), \hat{g}_x(x,\beta_i)\rangle \geq K_1\|x - x^*\|^2 - \gamma_i \tag{7.6.28}$$

$$\langle\hat{g}_x(x,\beta),\hat{g}_x(x,\beta)\rangle \leq K_2\|x - x^*\|^2 \tag{7.6.29}$$

$$0 < \alpha < 2K_1K_2^{-1} \tag{7.6.30}$$

$$\lim_{i\to\infty} \beta_i^{-2}N_i^{-1} = 0 \tag{7.6.31}$$

$$\lim_{i\to\infty} \gamma_i = 0, \qquad \gamma_i > 0 \tag{7.6.32}$$

$$E\|x_1\|^2 < \infty, \tag{7.6.33}$$

where $\|x\| = (\sum_{k=1}^{n} x_k^2)^{1/2}$ is the norm of x, and K_1 and K_2 are positive constants. Then process (7.6.27) converges in the mean square to the point x^*, that is, $\lim_{i\to\infty} E\|x_i - x^*\|^2 = 0$. If we replace condition (7.6.31) by

$$\sum_{i=1}^{\infty} \beta_i^{-2}N_i^{-1} < \infty \tag{7.6.34}$$

and conditions (7.6.32) by

$$\sum_{i=1}^{\infty} \gamma_i < \infty, \tag{7.6.35}$$

then process (7.6.27) converges, with probability 1, to x^*, that is,

$$P\left\{ \lim_{i\to\infty} \|x_i - x^*\| = 0 \right\} = 1.$$

Proof Without loss of generality we can set $x^* = 0$. Taking the conditional expectation of $\|x_{i+1}\|$, given $x_1, \ldots, x_i$, we obtain from (7.6.27)

$$E\left(\|x_{i+1}\|^2 | x_1, \ldots, x_i\right) \le \|x_i\|^2 - 2\alpha\langle x_i, E[\xi(x_i, \beta_i)]\rangle + \alpha^2 E\left[\langle \xi(x_i, \beta_i), \xi(x_i, \beta_i)\rangle\right]. \quad (7.6.36)$$

Substituting $E[\xi(x, \beta)] = \hat{g}_x(x, \beta)$ in (7.6.36), we obtain

$$E\left(\|x_{i+1}\|^2 | x_1, \ldots, x_n\right) \le \|x_i\|^2 - 2\alpha\langle x_i, \hat{g}_x(x_i, \beta_i)\rangle + \alpha^2 E\left[\langle \xi(x_i, \beta_i), \xi(x_i, \beta_i)\rangle\right]. \quad (7.6.37)$$

Now taking (7.6.26) through (7.6.29) into account, we obtain

$$\begin{aligned} E\left(\|x_{i+1}\|^2 \mid x_1, \ldots, x_i\right) &\le \|x_i\|^2 - 2\alpha K_1 \|x_i\|^2 + 2\alpha\gamma_i \\ &\quad + \alpha^2 N_i^{-1} \beta_i^{-2} \sigma^2 + \alpha^2 K_2 \|x_i\|^2 \\ &= \left(1 - 2\alpha K_1 + \alpha^2 K_2\right)\|x_i\|^2 + \alpha^2 N_i^{-1} \beta_i^{-2} \sigma^2 + 2\alpha\gamma_i. \end{aligned} \quad (7.6.38)$$

Taking the expectation of both sides of the last inequality, we obtain

$$\begin{aligned} E\|x_{i+1}\|^2 &\le \left(1 - 2\alpha K_1 + \alpha^2 K_2\right) E\|x_i\|^2 + \alpha^2 \beta_i^{-2} N_i^{-1} \sigma^2 + 2\alpha\gamma_i \\ &\quad + \left(1 - 2\alpha K_1 + \alpha^2 K_2\right)^i E\|x_1\|^2 \\ &\quad + \sum_{s=1}^{i} \left(\alpha^2 \beta_s^{-2} N^{-1} \sigma^2 + 2\alpha\gamma_s\right)\left(1 - 2\alpha K_1 + \alpha^2 K_2\right)^{i-s}. \end{aligned} \quad (7.6.39)$$

It follows from (7.6.30) that $1 - 2\alpha K_1 + \alpha^2 K_2 < 1$; therefore (7.6.39) can be rewritten as

$$E\|x_{i+1}\|^2 \le K_3^i E\|x_1\|^2 + \sum_{s=1}^{i} \left(\alpha^2 \beta_s^{-2} N_s^{-1} \sigma^2 + 2\alpha\gamma_s\right) K_3^{i-s}, \quad (7.6.40)$$

where

$$K_3 = 1 - 2\alpha K_1 + \alpha^2 K_2. \quad (7.6.41)$$

The first term in (7.6.38) converges to 0 as $i \to \infty$, since $K_3 < 1$ and $E\|x_1\|^2 < \infty$ (see (7.6.33)). Thus the theorem will be proven if we prove that

$$\lim_{i\to\infty} \sum_{s=1}^{i} (\alpha^2\beta_s^{-2}N^{-1} - 2\alpha\gamma_s)K_3^{i-s} = 0.$$

To prove this we assume that for any number ε we have chosen a number T such that, for all $s > T$, $\alpha^2\beta_s^{-2}N_s^{-1}\sigma^2 - 2\alpha\gamma_s$ is less than ε. Then

$$\sum_{s=1}^{i} (\alpha^2\beta_s^{-2}N_s^{-1}\sigma^2 + 2\alpha\gamma_s)K_3^{i-s} \le K_3^i \left[\sum_{s=1}^{T} (\alpha^2\sigma^2\beta_s^{-2}N_s^{-1} + 2\alpha\gamma_s)K_3^{-s} + \sum_{s=T+1}^{i} \varepsilon K_3^{-s} \right]. \tag{7.6.42}$$

In view of the fact that T is finite, the first term in (7.6.42) tends to zero as $i\to\infty$, since $K_3 < 1$. Using the formula for a geometrical progression, we obtain:

$$\lim_{i\to\infty} E\|x_i\|^2 \le \lim_{i\to\infty} \varepsilon K_3^i \sum_{s=T+1}^{i} K_3^{-s} \le \frac{\lim_{i\to\infty} \varepsilon K_3(1 - K_3^{i-T})}{1 - K_3}.$$

Since ε may be any positive number, we have $\lim_{i\to\infty} E\|x_i\|^2 = 0$.

This completes the proof of the first part of the theorem.

To prove the convergence of (7.6.27) with probability 1, it is sufficient to show that $\sum_{i=1}^{\infty} E(\|x_i\|^2) < \infty$. Summing both sides of (7.6.40), we have by (7.6.34) and (7.6.35)

$$\begin{aligned}
\sum_{i=1}^{\infty} E(\|x_{i+1}\|^2) &\le \frac{K_3}{1 - K_3} E(\|x_1\|^2) + \sum_{i=1}^{\infty} \sum_{s=1}^{i} (\alpha^2\beta_s^{-2}N_s^{-1}\sigma^2 + 2\alpha\gamma_s)K_3^{i-s} \\
&= \frac{K_3}{1 - K_3} E(\|x_1\|^2) + \sum_{s=1}^{\infty} \sum_{i=s}^{\infty} (\alpha^2\beta_s^{-2}N_s^{-2}\sigma^2 + 2\alpha\gamma_s)K_3^{i-s} \\
&= \frac{K_3}{1 - K_3} \left[E(\|x_1\|^2) + \sum_{s=1}^{\infty} (\alpha^2\beta_s^{-2}N_s^{-1}\sigma^2 + 2\alpha\gamma_s) \right] < \infty,
\end{aligned}$$

from which the result follows. Q.E.D.

Remark 1 The theorem remains valid for the deterministic optimization problem

$$\min_{x\in D\subset R^n} g(x) = g(x^*),$$

which is a particular case of problem (7.1.16′), when $W = 0$.

Remark 2 Condition (7.6.28), together with (7.6.32), allowed $g(x)$ to be nonconvex.

APPENDIX

Let Ξ be a random vector uniformly distributed over the surface of a unit n-dimensional sphere with its center at origin, and let R be any given unit vector issuing from the origin (see Fig. 7.A.1).

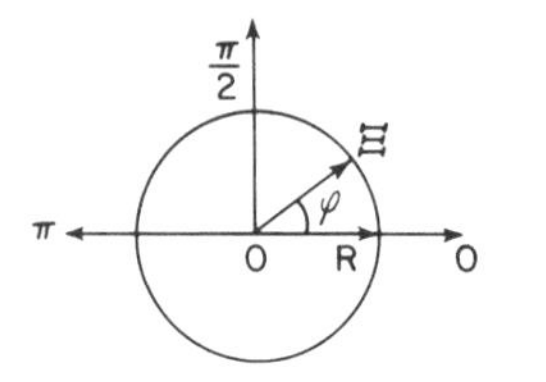

Fig. 7.A.1 Graphical representation of the random vector Ξ.

The p.d.f. of the random angle between Ξ and R is sought. For reasons of symmetry we confine ourselves to the semisphere $0 \leq \varphi \leq \pi$. The p.d.f. is then [28]:

$$h_n(\varphi) = \frac{\sin^{n-2}\varphi}{\int_0^{\pi} \sin^{n-2}\varphi \, d\varphi} = B_n \sin^{n-2}\varphi, \qquad 0 \leq \varphi \leq \pi,$$

where

$$B_n = \left(\int_0^{\pi} \sin^{n-2}\varphi \, d\varphi \right)^{-1} = \frac{\Gamma(n/2)}{\sqrt{\pi}\,\Gamma((n-1)/2)}.$$

The expected value of the r.v. φ is

$$E(\varphi) = \frac{\pi}{2},$$

from which it follows that on the average, R and Ξ are orthogonal.

It is readily verified that, as n increases, $h_n(\varphi)$ approaches to Dirac's δ function, that is,

$$h_n(\varphi) \underset{n\to\infty}{\to} \delta\left(\frac{\pi}{2} - \varphi\right).$$

Fig. 7.A.2 represents $h_n(\varphi)$ for different n.

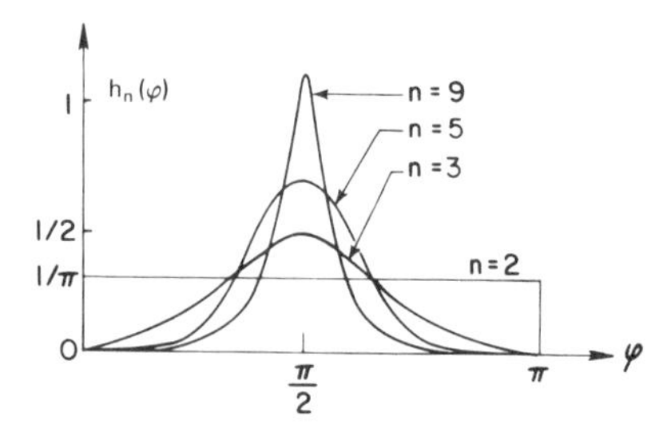

Fig. 7.A.2 The density function of φ for different n.

EXERCISES

1 Find the efficiency C_n (7.2.8) and var($\cos\varphi$) of Algorithm RS-2 analytically and of Algorithm RS-5 by simulation. For Algorithm RS-5 describe the random number generator and the flow diagram of your program.

2 Prove that for a linear function $g(x)$ the direction of $V_{i\Xi}$ in Algorithm RS-5 (see (7.1.13)) coincides, on the average, with that of the gradient of $g(x)$.

3 By analogy with algorithm RS-1 (see (7.1.18)) describe the nonlinear tactic Algorithm RS-2, the linear tactic Algorithm RS-3, the statistical gradient Algorithm RS-5 for solving problem (7.1.16).

4 Prove that, if $g(x)$ is convex in R^n and if the point x^* in which $g(x)$ attains its minimum value is unique, then $\hat{g}(x,\beta)$ (see (7.6.1)) is strictly convex.

5 Given a linear function $\langle c,x\rangle$ invariant for the convolution (7.6.1), that is, $\int h(\beta,x-u)\langle c,u\rangle\, du=\langle c,x\rangle$, prove that $\hat{g}(x,\beta)\geq g(x)$.

6 Prove (7.6.4) and (7.6.5).

7 Prove that, if $h_n(\varphi)=\frac{1}{2}B_n|\sin^{n-2}\varphi|$, $0\leq\varphi\leq 2\pi$, then $P_n(v)$, where $v=\cos\varphi$ is distributed according to (7.2.12).

8 Consider the following modification of Algorithm RS-1 (see (7.1.4)):

$$x_{i+1}=x_i+\frac{\alpha_i}{\beta_i}[g(x_k+\beta_i\Xi_i)-g(x_i)]\Xi_i,\qquad \alpha_1>0,\ \beta_i>0.$$

Find the efficiency C_n (7.2.8) and var($\cos\varphi$), assuming that $g(x)$ is a linear function.

REFERENCES

1 Archetti, F., A sampling technique for global optimization, in *Towards Global Optimization*, edited by L. C. W. Dixon and G. P. Szegö, North Holland, American Elsevier, New York, 1975.

2 Avriel, M., *Nonlinear Programming. Analysis and Methods*. Prentice-Hall, Englewood Cliffs, New Jersey, 1976.

3 Becker, R. W. and G. V. Lago, A global optimization algorithm, *Eighth Allerton Conference on Circuits and System Theory*, 1970, pp. 3–13.

4 Brooks, S. H., A discussion of random methods for seeking maxima, *Oper. Res.*, **6**, 1958, 244–251.

5 Chichinadze, V. K., Random search to determine the extremum of the function of several variables, *Eng. Cybern.*, **1**, 1967, 115–123.

6 Devroye, L. P., On the convergence of statistical search, *Inst. Elec. Electron. Eng. Trans. Syst., Man, Cybern.*, **6**, 1976, 46–56.

7 Dixon, L. C. W., Global optimization without convexity, Technical Report N85, The Hatfield Polytechnical Numerical Optimization Center, July 1977.

8 Dvoretzky, A., On stochastic approximation, in *Proceedings of the Third Berkeley Symposium on Mathematical Statistics and Probability*, Vol. 1, 1956, pp. 39–55.

9 Ermolyev, Yu. M., *Stochastic Programming Methods*, Nauka, Moscow, 1976 (in Russian).

10 Ermolyev, Yu. M., On the method of generalized stochastic gradients and quasi-Fejer-sequences, *Cybernetics*, **5**, 1969, 208–220.

11 Evtushenko, Yu. G. Numerical methods for finding global extrema (case of a non uniform mesh). *U.S.S.R. Comp. Math. Math. Phys.*, **11**, No. 6, 1971, pp. 38–55.

12 Gaviano, M., Some general results on the convergence of random search algorithms in minimization problems, in *Towards Global Optimization*, edited by L. C. W. Dixon and G. P. Szegö, North Holland, American Elsevier, New York, 1975.

13 Gladyshev, E. Y., On Stochastic Approximation. *Theory Prob. Appl.*, **1965**, No. 2, 272–275.

14 Gnedenko, B. V., Sur la distribution du terme maximum d'une serie aleatoire. *Ann. Math.*, **44**, 1943, 423–453.

15 Gomulka, J., Numerical experience with Törn's clustering algorithm and two implementations of Branin's method, in *Toward Global Optimization*, Vol. 2, edited by L. C. W. Dixon and G. P. Szegö, North Holland, American Elsevier, New York, 1977.

16 Hammersley, I. M. and D. C. Handscomb, *Monte Carlo Methods*, Wiley, New York; Methuen, London, 1964.

17 Katkovnik, V. Ya., *Linear Estimations and Stochastic Optimization Problems*, Nauka, Moscow, 1976 (in Russian).

18 Katkovnik, V. Ya. and Yu. Kulchitsky, Convergence of a class of random search algorithms, *Automat. Remote Control* **1972**, No. 8, 1321–1326.

19 Kiefer, J., and J. Wolfowitz, Stochastic estimation of the maximum of a regression function, *Ann. Math. Stat.*, **23**, 1952, 462–466.

20 Kleiza, V., On the modeling of nonlinearity by the sequence of Markov chains, *Lith. Math. J.*, **XV**, No. 4, 1975, 125–130.

21 Mangasarian, O. L., *Nonlinear Programming*, McGraw-Hill, New York, 1969.

22 Matyas, J., Random optimization, *Automat. Remote Control*, **26**, 1965, 246–253.

23 Meerkov, S. M., Deceleration in the search for the global extremum of a function, *Automat. Remote Control*, **1972**, No. 12, 129–139.

24 Metropolis, N., A. W. Rosenbluth, M. N. Rosenbluth, A. H. Teller, and E. Teller. Equations of state calculations by fast computing machines. *J. Chem. Physics*, **21**, 1953, 1087–1092.

25 Pincus, M., A closed form selection of certain programming problems,, *Oper. Res.*, **16**, 1968, 690–694.

26 Pincus, M., A Monte-Carlo method for the approximate solution of certain types of constrained optimization problems, *Oper. Res.*, **18**, 1970, 1225–1228.

27 Price, W. L., A controlled random search procedure for global optimization, in *Towards Global Optimization*, Vol. 2, edited by L. C. W. Dixon and G. P. Szegö, North Holland, American Elsevier, New York, 1977.

28 Rastrigin, L. A., *The Stochastic Methods of Search*, Nauka, Moscow, 1968 (in Russian).

29 Rastrigin, L. A., and Y. Rubinstein, The comparison of the random search and the stochastic approximation while solving the problem of optimization, *Automat. Control*, **2**, No. 5, 1969, 23–29.

30 Robins, H., and Monro, S., A stochastic approximation method, *Ann. Math. Stat.*, **22**, 1951, 400–407.

31 Rubinstein, Y., Convergence of the random search algorithm, *Automat. Control*, **3**, No. 1, 1969, 46–49.

32 Rubinstein, Y., Piece-wise-linear representation of function in situation of noise, *Automat. Control*, **2**, No. 5, 1968, 36–42.

33 Rubinstein, Y., Choice of the optimal search strategy, *J. Optimizat. Theory Appl.*, **18**, No. 3, March 1976, 309–317.

34 Rubinstein, Y. and J. Har-El, Optimal performance of learning automata in switched random environments, *Inst. Elec. Electron. Eng. Trans. Syst., Man, Cyber.*, **SMC-7**, 1977, 674–678.

35 Rubinstein, Y. and A. Karnovsky, Local and integral properties of a search algorithm of the stochastic approximation type, *Stochastic Processes Appl.*, **6**, 1978, 129–134.

36 Rubinstein, Y. and I. Weissman, The Monte-Carlo method for global optimization, *Cah. Cen. Étud. Rech. Opér.*, **21**, No. 2, 1979, 143–419.

37 Shubert, B. O., A sequential method for searching the global maximum of a function, *Soc. Indust. Appl. Math. J. Numer. Anal.*, **1972**, No. 9, 379–388.

38 Suitti, C., Convergence proof of minimization algorithms for nonconvex functions, *JOTA*, **23**, 1977, 203–210.

39 Strongin, R. G., Simple search algorithm for global extremum of function of several variables and its use in functions approximation problem, *Radiofizika*, **7**, No. 15, 1972, 1077–1085.

40 Törn, A., A search clustering approach to the global optimization problem, in *Towards Global Optimization*, Vol. 2, edited by L. C. W. Dixon and G. P. Szegö, North Holland, American Elsevier, New York, 1977.

41 Van Ryzin, J., On strong consistency of density estimations, *Ann. Math. Stat.*, **40**, 1969, 1765–1772.

42 Venter, H. J., On estimation of the mode. *Ann. Math. Stat.*, **38**, 1967, 1446–1455.

43 Wasan, M. T., *Stochastic Approximation*, Cambridge University Press, New York, 1969.

44 Wilde, D. J., *Optimum Seeking Methods*, Prentice-Hall, Englewood Cliffs, New Jersey, 1964.

45 Yakowitz, S. J. and L. Fisher, On sequential search for the maximum of unknown function, *J. Math. Anal. Appl.*, **41**, 1973, 234–259.

46 Zielinski, R., A Monte Carlo estimation of the maximum of a function, *Algorithms*, **VII**, No. 13, 1970, 5–7.

Index

About the author

Professor Reuven Y. Rubinstein is at the Faculty of Industrial Engineering and Management at the Technion-Israel Institute of Technology. He received his Ph.D. in Operations Research from Rigas Polytechnical Institute in 1969. In 1978-1979 he was a visiting professor at the University of Ilinois, Urbana and in 1980 a visiting scientist at IBM Thomas Watson Research Center. His main contribution is in the field of stochastic optimization, stochastic automata, Monte Carlo methods and simulation. Dr. Rubinstein has published over 40 articles and contributed to several books. In 1978 he was awarded Oded Lewin's Prize by the Operations Research Society of Israel.

Applied Probability and Statistics (Continued)

FLEISS • Statistical Methods for Rates and Proportions, *Second Edition*
GALAMBOS • The Asymptotic Theory of Extreme Order Statistics
GIBBONS, OLKIN, and SOBEL • Selecting and Ordering Populations: A New Statistical Methodology
GNANADESIKAN • Methods for Statistical Data Analysis of Multivariate Observations
GOLDBERGER • Econometric Theory
GOLDSTEIN and DILLON • Discrete Discriminant Analysis
GROSS and CLARK • Survival Distributions: Reliability Applications in the Biomedical Sciences
GROSS and HARRIS • Fundamentals of Queueing Theory
GUPTA and PANCHAPAKESAN • Multiple Decision Procedures: Theory and Methodology of Selecting and Ranking Populations
GUTTMAN, WILKS, and HUNTER • Introductory Engineering Statistics, *Second Edition*
HAHN and SHAPIRO • Statistical Models in Engineering
HALD • Statistical Tables and Formulas
HALD • Statistical Theory with Engineering Applications
HARTIGAN • Clustering Algorithms
HILDEBRAND, LAING, and ROSENTHAL • Prediction Analysis of Cross Classifications
HOEL • Elementary Statistics, *Fourth Edition*
HOLLANDER and WOLFE • Nonparametric Statistical Methods
JAGERS • Branching Processes with Biological Applications
JESSEN • Statistical Survey Techniques
JOHNSON and KOTZ • Distributions in Statistics
Discrete Distributions
Continuous Univariate Distributions—1
Continuous Univariate Distributions—2
Continuous Multivariate Distributions
JOHNSON and KOTZ • Urn Models and Their Application: An Approach to Modern Discrete Probability Theory
JOHNSON and LEONE • Statistics and Experimental Design in Engineering and the Physical Sciences, Volumes I and II, *Second Edition*
JUDGE, GRIFFITHS, HILL and LEE • The Theory and Practice of Econometrics
KALBFLEISCH and PRENTICE • The Statistical Analysis of Failure Time Data
KEENEY and RAIFFA • Decisions with Multiple Objectives
LANCASTER • An Introduction to Medical Statistics
LEAMER • Specification Searches: Ad Hoc Inference with Nonexperimental Data
McNEIL • Interactive Data Analysis
MANN, SCHAFER and SINGPURWALLA • Methods for Statistical Analysis of Reliability and Life Data
MEYER • Data Analysis for Scientists and Engineers
MILLER • Survival Analysis
MILLER, EFRON, BROWN, and MOSES • Biostatistics Casebook
OTNES and ENOCHSON • Applied Time Series Analysis: Volume I, Basic Techniques
OTNES and ENOCHSON • Digital Time Series Analysis
POLLOCK • The Algebra of Econometrics
PRENTER • Splines and Variational Methods
RAO and MITRA • Generalized Inverse of Matrices and Its Applications
RIPLEY • Spatial Statistics
SCHUSS • Theory and Applications of Stochastic Differential Equations
SEAL • Survival Probabilities: The Goal of Risk Theory
SEARLE • Linear Models
SPRINGER • The Algebra of Random Variables
UPTON • The Analysis of Cross-Tabulated Data
WEISBERG • Applied Linear Regression
WHITTLE • Optimization Under Constraints